NANOTECHNOLOGY SCIENCE AND TECHNOLOGY

PHOTOCATALYSIS

PERSPECTIVE, MECHANISM, AND APPLICATIONS

NANOTECHNOLOGY SCIENCE AND TECHNOLOGY

Additional books and e-books in this series can be found on Nova's website under the Series tab.

NANOTECHNOLOGY SCIENCE AND TECHNOLOGY

PHOTOCATALYSIS

PERSPECTIVE, MECHANISM, AND APPLICATIONS

PREETI SINGH
M. M. ABDULLAH
MUDASIR AHMAD
AND
SAIQA IKRAM
EDITORS

nova
science publishers
New York

Library of Congress Cataloging-in-Publication Data

ISBN: 978-1-53616-044-4

Published by Nova Science Publishers, Inc. † New York

CONTENTS

PREFACE

Photocatalysis is becoming one of the most active research and development areas among the scientific communities. These reactions are finding applications in industries, pharmaceutics and in environmental chemistry stressing on the sustainable developments such as antifouling agents, controlling air pollution by degrading contaminants present in the environment, wastewater treatments parallel with inactivation of microorganisms and hydrogen generation etc. Though, many books are devoted to the description of Photocatalysis technique/process for the research and industrial production but there is no book exclusively relevant to intrigue topic of Nano-materials for photocatalysis. The main aim for compiling this book is to present, an easy-to-follow sequence, a description of types of photocatalytic process which can be used to estimate and modify its physical and chemical properties. This book offers a description about photocatalyses, starting from their fundamentals to the mechanism involved, followed by their sustainable environmental applications by using Biopolymer in the field of environmental studies and water treatment. This book also summarizes artificial photosynthesis techniques for clean hydrogen and oxygen production due to their utmost importance to industry and academia.

We hope this book will assist all level of readers. It is dedicated not only to academia but to the researchers and to the industrialist who will find this book to be a source of knowledge as well as a launching pad for novel ideas and inventions. In particular this book is expected to be of interest to the people involved in nanomaterials, water treatment, environmental industries and engineers. Potential readers also include both professionals and dedicated non-professional environmentalist, and those working on the development of novel- innovative work on Photocatalyses. In the end, it is expected that this book will find a prominent place in the traditional universities and research institutions libraries where chemistry, physics and/or material science, environmental science and technology are being pursued among core studies in the designing and operating Photocatalytic devices for generating clean energy to greener environmental cleaning.

ACKNOWLEDGMENTS

It is our great pleasure to write the acknowledgement for the book entitled *Photocatalysis: Perspective, Mechanism, and Applications.* We are indebted to all the contributing authors for providing their significant articles as per the Publisher's requirement with static dedication. In addition, a sincere gratitude goes to the editors for their vital support, guidance and encouragements without whom it was impossible to accomplish the end task. Our sincere thanks also goes to Nova Science Publishers who had given us the opportunity to edit the present book and subsequently undertaken the responsibility in its compilation, printing, binding, publication etc., in the best possible way.

We express our gratitude towards the support and love of our family members, parents, family, friends, and colleagues. At last, we are highly thankful to the readers of this book.

Editors

In: Photocatalysis
Editors: P. Singh, M. M. Abdullah, M. Ahmad et al.
ISBN: 978-1-53616-044-4

Chapter 1

PHOTOCATALYSIS: ESSENTIALS, MECHANISM, AND EFFECTIVE PARAMETERS

***Preeti Singh*[1], *M. M. Abdullah*[2], *Suresh Sagadevan*[3] and *Saiqa Ikram*[1,*]**

[1]Bio/Polymers Research Laboratory, Department of Chemistry, Jamia Millia Islamia, New Delhi, India

[2]Promising Center for Sensors and Electronic Devices (PCSED), Department of Physics, Faculty of Science & Arts, Najran University, Najran, Saudi Arabia

[3]Nanotechnology & Catalysis Research Centre, University of Malaya, Kuala Lumpur, Malaysia

ABSTRACT

Contamination/pollutants in the environment has been influencing/damaging the plants, earth, seas, living organism, water, air, and various other environmental species over the globe. Water sullying assumes to be a significant factor in ecological contamination. The dyes get decomposed in water and form the harmful compounds. Therefore, there is a necessity for a technique to remove the contaminants. Photocatalysis is one of the advanced techniques for the removal of dyes from the waste water. In addition, it is a remarkable technique for rectifying energy and ecological issues such as, degradation of poisonous natural contaminations in wastewater, harmful gases and microscopic organisms in their diverse media, the creation of hydrogen, air decontamination, soil remediation and antibacterial action. In view of its importance, we have discussed the basics, principles, mechanism, and operating parameters of

[*]Corresponding Author's E-mail: sikram@jmi.ac.in; aries.pre84@gmail.com

photocatalysis in detail. The roles of various operating parameters of the photocatalytic process were found to influence the method. A possible study of photocatalytic effective parameters and their theoretical interpretation demonstrate the significant features of this technique, and its importance to the environment and living beings.

Keywords: contamination, waste treatment methods, photocatalyst, mechanism, oxidation, reduction, effective parameters

1. INTRODUCTION

Inclusion of pollutants into the environment causes adverse effectsto living as well as non living beings. After air, water is an essential requirement for creatures. Living creatures need clean water for drinking, cooking, cleaning, horticulture, etc. Water is becoming polluted due to many factors such as large deforestations, decreased agricultural lands, extensive use of chemical pesticides, industrial discharge of chemicals etc. These factors have disrupted the existing natural processes of recycling of waste-water into clean-water. Due to this, most of the natural resources of drinking water are found to be contaminated with various poisonous/toxic/hazardous materials and pathogenic microorganisms (Baruah et al. 2010). Dumping of untreated hazardous materials/pollutants into water disturbs the ecological safety and quality of life (Pelaez et al. 2012). In addition, water pollutants exist in various unsafe squanders like pharmaceutical squanders, pesticides, herbicides, textiles, gums and phenolic compound (Chong et al. 2010). Some potential sources of water pollution or contamination are shown in Figure 1.

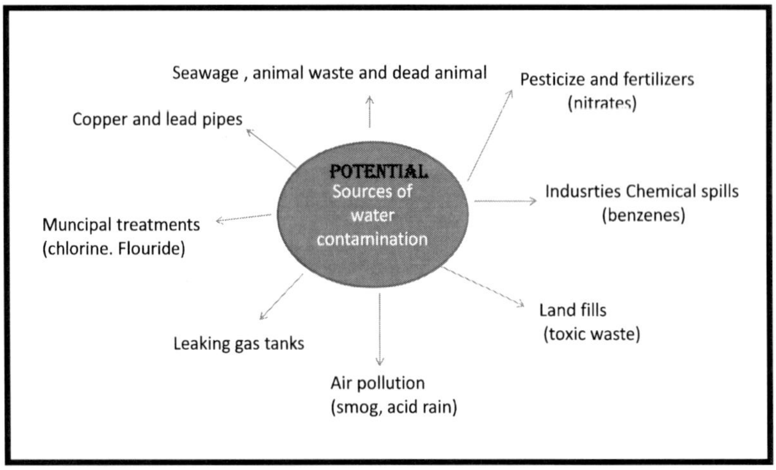

Figure 1. Represents the potential sources of water contamination.

Removing pollutants from water has become a universal need. In current scenario, it becomes very essential to decrease/remove this spectrum of pollutant to maintain a high-quality eco-friendly nature/atmosphere. To remove the pollutants, various industries have employed natural treatment, turn around assimilation, ozonation, filtration, adsorption on strong stages, burning, and coagulation for water preservationist treatment forms. However, every treatment methodology comprises of their own advantages with certain limitations such as incineration method can result in deadly venomous volatile products. Conventional biological treatment processes are also not so flourishing due to their intractable nature of synthetic dyes and the high salinity of wastewater containing dyes. It further consumes prolonged time for treatment of waste resulting in an unpleasant smell. Chlorination and ozonation are also not sufficiently appropriate owing to their high expenses. Although, ozonation can be considered as a viable path for the treatment of natural profluent from contaminated water, yet the stability and solidness of ozone effectively get affected by the presence of salts, pH, and the temperature. Some physical methods, for example, adsorption on enacting carbon, ultra-filtration, turn around assimilation, coagulation by concoction specialists, particle trade on engineered adsorbent pitches, and so on have been utilized for the expulsion of dye from polluted water. These techniques do not reduce the organic molecules completely, but only breaks the organic compound from water to another phase and thus creates a secondary pollutant, which further requires treatment of solid wastes and restoration of the adsorbent, subsequently making the process expensive and time consuming (Gupta et al. 2012; Khan et al. 2017). Compared to these classical methods, it has been observed that photocatalysis is one of the distinctive and effective methods for the removal of water contaminants. In 1972, Fujishima and Honda discovered the phenomenon of photocatalytic splitting of water on a TiO_2 electrode under ultraviolet (UV) light (Fujishima and Honda.1972).

In the present scenario, advanced oxidation processes (AOPs) are a cost-effective approach to the treatment of organic effluent from contaminated water. AOPs can be classified into two forms: Non-photochemical and Photochemical AOPs (homogeneous and heterogeneous photocatalytic processes). Researchers explore that AOPs is dependent on advanced photocatalytic strategies and thus benefitting our atmospheric conditions by complete oxidation of natural organic molecules into non-harmful components CO_2 and H_2O. In addition to these AOPs, also generates hydroxyl free radicals (·OH). These ·OH radicals are non- selective in nature, act as strong oxidizing agents and readily react with contaminants to destroy and remove them from the water. Also, it completely degrades non-biodegradable organic compounds and makes water reusable (Rehman et al. 2009). Thus, photocatalytic phenomenon practices over the classical techniques, for example, quick oxidation, no development of poly-cyclic compounds, complete oxidation of contaminants and high productivity.

Application of nanotechnology provides cost-effective remediation techniques imposing a major challenge for the researchers for developing an adequate water remediation method. Nanomaterials are most rapidly used and lucrative approach to eradicate wastes from water because of its high surface to volume ratio. Thus, nanotechnology has a significant role in minimizing the harmful wastes from water to provide safe water (Mansoori et al. 2008). Hence, Photocatalytic degradation emerged as one of the foremost rich procedures for the treatment of toxic organic effluents by utilizing nanomaterials, such as, TiO_2, ZnO, ZnS etc. which act as a catalyst (Ganaprakasam et al. 2015). Therefore, the main core of this chapter is to address the basics, principles, mechanism, and essentials of photocatalytic processes and to evaluate the effective parameters of photocatalytic method.

2. Principle and Mechanism

A photocatalyst is defined as a material which produces electron-hole (e^-/h^+) pairs on absorbing photons from the source to enable the chemical transformations of the reactants and regenerate their chemical composition after each rotation of e^-/h^+ pair interactions (Diuresis et al. 2014). Catalytic materials such as TiO_2, ZnO, ZnS, etc., in nano size are found efficient compared to their bulk counterpart and is termed as nanocatalyst. A significant feature of photocatalysis is that it requires sunlight or UV light as a source of irradiation. Photocatalysis reaction potentially depends on the wavelength of light (photon). Also, the degradation of dye is identified with the redox capabilities or potential of dyes and the energy band gap of the material used. The light can be irradiated directly or indirectly along with the catalyst on the dye. Thus, the photocatalytic mechanism involves two kinds of kinetics: Direct and Indirect. In general, the direct photocatalysis occur under visible light, whereas the indirect photocatalysis happen under UV-irradiation (Khan et al. 2015). The detail mechanism of these two photocatalytic processes is as follows:

2.1. Indirect Mechanism

In principle, the degradation of dye mainly depends on oxidation and reduction of the photocatalyst. Its mechanism can be summarized as follows in following four main steps:

2.1.1. Photo-Excitation

First basic step of photocatalysis is the interaction of aphoton with the surface of nanocatalyst. Photocataltic process starts with the irradiation of light on nanocatalyst. The valence band electron of catalyst gets excited to the conduction band by absorbing the photon of energy (hv), which further creates the hole (positive openings (h^+)) in the valence band, and thus generates a pair of electron-hole (e^-/h^+). This mechanism is represented in the equation (1) below.

$$\text{Nanocatalyst} + h\nu \rightarrow \text{Nanocatalyst}\,(e^-(CB) + h^+(VB) \quad (1)$$

2.1.2. Ionization of Water Molecules

Nanocatalyst surface contains contaminated water or dye solution. Due to light illumination, the electron from the valence band of the nanocatalyst gets excited to the conduction band and thus creates holes in the valence band. Furthermore,the contaminated water gets oxidized by the holes (h^+) generated in the valence band, consequently clearing a path for the arrangement of hydroxyl (OH·) radicals, or specialists having colossally solid oxidative disintegrating power represented in equation (2):

$$H_2O + h^+(VB) \rightarrow OH^- + H^+ \quad (2)$$

OH radical being non-selective in nature, they get easily absorbed by the organic molecules present at the surface of nanocatalyst, resulting in degrading or mineralizing their complex structure into simpler. It also readily attacks the microorganisms for decontamination.

2.1.3. Ionosorption of Oxygen

Photogenerated hole (h^+) reacts with surface-bound water to produce the hydroxyl radical. This is because of the fact that the electron in the conduction ($e^-{}_{(CB)}$) is taken up by the oxygen. Furthermore, oxygen reacts with the intermediate radical present in the organic compound experiences a radical chain reaction resulting to generate anionic superoxide radical (O_2-). Reduction of oxygen happens as an option in contrast to hydrogen generation because of the way that oxygen is an effortless reducible substance. The conduction band electrons respond with dissolved oxygen species to shape superoxide anions (equation (3))

$$O_2 + e^-(CB) \rightarrow O_2^- \quad (3)$$

The superoxide (O_2^-) particle is the main force behind the oxidation procedure, shaping peroxide or changing to hydrogen peroxide and afterward to water. The higher concentration of organic matter leads to increase the number of positive holes. This decreases the probability of carrier recombination of the electron-hole pair. This maintains the electron neutrality within the nanocatalyst, and thus upgraded the photocatalytic process.

2.1.4. Protonation of Superoxide

The superoxide (O_2^-) produced gets protonated to form hydroperoxyl radical ($HO_2^{\cdot}$) and consequently transforms to H_2O_2 which readily dissociates into highly reactive species hydroxyl radicals ($OH^{\cdot}$). The response of biodegradable intermediates with oxidants is alluded to as mineralization which is the creation of water, carbon dioxide, and inorganic particles which are biodegradable intermediates represented in equations (4) to (9):

$$O_2^- + H^+ \rightarrow HOO^* \quad (4)$$

$$2HOO^{\cdot} \rightarrow H_2O_2 + O_2 \quad (5)$$

$$H_2O_2 \rightarrow 2OH^{\cdot} \quad (6)$$

$$\text{Dye} + OH^{\cdot} \rightarrow CO_2 + H_2O \quad (7)$$

$$\text{Dye} + h^+(VB) \rightarrow \text{Products from Oxidation} \quad (8)$$

$$\text{Dye} + e^-(CB) \rightarrow \text{Product from reduction} \quad (9)$$

Both oxidation and reduction processes commonly take place on the surface of the photoexcited semiconductor photocatalyst (Ajmal et al. 2014). The complete process has been represented in Figure 2.

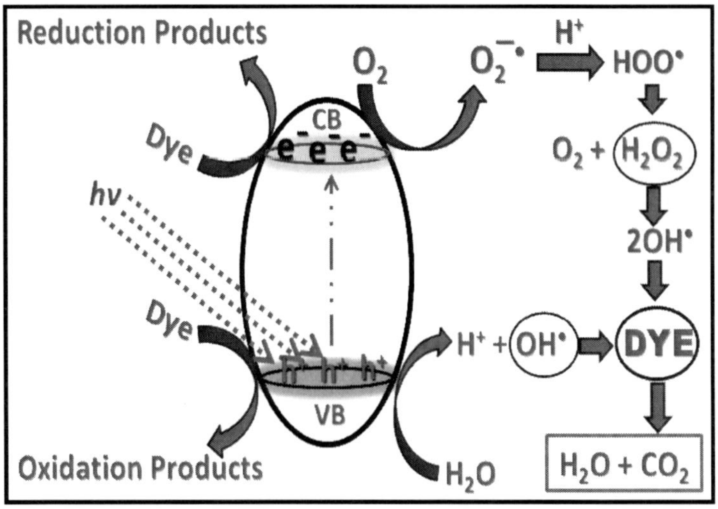

Figure 2. Schematic representation of indirect photocatalysis process.

2.2. Direct Mechanism

The mechanism involves the excitation of dye from ground state to the triplet excited state under visible light photons of wavelength $\lambda > 400$ nm. This excited state of original dye species is converted into a semi-oxidized radial cation (dye^+), because the excited electron of the dye molecule transfers to the conduction band of the nanocatalyst which ionizes the nanocatalyst as anion. The kinetics of the reactions are represented in equation (10) and (11).

$$Dye(GroundState) + hv(VisibleLight) \rightarrow Dye^*(TripletState) \tag{10}$$

$$Dye^* + Nanocatalyst \rightarrow Dye^+(Cation) + Nanocatalyst^-(Anion) \tag{11}$$

These transferred excited electrons of dye molecule dissolves the oxygen in the system and produces the superoxide radical anions ($O_2^{\cdot-}$), which further protonated into hydroperoxyl radical ($HO_2^{\cdot}$) and consequently transforms to H_2O_2, and finally dissociates into hydroxyl radicals (OH•) (Galino et al. 2000). This OH• radical oxidizes the dye molecules as shown in equation (6) to (9) and represented in the schematic diagram (Figure 3). In general, the indirect mechanism is dominant over direct mechanism (Chen et al. 2001). Dye degradation via UV-irradiation is more prominent than the visible light illumination.

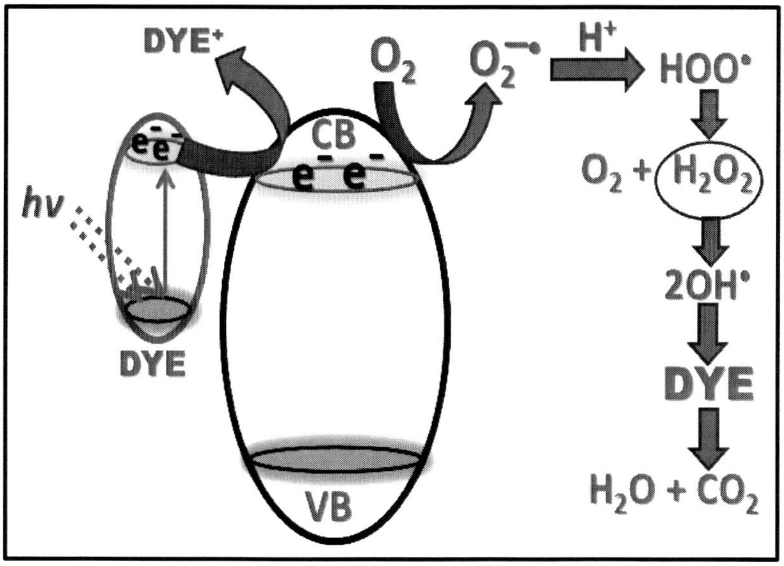

Figure 3. Schematic representation of indirect photocatalysis process.

3. Effective Parameters of Photocatalysis

The process of photocatalysis mainly depends on the particle size and structure of catalyst, the amount of the catalyst loaded, concentration of pollutant (dye), pH value, free oxygen in solution, oxidants, intensity and wavelength of light, reaction temperature, etc (Fujishima et al. 2000). These effective parameters have been discussed below in detail:

3.1. Catalyst: Size, Surface Area, Structure and Morphology

Nanocatalysis is considered as the main bridge of the photocatalytic system. Nanocatalyst having capacious surface territory and smaller size were found more efficient as photocatalyst (Han G et al. 2014). As the size (nano) of the catalyst is low as a result more number of active sites are available to produce higher interfacial charge bearer exchange rates, and thus enhances the catalytic activity (Cernuto et al. 2011). At the same time, due to the insolubility of nanocatalyst in the reaction solvent, it can be easily separated from the reaction mixture. The photocatalytic redox response predominantly happens at thesurface of the nanocatalyst shows that the surface properties altogether impact the productive efficiency of catalyst (Khan et al. 2015b). Thus, nanocatalysts provide an exclusive photocatalytic activity with high selectivity (Ameta et al. 2009).

Structure of nano-catalyst assumes a key job in accomplishing prevalent photocatalytic action. For example, TiO_2 material can be synthesized in three different phases such as anatase, rutile, and brookite. Among these three phases, anatase is observed to be a most delicate and alluring phase because of its soundness stability, the situation of the conduction band, high level of hydroxylation and adsorption control (Gnanasekaran et al. 2015) as compared to others. Morphology (shape) of nanocatalyst plays a potential factor which can also influence the final degradation efficiency of organic compound in the photocatalysis process as reported earlier such as spherical-shaped ZnO samples show higher efficiency to dye degradation by using photocatalyst process as contrasted with the spindle-and rod-shaped ZnO samples because of its expansive surface zone (Saravanan et al. 2013a).

3.2. Amount of Catalyst Loaded

Photocatalysis process is directly influenced by the concentration of nanocatalyst. At low concentration of catalyst, the degradation of an organic molecule is also very low, because a large amount of light can be transmitted throughout the reactor, leaving behind the less number of transmitting radiation for the reaction (Mai et al. 2008). But when an optimum amount of catalyst has been loaded into the solution, the solution becomes opaque and in such situation, no more light radiation can enter into the activate sites of catalyst particles, which subsequently decreases the rate of dye degradation (Pouretedal et al. 2009). The opacity obtained at high concentrations of nanocatalyst is due to the agglomeration of nanoparticles on the surface of the solution, which as a result decreases the active-surface-sites available for the degradation (Sohrabnezhad et al. 2009). These blockage of active sites ultimately, decreases the light penetration depth into the solution and thus decreases the degradation rate. So this is a significant finding, to evaluate the optimum amount of nanocatalyst to be used for optimum degradation of dye in a given solution (Nappolian et al. 2002).

3.3. A Concentration of Pollutants (or Dye)

Photocatalysis is a method used for dye degradation. Dye (or Pollutant) is removed from the waste water by getting adsorbed on the surface of nanocatalyst. In this degradation process the dye is to be added to certain time constraints to get absorbed as well get adsorbed but the whole of the dye molecule does not degrade. If the dye concentration is increased the number of dye molecule in the solution is increased which affect the degradation rate. Hence, for the optimum degradation rate, the concentration of dye should not be increased after a certain limit. Thus an optimum dye concentration is to sort out during the experiment is very necessary for successful results at the end. Also, to find out the degradation rate, pollutant type, and their concentration is the main factor to observe carefully (Rajeshwar et al. 2008; Rehman et al. 2009). Kiriakidou et al., have already reported the degradation of Acid Orange 7 dye by using TiO_2 as nanocatalyst. They perform the experiment using TiO_2 as a catalyst under comparable working conditions by utilizing a various concentration of Acid Orange 7 (25– 600 mg/L) and saw that entire degradation rate was accomplished at (25– 100 mg/L). After a specific time, the degradation rate decreases at higher concentration of dye (200– 600 mg/L). Complete decolorization of the solution occurred in under an hour for generally low concentration values (25– 100 mg/l), while this was not the situation for higher initial dye concentrations (200–600 mg/l) (Kiriakidou et al. 1999). Similar observations were also reported by Reza et al., who found that with increasing the initial concentration of the dye the percentage of the degradation decreases. This all is because of development of $OH^{\cdot}$ radicals on the catalyst surface, which promptly responds with dye particles and thus substantiates the degradation rate. Increasing the initial dye concentration, progressively increases the number of dye atoms, which are available for excitation and energy exchange. This energy exchange prompts the arrangement of a few monolayers of adsorbed dye at nanocatalyst surfaces especially at high concentration of dye. Thus completely covered catalyst surfaces keeps the reaction rates constant.

Also, a significant amount of UV is consumed by the dye atoms as opposed to the nanocatalyst particles can likewise be the explanation for the decline in degradation efficiency with the increase in dye concentration. This unreveals the fact that the infiltration of light at the surface of the catalyst diminishes because of the production of hydroxyl radicals on the active sites which are adsorbed by the dye. This adsorbed dye on the photocatalyst surface hinders the response of adsorbed atoms with the photo-induced positive holes or hydroxyl radicals since there is no immediate contact of the semiconductor with them. As a result, the path length of the photons entering the solution gets reduced by a high concentration of dye. Again, as the initial concentration of the dye increases, the requirement of catalyst surface needed for the degradation also increases. Since the illumination time and the amount of catalyst are consistent, the OH- radical (essential oxidant) shaped on the surface of nanocatalyst is additionally steady. So the overall number of free radicals attacking the dye particles diminishes with expanding measure of the dye (Reza et al. 2017).

3.4. Effect of pH

The degradation rate of dye concentration is also affected by the pH value of the solution because of the charges generated at the surface of nanocatalyst. The surface charge gets protonated or deprotonated with the changing pH value. Thus, pH affects act on the efficiency of degradation of dye can be interpreted in provisions based on the presence of electrostatic charge interactions exists between charged particles and the contaminants. This electrostatic interaction greatly influences the adsorption and subsequently the surface properties of nanocatalyst. Hence, pH value has a great impact on the adsorption and separation of the organic molecule (He et al. 2010) surface charge of nanocatalyst and the oxidation potential of the valence band (Venkatachalam et al. 2007). When nanocatalyst surface is predominantly filled with negative charge, the pH value of solution gets increased beyond the isoelectric point of nanocatalyst used.On increasing the pH value, the number of positive charges on the surface of nanocatalyst is also increasing as the functional groups get protonated. At a point, when the surface of nanocatalyst will be negatively charged, which as a result increases the adsorption cationic molecules at higher pH value and vice-versa in the reverse situation it would adsorb anionic molecules very easily (Rajabi et al. 2013). Whereas, on increasing the pH value (pH >12), the degradation rate is repressed by the presence of the hydroxyl radicals because these hydroxyl ions contend with contaminants for the adsorption on the surface of the catalysts (Mai et al. 2008). On the contrary, at low pH, the adsorption of cation particle on the photocatalyst surface is reduced because of the positively charged surface of the photocatalyst, which results in the decline in the adsorption of cations. Therefore, the degradation efficiency is declined in lower pH or acidic solution. For example, Neppolian, 2002 observed the photocatalytic degradation of TiO_2 material for reactive blue 4 under distinctive pH conditions. In an acidic medium where pH < 5, they observed the low degradation efficiency due to the presence of high concentration of proton. However, when the experiment was conducted in alkaline medium pH > 10, the existence of hydroxyl ions defuses the acidic end products that are produced by the photo-degradation reaction. Furthermore, an unexpected drop of degradation has been detected in the alkaline range (pH 11–13) because of rapidly scavenged hydroxyl radicals (OH) (Neppolian et al. 2002). The impact of pH on the rate of reaction can be interpreted related to the electrostatic associations between charged particles and the contaminants. This impacts the adsorption and the surface properties. On the contrary, at low pH, the adsorption of the cationic organic molecule on the photocatalyst surface is reduced. This is due to the positively charged surface of the photocatalyst which reduces the adsorption of cationic atoms. Hence, the degradation efficiency is declining in lower pH or acidic solution (Sohrabnezhad et al. 2009).

3.5. Free Oxygen in Solution

The degradation rate is also affected by the production of free radical oxygen via ionsorption of oxygen. From O_2 molecule, solution accepts the electron (e-) from the conduction band leads to the formation of stabilizing O_2• radicals, which decreases the rate of e-/h+ recombination pair. O_2• radicals oxidize the dye molecule which readily starts degrading as O_2 and thus O_2• radicals does not get adsorb on the surface of nanocatalyst (Ullah et al. 2012). These free O_2 molecule present in solution stabilizes intermediate radicals, which are chiefly responsible for the mineralization (degradation) of the dye molecule. It also induces the cleavage mechanism of aromatic bonds inorganic which pollute water (Fujishima et al. 2008).

3.6. Effect of Oxidant

Degradation of dye efficiency is greatly enhanced by the addition of oxidant during the reaction.The electron-hole pair recombination of charges in nanocatalyst can be diminished by adding some irreversible electron acceptors such as H_2O_2, $(NH_4)_2S_2O_8$, $KBrO_3$, and $K_2S_2O_8$ to the solution (Sobana et al. 2008). In general, dihydrogen peroxide (H_2O_2) is used to improve the photocatalytic action of the arranged photocatalyst. The mechanism in which H_2O_2 alters the photocatalytic degradation is given as follows in equations (12) to (14):

$$H_2O_2 + O_2^{\cdot -} \rightarrow \cdot OH + OH^- + O_2 \quad (12)$$

$$H_2O_2 + hv \rightarrow 2(\cdot OH) \quad (13)$$

$$H_2O_2 + e^-(CB) \rightarrow \cdot OH + OH \quad (14)$$

Thus, by observing the above chemical reaction numbers from 13 to 15, we can conclude that addition or increase in the concentration of O_2 on the surface enhances the degradation rate of the organic molecules. As the light energy strikes the nanocatalyst surface, photolysis of H_2O_2getsactivated producing the OH radicals and O_2 on the surface of nanocatalyst. The product formed by photolysis helps in the trapping of electrons, and thus preventing the recombination of electron-hole pairs, in consequence, increasing the rate of photocatalyst degradation of dye. But beyond a certain limit, when the concentration of H_2O_2 reaches to its optimum value, quenching of $OH^{\cdot}$ radical starts taking place at the surface nanocatalyst whichacts as a hole scavenger to photo produced holes consequently decreases the degradation rate of the dye (Sobana and Swaminathan. 2007; Neppolian et al. 2003).

3.7. Effect of Intensity and Wavelength of Light

The wavelength and intensity of UV light used as irradiation sources during the photocatalysis also affect the rate of photodegradation of pollutant (dye) by altering the light intensity used at the surface of nanocatalyst and also the irradiation time of light (Herrmann J.M. 2010). Longer exposure to light (hv) causes the byproducts to be accumulated on the active sites of nanocatalyst which subsequently deactivated the photocatalyst, thus the degradation rate is affected by irradiation time also (Ullah et al. 2012). Degradation of Azo dye is done in aqueous solution by using TiO_2 catalyst powder in photocatalytic reactor, where artificial UV light is used as a source of illumination. The artificial UV irradiation used is found to be more efficient for the degradation of dye as compared to natural sunlight (Konstantinou and Albanis, 2004).

At low light intensities (0–20 mW/cm2), the degradation of dye rate would increase linearly with increasing light intensity. As we increase the light intensity say an intermediate light intensities beyond a certain value (approximately 25 mW/cm^2), the degradation rate would depend on the square root of the light intensity (Ollis et al. 1992). The rate of dye degradation becomes independent of light intensities at higher light intensity. This may be due to the number of photons produced per unit time per unit area at the nanocatalyst surface in aqueous medium reaches the optimum photocatalytic power. Also, the number of active sites remains constant with increasing lightintensity, achieves a specific reaction rate even though the light intensity continues to increase

(Reza et al. 2017). In addition, the wavelength of the light can effectively influence the efficiency of the photocatalyst. It is observed that electron-hole generation is promoted more by irradiation of nanocatalyst by a shorter wavelength which consequently enhances the efficiency of the catalyst (Nguyen et al. 2014).

3.8. Effect of Reaction Temperature

The photocatalytic degradation efficiency of organic molecules consistently increases as the temperature of the solution is raised. In general, when the temperature is increased, it forms the bubbles at the surface of the solution resulting in the generation of free radicals. Furthermore, the increase in temperature encourages the degradation reaction to beat the recombination of an electron-hole pair. Other than that, raise in temperature may upgrade the oxidation rate of organic particles at the interface (L. Karimi et al. 2014). On the other hand, the photocatalytic reaction (e.g., TiO_2 material at a temperature over 80 °C) at higher temperature generally improves the pair recombination and desorption procedure of adsorbed reactant species, which causes the decline in photocatalytic activity. The dependency of degradation rate on temperature is reflected by the low activation energy (5–20 kJ mol^{-1}) as compared with ordinary thermal reactions. Due to the presence of photonic activation energy, photocatalysis systems do not require heat and can work in ambient conditions also. The ideal reaction temperature for photocatalytic activity of TiO_2 material is observed to be in the range of 20–80 °C. This ideal optimum range mostly relies upon the activation energy of the material in the photocatalytic reaction (Mozia et al. 2009).

Conclusion

Waste water treatment by photocatalysis using nanocatalyst was found efficient for the degradation of pollutant at a low cost. The mechanism of photocatalysis basically depends on the excitation of photons in nanocatalyst, ionization of the water molecule, ionosorption of oxygen, and protonation of superoxides. The added advantage of this technique is that the obtained end products such as water, carbon dioxide, and inorganic ions are environmentally friendly. Furthermore, it has been found that certain parameters such as, size and shape of nanocatalyst, catalyst loading, pH values, oxidants, light intensity, and temperature influences the photocatalytic results. Optimization of these operating parameters enhances the degradation efficiency, which advocates the potential application of this method. In this way, the photocatalysis technique could be used industrially as a modern technique in an ecological framework to remove pollutants from the environment.

REFERENCES

Ajmal A., Majeed I., Naseem M.R., Idriss H. and Nadeem M.A. 2014 "Principles and mechanisms of photocatalytic dye degradation on TiO_2based photocatalysts: a comparative overview" *Royal Society of Chemistry Advances* 4:37003-37026.

Ameta J., Kumar A., Ameta R., Sharma V. K. and Ameta S. C. 2009 "Synthesis and Characterization of $CeFeO_3$ Photocatalyst used in Photocatalytic Bleaching of Gentian Violet" *Journal of the Iranian Chemical Society*6: 293-299.

Baruah S., Samir K.P. and DuttaJ. 2010 "Nanostructured Zinc Oxide for water treatment" *Nanoscience & Nanotechnology-Asia* 2:90-102.

Cernuto G., Masciocchi N., Cervellino A., Colonna G.M. and Guagliardi A. 2011 "Size and shape dependence of the photocatalytic activity of TiO_2 nanocrystals: a total scattering Debye function study" *Journal of American Chemical Society* 133:3114–3119.

Chen F., Xie Y., Zhao J. and Lu G. 2001 "Photocatalytic degradation of dyes on a magnetically separated photocatalyst under visible and UV irradiation" *Chemosphere* 44:1159–1168.

Chong M.N., Jin B., Chow C. W. K. and Saint C., 2010 "Recent developments in photocatalytic water" *Water Research* 44:2997-3027.

Djurisic A.B., Leung and Ng A.M.C., 2014 "Strategies for improving the efficiency of semiconductor metal oxide photocatalysis" *Materials Horizon* 1:400–410.

Fujishima A. and Honda K. 1972 "Electrochemical photolysis of water at a semiconductor electrode" *Nature* 238:37–38.

Fujishima A., Rao T.N. and Tryk D.A. 2000 "Titanium Dioxide Photocatalysis" *Journal ofPhotochemistry and Photobiology C* 1:1–21.

Fujishima A., Zhang X. and Tryk D.A., 2008 "TiO_2 photocatalysis and related surfacephenomena" *Surface Science Reports* 63:515–582.

Galindo C., Jacques P. and Kalt A. 2001"Photochemical and photocatalytic degradation of an indigoid dye: a case study of acid blue 74 (AB74)"*Journal of Photochemistry and Photobiology A: Chemistry* 141:47-56.

Gnanaprakasam A., Sivakumar V.M., and Thirumarimurugan M. 2015 "Influencing Parameters in the Photocatalytic Degradation of Organic Effluent via Nanometal Oxide Catalyst: A Review" *Indian Journal of Materials Science* 2015:1-16.

Gnanasekaran L., Hemamalini R. andRavichandran K. 2015 "Synthesis and characterization of TiO_2 quantum dots for photocatalytic application" *Journal of Saudi Chemical Society* 19:589–594.

Gupta V.K., Ali I., Saleh T.A., Nayak A. and Agarwal S. 2012 "Chemical treatment technologies for waste-water recycling-an overview" *Royal Society of Chemistry Advances* 2:6380-6388.

Han G., Wang L., Pei C., Shi R., Liu B., Zhao H., Yang H. and Liu S. 2014 "Size-dependent optical properties and enhanced visible light photocatalytic activity of wurtzite CdSe hexagonal nanoflakes with dominant 001 facets" *Journal of Alloys and Compound* 610:62–68.

He H.Y., Huang J. F., Cao L. Y., and Wu J. P.2010 "Photodegradation of methyl orange aqueous on MnWO4 powder under differentlight resources and initial pH" *Desalination* 252:66–70.

Herrmann J. M., 2010 "Photocatalysis fundamentals revisited to avoid several misconceptions" *Applied Catalysis B: Environmental* 99: 461–468.

Karimi L., Zohoori S., and Yazdanshenas M. E.2014 "Photocatalytic degradation of azo dyes in aqueous solutions under UV irradiation using nano-strontium titanate as the nanophotocatalyst," *Journal of Saudi Chemical Society*18:581–588.

Khan M. M., Pradhan D. and Sohn Y. (Eds.) 2017 "Nanocomposites for Visible Light-induced Photocatalysis", Springer Series on *Polymer and Composite Materials* chapter.

Khan M.M., Adil S.F. and Al-Mayouf A. 2015 "Metal oxides as photocatalyst" *Journal of Saudi Chemical Society* 19:462–464.

Kiriakidou F., Kondarides D.I. and Verykios X.E. 1999 "The effect of operational parameters and TiO_2- doping on the photocatalytic degradation of azo-dyes" *Catalysis Today* 54:119–130.

Konstantinou I. K. and Albanis T. A. 2004 "TiO_2-assisted photocatalytic degradation of azo dyes in aqueous solution: kinetic and mechanistic investigations: a review" *Applied Catalysis B* 49:1–14.

Mai F. D., Lu C. S., C. W., Wu Huang C. H., Chen J. Y., and Chen C. C. 2008 "Mechanisms of photocatalytic degradation of Victoria Blue R using nano-TiO_2," *Separation and Purification Technology* 62:423–436.

Mai F. D., Lu C. S., Wu C. W., Huang C. H., Chen J. Y. and Chen C. C. 2008 "Mechanisms of photocatalytic degradation of Victoria Blue R using nano-TiO_2" *Separation and Purification Technology* 62:423–436.

Mansoori G.A., Bastami R.T., Ahmadpour A., and Eshaghi2008 "Environmental application of nanotechnology" *Annual Review of Nano Research* 2:1-73.

Mozia S., Morawski A.W., Toyoda M. and Inagaki M. 2009 "Application of anatase-phase TiO_2 for decomposition of azo dye in a photocatalytic membrane reactor" *Desalination* 241:97–105.

Neppolian B., Choi H. C., Sakthivel S., Arabindoo B., and Murugesan V. 2002 "Solar/UV-induced photocatalytic degradation of three commercial textile dyes," *Journal of Hazardous Materials* 89:303–317.

Neppolian B., Choi H.S., Sakthivel S., Arabindoo B. and Murugesan V. 2002 "Solar light induced and TiO_2 assisted degradation of textile dye reactive blue 4" *Chemosphere* 46:1173–1181.

Neppolian B., Kanel S. R., Choi H. C., Shankar M. V., Arabindoo B. and Murugesan V.2003 "Photocatalytic degradation of reactive yellow 17 dye in aqueous solution in the presence of TiO_2 with cement binder" *International Journal of Photoenergy* 5:45–49.

Nguyen V.H., Shawn D.L., Wu J.C. S. and Bai H. 2014 "Artificial sunlight and ultraviolet light induced photo-epoxidation of propylene over V-Ti/MCM-41 photocatalyst" *Journal of Nanotechnology* 5:566–576.

Ollis D. F., Pelizzetti E, Serpone N.1992 "Photocatalyzed destruction of water contaminants" *Environmental Science and Technology* 25:1522–1529.

Pelaez M., Nolan N.T., Pillai S.C., Seery M.K., Falaras P., Kontos A.G., Dunlop P.S., Hamilton J.W., Byrne J., O'Shea K., Entezari M.H. andDionysiou D.D. 2012 "A review on the visible light active titanium dioxide photocatalysts for environmental applications" *Applied Catalysis B* 125:331–349.

Pouretedal H. R., Norozi A., Keshavarz M. H., and Semnani A. 2009 "Nanoparticles of zinc sulfide doped with manganese, nickel and copper as nanophotocatalyst in the degradation of organic dyes," *Journal of Hazardous Materials*162:674– 681.

Rajabi H. R., Khani O., Shamsipur M. and Vatanpour V. 2013 "Highperformance pure and Fe3+-ion doped ZnS quantum dots as green nanophotocatalysts for the removal of malachite green under UV-light irradiation" *Journal of Hazardous Materials* 250-251:370–378.

Rajeshwar K., Osugi M.E., Chanmanee W., Chenthamarakshan C.R., Zanoni M., Kajitvichyanukul P. andKrishnan-Ayer R. 2008 "Heterogeneous photocatalytic treatment of organic dyes in air and aqueous media" *Journal of Photochemical Photobiology C* 9:171–192.

Rehman S., Ullah R., Butt A.M. and Gohar N.D. 2009 "Strategies of making TiO_2 and ZnO visible light active" *Journal of Hazardous Material* 170:560–569.

Rehman S., Ullah R., Butt A.M. and Gohar N.D. 2009 "Strategies of making TiO_2 and ZnO visible light active" *Journal of Hazardous Material* 170:560–569.

Reza K.M., Kurny A.S. and Gulshan F. 2017 "Parameters affecting the photocatalytic degradation of dyes using TiO_2: a review" *Applied Water Science* 7:1569-1578.

Reza M.K., Kurny A. S.W. and Gulshan F. 2017 "Parameters affecting the photocatalytic degradation of dyes using TiO_2: a review" *Applied Water Science* 7:1569–1578.

Saravanan R., Gupta V.K., Narayanan V., Stephen A. 2013 "Comparative study on photocatalytic activity of ZnO prepared by different methods" *Journal of Molecular Liquids* 181:133–141.

Sobana N. and Swaminathan M., 2007"The effect of operational parameters on the photocatalytic degradation of acid red 18 by ZnO" *Separation and Purification Technology* 56:101–107.

Sobana N., Selvam K. and Swaminathan M. 2008 "Optimization of photocatalytic degradation conditions of Direct Red 23 using nano-Ag doped TiO_2" *Separation and Purification Technology* 62:648–653.

Sohrabnezhad S., Pourahmad A., and Radaee E. 2009 "Photocatalytic degradation of basic blue 9 by CoS nanoparticles supported on AlMCM-41 material as a catalyst," *Journal of Hazardous Materials* 170:184–190.

Sohrabnezhad, A. Pourahmad, and E. Radaee, 2009 "Photocatalytic degradation of basic blue 9 by CoS nanoparticles supported on AlMCM-41 material as a catalyst" *Journal of Hazardous Materials*170:184–190.

Ullah I., Ali S., Hanif M. A. and Shahid S.A., 2012 "Nanoscience for environmental remediation: A Review" *International Journal of Chemical and Biochemical Sciences*, 2:60-77.

Venkatachalam N., Palanichamy M., Arabindoo B., and Murugesan V. 2007 "Enhanced photocatalytic degradation of 4-chlorophenol by Zr4+ dopednanoTiO_2," *Journal of Molecular Catalysis A: Chemical* 266:158–165.

In: Photocatalysis
Editors: P. Singh, M. M. Abdullah, M. Ahmad et al.
ISBN: 978-1-53616-044-4

Chapter 2

PHOTOCATALYSIS: FUNDAMENTAL CLASSIFICATIONS

Nida Qutub***
Department of Chemistry, Jamia Millia Islamia University,
New Delhi, India

ABSTRACT

Photocatalysis is emerging as a promising way out towards challenges associated with the intermittent nature of sunlight which is considered as a renewable and ultimate energy source to power activities on Earth. Photocatalysis is based upon the electronic properties of the catalyst and may involve an oxidation reaction or reduction reaction. A photocatalyst can get excited for a photoreaction by absorbing the light itself directly or with the help of a sensitizer. Nanocomposite photocatalysts can be differentiated on the basis of the band gap of various nanoparticles and their combinations, like sand-witch or core-shell. This chapter's aim is to provide a broad overview of the reactions involved in photocatalysis along with types of nanocomposites photocatalysis.

Keywords: photocatalytic reactions, nanocomposites

---

* Corresponding Author's E-mail: drnidaqutub@gmail.com.

1. INTRODUCTION

Photocatalysis is the segment of catalysis, which includes the collection of the reactions proceeding under the influence of light. It involves phenomena like catalysis of photochemical reactions, photo-activation of catalysts, and photochemical activation of catalytic processes. IUPAC defined the term "*photocatalysis" as follows "Photocatalysis is the catalytic reaction involving light absorption by a catalyst or a substrate*" (Reithmaier.2009). A more detailed definition may be written as "*Photocatalysis is a change in the rate of chemical reactions under the action of light in the presence of the substances (photocatalysts) acting as a catalyst after absorbing light quanta and are involved in the chemical transformations of the reactants, and regenerating their chemical composition after each cycle of such interactions*"(Reithmaier 2009). Usually, the most typical processes that are covered by "photocatalysis" are the photocatalytic decomposition (PCD) and the photocatalytic oxidation (PCO) of substrates, which most often belong to the organic class of compounds. The former takes place in the absence of O_2, while the later process employs the use of gas-phase oxygen as a direct participant to the reaction. In Photocatalysis, semiconductor materials are used as catalysts. The catalyst can be homogenous or heterogeneous to the degrading substance. The photoexcitation of semiconductor catalyst particles generates electron-hole pairs due to the adsorption of UV or visible light that subsequently react with species near or on the surface of the nanomaterial. The chemical reactions involving the photogenerated electrons are photoreduction reactions, while reactions involving photogenerated holes are photooxidations (Conde et al. 2003)(Murray et al. 1993). A large percentage of the initially created charge carriers are quickly trapped by surface trap states (e_t and h_t) (on the timescale of a few hundreds of Fermi seconds to a few tens of Pico seconds). Both trapped and free carriers can participate in reactions with species on or near the surface. The trapped carriers are less energetic than free carriers. Electron or hole transfer across the interface region is a critical step in the overall reaction processes. Trapping and transfer of free electrons are competing processes and often occur on ultrafast timescales. Another competing process is electron-hole recombination. Electron transfer can take place following trapping as well, but on longer timescales, nano-second or longer. Similar events take place for the hole. However, the timescale for hole transfer and trapping can be different from that for the electron (Murray et al. 1993). As shown schematically in Figure 1, the different processes involving photoexcited charge carriers are illustrated. The figure showed the (1) electronic cooling within the Conduction band (CB), (2) trapping of electrons (e_t) by trap states, (3) electron-hole recombination at band-edges, (4) electron-hole recombination of trapped electrons, (5) electron transfer and reduction reaction with an electron acceptor at conduction band and (6) hole transfer and oxidation reaction with a hole acceptor or an electron donor at valence band (VB).

If the semiconductor is excited by absorbing the energy coming from solar radiation, the process is called solar photocatalysis (Pandian et al. 2011)(Kozhevnikova & Vorokh.2010). Semiconductor photocatalysis has received much attention during the last three decades as a promising solution for both energy generation and environmental problems. Semiconductor photocatalysis is becoming a hopeful way to solve the current environmental and energy problems using the abundant solar light (Pandian et al. 2011). It can decompose harmful organic and inorganic pollutants present in air and water and can also split water to produce clean and recyclable hydrogen energy. Up to now, a lot of photocatalysts, such as Ag_3PO_4, Ag_2S, $Bi_{12}TiO_{20}$, TiO_2, WO_3, WS_2, ZnO, Fe_2O_3, V_2O_5, CeO_2, CuS, CdS, and ZnS (Guo et al. 2005)(Zhang et al. 2007)(Li et al. 2010)(Pandian et al. 2011)(Cui et al. 2001)(Patolsky et al. 2006)(Hashimoto et al. 2005) have been prepared and demonstrated to be able to produce hydrogen and decompose pollutants under UV or visible light irradiation (Liu et al. 2006)(Manzoor et al. 2004)(Moore.1965)(Moore.1995).

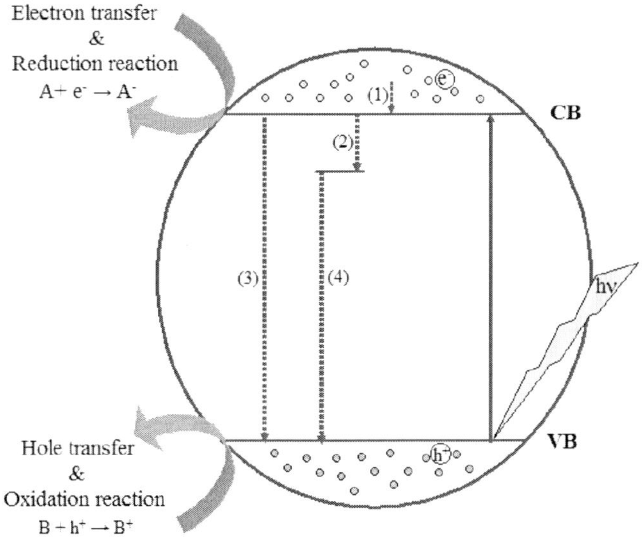

Figure 1. Schematic representation of different relaxation, transfer and reaction pathways of photogenerated charge carriers (electrons and holes) in semiconductor nanoparticles.

2. Types of Photocatalytic Reactions

As mentioned above, in order to show photocatalytic properties, an inorganic solid compound must have electronic properties described by a semiconducting behavior (Sant and Kamat.2002). Basic understanding of the nanostructured semiconductor systems is also important for establishing their practical use in various photochemical processes like photochemical solar cells, electrochromic devices, sensors, photocatalytic hydrogen production and photocatalytic degradation of organic contaminants (Rao et al. 2006)(Zhao et al. 2010)(Shen et al. 2010)(Lin et al. 2002). Semiconductor photocatalysis is initiated by photoexcited electron-hole pairs (e^-/h^+) after bandgap excitation, as an effect of interaction with photons having an energy comparable (equal or higher) to the bandgap. When a photocatalyst is illuminated by appropriate energy the valence band electrons are excited to the conduction band, leaving a positive hole in the valence band (eq.1):

$$\text{Photocatalyst} \xrightarrow{h\nu} e^-_{CB} + h^+_{VB} \quad (1)$$

The excited electron-hole pairs can recombine, releasing the input energy in the form of heat, without any chemical effect. The high efficiency of recombination of the photogenerated electrons and holes create an obstacle to the practical application of the semiconductor nanoparticles in photocatalysis. However, if the electrons (or holes) gets migrated to the surface of the semiconductor prior to recombination, they can participate in various oxidation and reduction reactions (Choi.2010). The participation of semiconductor nanoparticles in a photocatalytic reaction can either be direct or indirect as illustrated in Figure 2 and Figure 3, respectively (Choi.2010)(Rao et al. 2006).

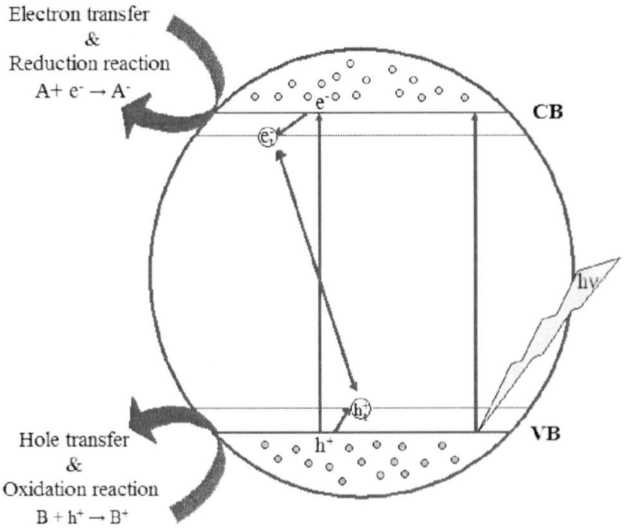

Figure 2. Simplified mechanism of semiconductor photocatalytic process, involving photoinduced charge transfer in semiconductor nanoclusters under bandgap excitation (CB & VB refer to conduction and valence bands and e_t & h_t refer to trapped electrons and holes).

In direct participation, the photogenerated electrons and holes are capable of oxidizing or reducing the adsorbed species such as water, oxygen, and other organic or inorganic species. These oxidation and reduction reactions are the basic mechanisms of photocatalytic water/air purification and photocatalytic hydrogen production, respectively (Figure 2). Alternatively, in the indirect participation, the semiconductor nanoclusters promote a photocatalytic reaction by acting as a mediator for the charge transfer between two adsorbed molecules (Figure 3). This process, which is commonly referred to as photosensitization, is extensively used in various applications like in photo-electrochemistry and imaging science. In the direct participation, the bandgap excitation of a semiconductor particle is followed by charge transfer at the semiconductor/electrolyte interface. However, in the indirect participation, the semiconductor nanoparticle quenches the excited state by accepting an electron and then transfers the charge to another substrate (e.g., adsorbed oxygen) or to a collecting electrode surface to generate photocurrent. The energy of the CB and VB of the semiconductors and the redox potential of the adsorbed molecules control the reaction pathway of the surface photochemical reactions (Rao et al. 2006)(Choi.2010)(Fendler.2008).

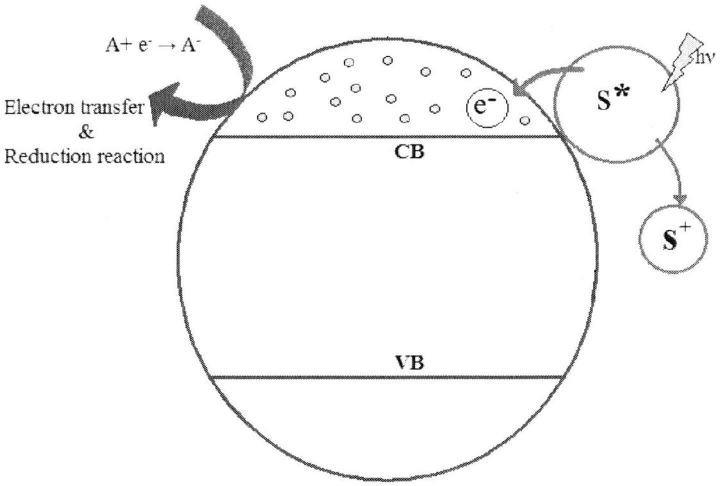

Figure 3. Simplified mechanism of semiconductor photocatalytic process involving photoinduced charge transfer in semiconductor nanoclusters under sensitized charge injection by exciting adsorbed sensitizer (S), (CB & VB refer to the conduction and valence bands).

3. Types of Nanocomposites Photocatalysis

In this connection, the semiconductor nanocomposites, composed of different semiconductors with different electronic level structure, are of considerable interest. With these combinations, it is possible to attain high quantum yields of charge separation. The nanoparticles schematically shown in Figure 4 belong to this class of heterostructures (Khairutdinov.1998). For Sandwich type nanocomposites, like in case of CdS/TiO_2(Sant and Kamat.2002) (see Figure 4a), photoexcitation into the absorption band of CdS results in the electron transfer from CdS to TiO_2, while the hole remains in CdS leading to increased charge separation. These structures are also of interest for molecular electronics as nanoscale rectifier units. The structure analogous to that shown in Figure 4a is also known for CdS/ZnS(Zhanga et al. 2003), CdS/TiO_2(Zhao et al. 2010), PANI-SnO_2–TiO_2(Nur et al. 2007) and AgI/Ag_2S (Khairutdinov 1998). The decrease in recombination and increase in charge separation can also be achieved by building nanocomposites of the core-shell type (see Figure 4 (b) & (c)) as demonstrated by the CdSe/ZnSe(Rawalekar et al. 2012), CdS/CdSe(Tschirner et al. 2012), ZnS/CdSe(Hines and Guyot-Sionnest.1996), CdTe/CdSe(Chuang et al., 2010), CdS/ZnS and ZnS/CdS Nanoparticles (Hassan and Ali.2008). These nanoparticles are obtained by the controlled precipitation of semiconductor molecules of one type (shell) on the pre-synthesized nanoparticles of another type (core). These nanocomposites can, in turn, be coated with a layer of one more semiconductor e.g., CdSe/CdS/ZnS, ZnSe/CdS/ZnSe and CdSe/ZnSe/ZnS Core-Shell-Shell Nanocrystals(Khairutdinov 1998)(Talapin et al., 2004)(Hewa-Kasakarage et al., 2010).

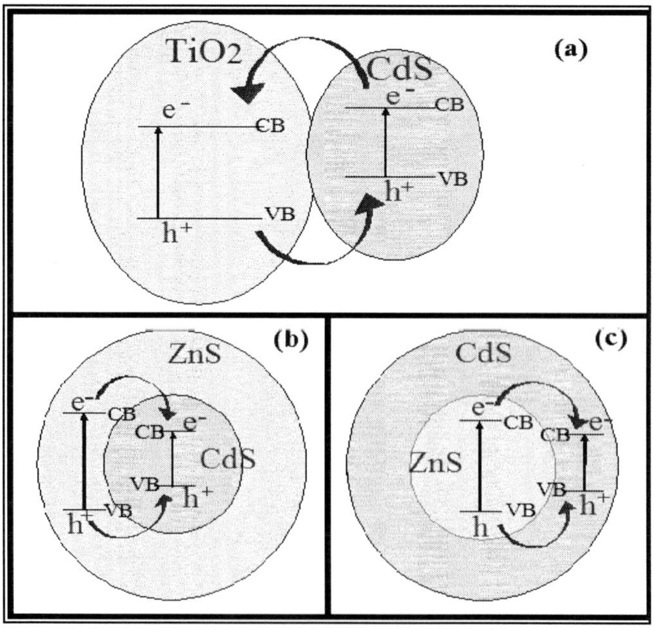

Figure 4. Schematic view of hetero-nanoparticles of different structure. (a) sandwich type hetero-nanoparticles, (b) core-shell type-I and (c) core-shell type-IV CB stands for conduction band; VB stands for valence band.

Conclusion

In the current scenario, photocatalytic wastewater treatment has become a promising technology for degradation of pollutant. Having a long history of active investigation among the researchers, the topic still retains its importance in current research; improvements have been made to enhance the treatment process. The provided evidence proves that semiconductor photocatalysis is emerging as an innovative and green method for achieving chemical transformations of interest in synthesis. Semiconductor photocatalysis technology has been used in wastewater treatment during the last decade and has shown increased efficiency ever since. The classifications covered in this chapter can be applied to all nanoparticles for carrying photocatalysis in general. The promising arrangement of electronic structure, light absorption properties, and charge transport characteristics of most of the nanomaterials has made possible for them to act as photocatalysts.

References

Choi J. 2010 *Development of Visible-Light-Active Photocatalyst for Hydrogen Production and Environmental Application,* Thesis, California Institute of Technology.

Chuang C.H., Lo S. S., Scholes G. D., and Burda C. 2010 "Charge Separation and Recombination in CdTe/CdSe Core/Shell Nanocrystals as a Function of Shell Coverage: Probing the Onset of the Quasi Type-II Regime" *The Journal of Physical Chemistry Letters* 1: 2530–2535.

Conde O., Rolo A. G., Gomes M. J. M., Ricolleau C. and Barber D. J. 2003 "HRTEM and GIXRD Studies of CdS Nanocrystals Embedded in Al2O3 Films Produced by Magnetron RF-Sputtering" *Journal of Crystal Growth*247: 371–380.

Cui Y., Wei Q., Park H. and Lieber C. M. 2001 "Nanowire Nanosensors for Highly Sensitive and Selective Detection of Biological and Chemical Species" *Science* 293:1289–1292.

Fendler J. H. 2008 "*Nanoparticles and Nanostructured Films"* Wiley VCH, Federal Republic of Germany.

Guo Y., Zhang H., Wang Y., Liao Z.-L., Li G.-D. and Chen J.-S. 2005 "Controlled Growth and Photocatalytic Properties of CdS Nanocrystals Implanted in Layered Metal Hydroxide Matrixes" *The Journal of Physical Chemistry B* 109: 21602–21607.

Hashimoto K., Irie H. and Fujishima A. 2005 "TiO_2 Photocatalysis: A Historical Overview and Future Prospects" *Japanese Journal of Applied Physics Part 1, AAPPS Bulletin* 44:12–28.

Hassan M. L. and Ali A. F. 2008 "Synthesis of Nanostructured Cadmium and Zinc

Sulfides in Aqueous Solutions of Hyperbranched Polyethyleneimine" *Journal of Crystal Growth* 310:5252–5258.

Hewa-Kasakarage N. N., El-Khoury P. Z., Tarnovsky A. N., Kirsanova M., Nemitz I., Nemchinov A. and Zamkov M. 2010 "Ultrafast Carrier Dynamics in Type II ZnSe/CdS/ZnSe Nanobarbells" *ACS Nano* 4:1837–1844.

Hines M. A. and Guyot-Sionnest P. 1996 "Synthesis and Characterization of Strongly Luminescing ZnS-Capped CdSe Nanocrystals" *The Journal of Physical Chemistry* 100:468–471.

Khairutdinov R. F. 1998 "Chemistry of Semiconductor Nanoparticles" *Russian Chemical Reviews* 67:109–122.

Kozhevnikova N. S. and Vorokh A. S. 2010 "Preparation of Stable Colloidal Solution of Cadmium Sulfide CdS Using Ethylenediaminetetraacetic Acid" *Russian Journal of General Chemistry* 80:391–394.

Li F., Wu J., Qin Q., Li Z., and Huang X. 2010 "Controllable Synthesis, Optical and Photocatalytic Properties of CuS Nanomaterials With Hierarchical Structures" *Powder Technology* 198:267–274.

Lin C. F., Shih S. M. and Su W. F. 2002 "CdS Nanoparticle Light-Emitting Diode on Si" *Symposium on Integrated Optoelectronic Devices*: 102–110, International Society for Optics and Photonics.

Liu Y., Yan W. and Liu H. 2006 "A Composite Visible-Light Photocatalyst for Hydrogen Production" *Journal of Power Sources* 159: 1300–1304.

Manzoor K., Vadera S. R., Kumar, N. and Kutty T. R. N. 2004" Multicolor Electroluminescent Devices Using Doped ZnS Nanocrystals" *Applied Physics Letters* 84: 284–286.

Moore G. E. 1965 "Cramming More Components Onto Integrated Circuits" *Electronics* 38:114–117.

Moore G. E. 1995 "Lithography and the Future of Moore's law" *SPIE's 1995 Symposium on Microlithography*: 2–17. International Society for Optics and Photonics.

Murray C. B., Norris D. J. and Bawendi M. G. 1993 "Synthesis and Characterization of Nearly Monodisperse CdE (E= sulfur, selenium, tellurium) Semiconductor Nanocrystallites" *Journal of the American Chemical Society* 115:8706–8715.

Nur H., Misnon I. I. and Wei L. K. 2007 "Stannic Oxide-Titanium Dioxide Coupled Semiconductor Photocatalyst Loaded with Polyaniline for Enhanced Photocatalytic Oxidation of 1-Octene" *International Journal of Photoenergy* 1–6.

Pandian S. R. K., Deepak V., Kalishwaralal K. and Gurunathan S. 2011 'Biologically Synthesized Fluorescent CdS NPs Encapsulated by PHB" *Enzyme and Microbial Technology* 48:319–325.

Patolsky F., Zheng G. and Lieber C. M. 2006 "Nanowire-Based Biosensors" *Analytical Chemistry* 78:4260–4269.

Rao C. N. R., Müller A. and Cheetham A. K. 2006 "*The Chemistry of Nanomaterials*

(Vol. 1). Wiley VCH, Federal Republic of Germany.

Rawalekar S., Raj M. V. Ni. and Ghosh H. N. 2012 "Synthesis and Optical Properties of Type I CdSe/ZnSe Core-Shell Quantum Dot" *Science of Advanced Materials* 4:637–642.

Reithmaier J. P. 2009 "*Nanostructured Materials for Advanced Technological Applications*. Springer, Netherlands.

Sant P. A. and Kamat P. V. 2002 "Interparticle Electron Transfer Between Size-Quantized CdS and TiO2 Semiconductor Nanoclusters" *Physical Chemistry Chemical Physics* 4:198–203.

Shen S., Guo L., Chen X., Ren F. and Mao S. S. 2010 "Effect of Ag2S on Solar-Driven Photocatalytic Hydrogen Evolution of Nanostructured CdS" *International Journal of Hydrogen Energy* 35:7110–7115.

Talapin D. V, Mekis I., Götzinger S., Kornowski A., Benson O. and Weller H. 2004" CdSe/CdS/ZnS and CdSe/ZnSe/ZnS Core-Shell-Shell Nanocrystals" *The Journal of Physical Chemistry B* 108:18826–18831.

Tschirner N., Lange H., Schliwa A., Biermann A., Thomsen C., Lambert K. and Hens Z. 2012 "Interfacial Alloying in CdSe/CdS Heteronanocrystals: A Raman Spectroscopy Analysis" *Chemistry of Materials* 24:311–318.

Zhang H., Chen X., Li Z., Kou J., Yu T. and Zou Z. 2007 "Preparation of Sensitized ZnS and its Photocatalytic Activity Under Visible Light Irradiation" *Journal of Physics D: Applied Physics* 40:6846–6849.

Zhang Y., Wang X., Fu D., Zhang H., Gu N., Liu J. and Xu L. 2003 "Hyper-Rayleigh Scattering and Fluorescence of CdS-ZnS Nanoparticle Composites" *Fall 2002-Symposium E: Physics and Technology of Semiconductor Quantum Dots,* Materials Research Society. 737: 9-13.

Zhao W., Bai Z., Ren A., Guo B., and Wu C. 2010 "Sunlight Photocatalytic Activity of CdS Modified TiO_2 Loaded on Activated Carbon Fibers" *Applied Surface Science* 256:3493–3498.

In: Photocatalysis
Editors: P. Singh, M. M. Abdullah, M. Ahmad et al.
ISBN: 978-1-53616-044-4

Chapter 3

ARTIFICIAL PHOTOSYNTHESIS: CLASSICAL APPROACH OF PHOTOCATALYST

Neelu Chouhan[1,*] and Hari Shanker Sharma[2]
[1]Department of Pure and Applied Chemistry, University of Kota, Kota (Rajasthan), India
[2]Department of Chemistry, Government Science College Kota, Kota (Rajasthan), India

ABSTRACT

Clean and cheap energy is the basic driving force to boost all type of development in livelihood. Natural photosynthesis process occurring in plants inspired us to harvest solar power to generate oxygen and hydrogen/hydrogenated compounds by splitting of water. In the light of the current energy scenario and available energy sources, we herewith discussed the artificial photosynthesis process, which is divided into two parts i.e., photocatalytic water splitting for production of clean hydrogen and oxygen fuel and reduction of carbon dioxide using water into C_1 (single carbon containing compounds) or C_n (higher carbon containing compounds) in presence of photocatalyst under sunlight. In this chapter, we brief the overall photocatalytic water splitting phenomenon that discussed along with the hydrogen and oxygen producing photocatalysts. Chapter ends with the future prospective of the artificial photosynthesis process.

Keywords: artificial photosynthesis, energy scenario, water splitting, hydrogen generation, oxygen generation

* Corresponding Author's E-mail: neeluchouhan@uok.ac.in.

1. Introduction

To underline the importance of solar power, Thomas A. Edison (1931) state that I'd put my money on the sun and solar energy. What a source of power! I hope we don't have to wait until oil and coal run out before we tackle that. We think he was right as the energy is the main driving force behind the development of human race and the conventional fuels (oil and coal) are depleting day by day. Therefore, there is a need to search some other powerful energy sources than traditional. The Earth, the Sun, the Galaxy and the Universe have ample energy to power our civilization for the next decades, centuries, and millennia but to harvest that energy, we need to use the proper technology. Beyond the current scientific understanding and knowledge, the Russian astrophysicist N. Kardashev (Kardashev, 1984) categorized our civilization in seven parts on the basis of their energy harness capability. Where, Type-I civilization can use the power available on a single planet (like Earth ~ 174 × 1015 W), Type-II civilization can use the energy of a single star (in our case, Sun ~386 × 1024 W). Similarly, the Type-III civilization can be able to exploit the power of a single galaxy (in our case: Milky Way ~5 × 1036 W), Type-IV civilization can be able to harvest the energy of a galactic supercluster (~1046 W) and Type-V civilization can be able to make use of the energy of the entire Universe (~1056 W). Frank J. Tipler's "Omega point (as per spiritual belief and a scientific speculation, it is an end point of divine unification.)" (Castillo, 2012) would occupy this level. Some science fiction writers talked about a Type-VI civilization that can tap the energy of multiple Universes (technically infinite power) and a Type-VII civilization is a status of a deity (able to create Universes at will, using them as an energy source). According to Kardashev, currently our civilization acquired a Type-0 stage that might be reached at Type-I stage by twenty-second century. Type-I stage can be achieved by use of a variety of alternative resources to conventional that can create a diversified energy matrix based on a mixture of energy alternatives. But our energy is mostly powered by fossil fuels, which are depleting and generate a huge amount of pollutant that is not good for the environmental health. Therefore, a considerable amount of attention from the researchers was attracted to develop the energy without harming environment by using the more sustainable and recyclable alternatives (Kardashev, 1984). Some of the most popular energy alternatives to conventional fuel are solar-, geothermal-, wind-, tidal- or hydroelectric-power plants but they had the major disadvantage of them is that they provide the energy in the form of electricity. The main difficulty associated with the electricity generating via renewable power is that the output electricity either needs to be immediately used or stored but their storage and transport is not easy. But the copious amount (17 Terawatts) of the solar radiation reaching to the earth's surface every day and if we can convert and store a few percent of this enormous- and no-cost- energy reserve, we need not to struggle for satisfying our energy needs. There are two major approaches to tap solar energy: first is the solar to chemical conversion and second is the solar energy to 'electricity' conversion. As the electricity can't be easily stored and transported. Therefore, first option is more viable because if the energy saved in form of chemical bonds (high energy fuels such as molecular hydrogen, obtained from water), it can easily release energy on demand. By utilizing the solar energy, hydrogen can be generated by splitting of the water. Hydrogen is a good choice, as it is a highest energy efficient material and can be easily transported and stored. Natural photosynthesis is the perpetual machine of nature, which is working fine from years of providing food and energy for the good health of the planet and its inhabitants. Natural photosynthesis (Figure 1a) (Lewis, 2014) is a combination of the many useful but complex reactions (light absorption, energy transfer, electron transfer, redox, catalysis) that occurred under the sunlight including the oxidation of water in to molecular oxygen and reducing equivalents, which appear as Nicotinamide adenine dinucleotide phosphate (NADPH). In photosystem -I, the reducing equivalents in NADPH are used to fix CO_2 in the various form of sugars (methane, methanol, formaldehyde, formate, carbon monoxide, oxalate, reduced the proton to hydrogen) as given in Equation 1 (Kalyanasundaram and Graetzel, 2010). This process involves a sequence of many light-induced multielectron-transfer reactions take place in chlorophylls with the help of different enzymes.

$$6H_2O + 6CO_2 + 48h\nu \rightarrow C_6H_{12}O_6 + 6O_2 \quad (1)$$

Researchers developed the artificial photosynthesis (Figure 1b) (Lewis, 2014) process to harness the energy of the sun to drive high-energy small-molecule reactions such as breaking of water (Equation 2) and reduction of carbon dioxide (Equations 3-8 and Figure1c) (Kalyanasundaram and Graetzel, 2010) by mimicking the natural photosynthesis processes but by using simpler chemicals. Further details are discussed in section 4 of the chapter.

$$2H_2O + 4h\nu \rightarrow 2H_2 + O_2 \ E = -1.23V \quad (2)$$

$$2H_2O + 2CO_2 + 4h\nu \rightarrow 2HCOOH + O_2 \ E = -1.43 \ V \quad (3)$$

$$H_2O + 2CO_2 + 4h\nu \rightarrow 2CO + O_2 + H_2O \ E = -1.35 \ V \quad (4)$$

$$2H_2O + 2CO_2 + 4h\nu \rightarrow (HCOOH)_2 + O_2 \ E = -1.31 \ V \quad (5)$$

$$2H_2O + 2CO_2 + 4h\nu \rightarrow 2HCHO + 2O_2 \ E = -1.30 \ V \quad (6)$$

$$4H_2O + 2CO_2 + 4h\nu \rightarrow 2CH_3OH + 3O_2 \ E = -1.20 \ V \quad (7)$$

$$4H_2O + 2CO_2 + 4h\nu \rightarrow 2CH_4 + 4O_2\ E = -1.06\ V \quad (8)$$

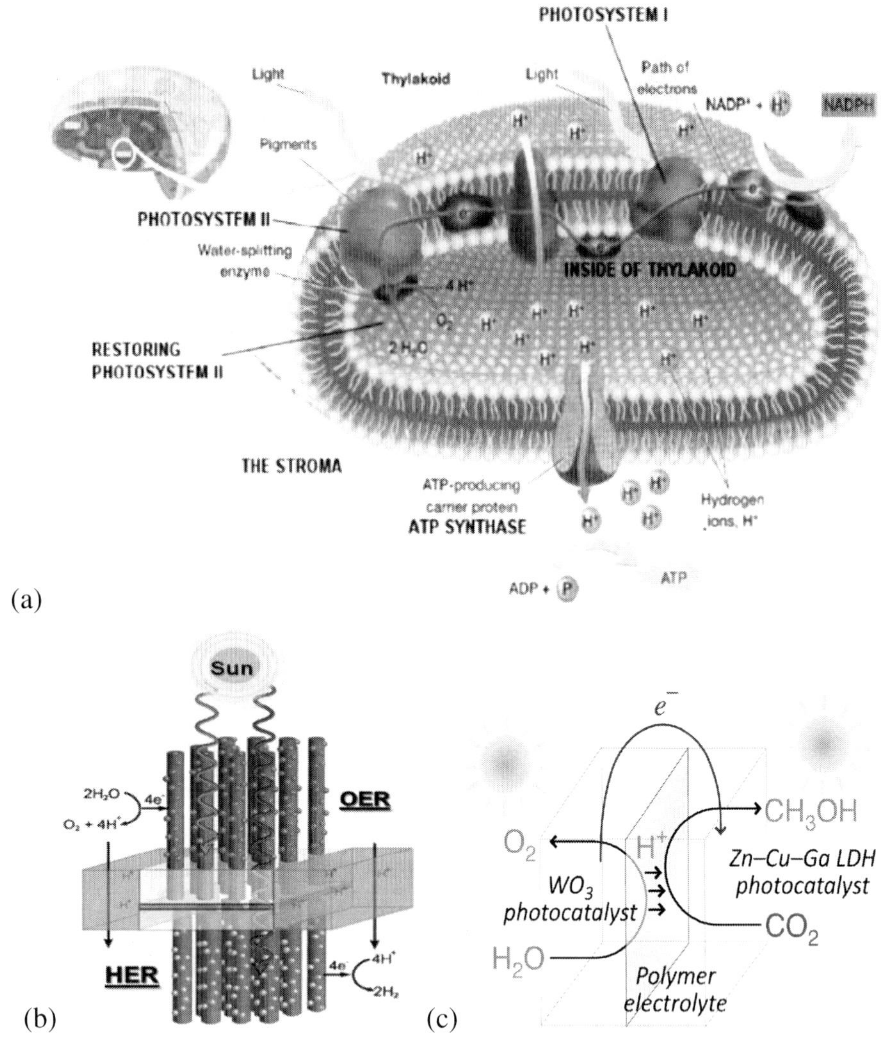

Figure 1. Represents the (a) Natural photosynthesis process (Lewis, 2014) and the artificial photosynthesis process, which involves two steps i.e., (b) Water splitting and (Lewis, 2014) (c) Reduction of carbon dioxide and water (Morikawa et al. 2014). Figure 1a and 1b are taken with permission as they are from open access but Figure 1c (Morikawa et al. 2014) – reproduced by permission of *The Royal Society of Chemistry*.

Artificial photosynthesis (Figure 1b) may either deliver chemical energy or electrical energy during the photocatalytic water splitting via H_2 and O_2 generation or photoelectrochemical solar cells for the direct conversion of sunlight to electricity, respectively. Decomposition of water to its constituents in gaseous form as mentioned by reaction (Equation 2), is a holy grail of chemistry because i) is a uphill thermodynamic reaction; ii) strong oxidants and reductants are required; and iii) involved multi electron–transfer processes to yield molecular gases products i.e., H_2 and O_2 under the sunlight exposure to decompose the water (iv) separation of the both gaseous products (v) apt photocatalytic material with required band gaps. Beside above mentioned problems, the searching of the appropriate photocatalytic materials of high solar efficiency to chemical/electrical energy conversion efficiency and easy H_2/O_2 storage, are needed to encash the energy of Sun. In looking to the above mentioned materials following aspects are need to be addressed: i) suitable molecular architectures that can mimic the antennal chlorophyll function of photosynthesis; ii) suitable molecular redox catalyst with suitable band gap and band position that permit formation of hydrogen (H_2) and oxygen (O_2) in the presence of suitable oxidants and reductants; iii) suitable scavengers for breaking down the water in to H_2 and O_2 by using sunlight; iv) or direct conversion of the sunlight in to electricity; and v) reduction of aqueous CO_2 in to various C_1 compounds that can be used as fuels or raw materials for industries and research purpose.

(a)

(b)

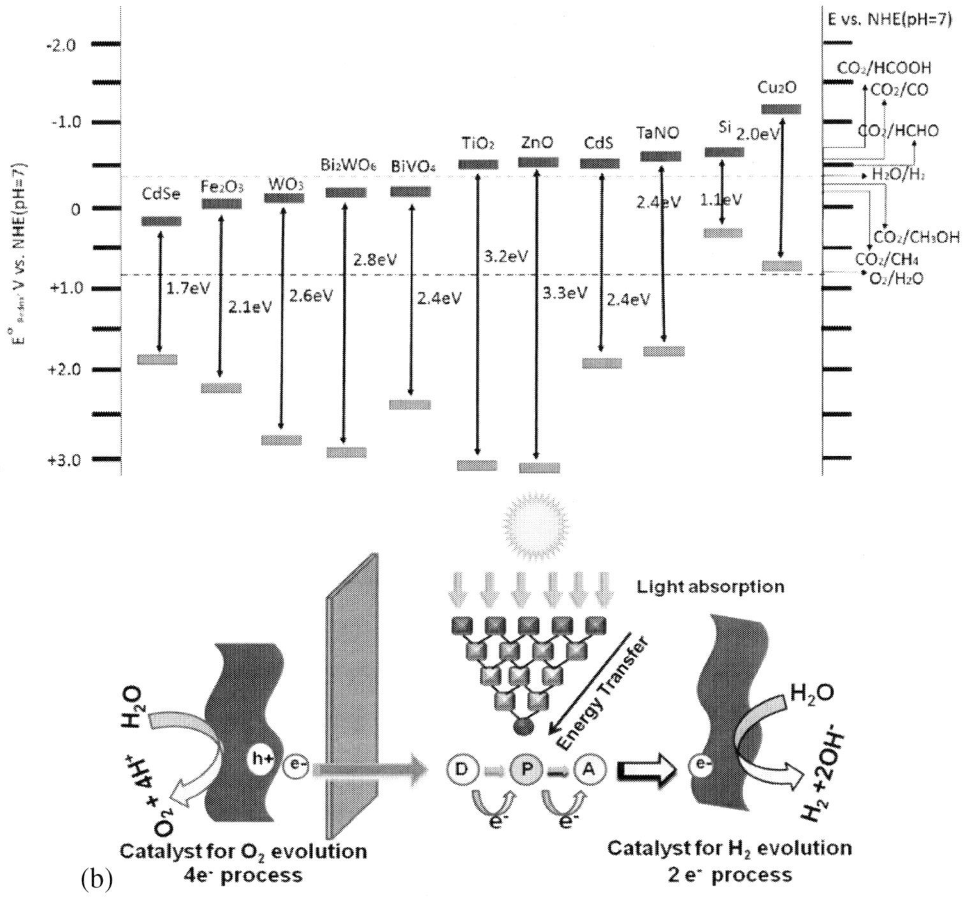

Figure 2. (a) Schematic representation of the various semiconductor photocatalysts with their conduction band, valence band positions, and band gap energies with respect to their relative redox potentials for CO_2 reduction at pH = 7 and water splitting; (b) Elements of the photocatalytic material used in water splitting for O_2 and H_2 generation (Nahar et al. 2017). Figure 2a and 2b reproduced by permission of *The Royal Society of Chemistry*.

Artificial photosynthesis includes four main steps that to mimic the natural photosynthesis process:

1. *Light harvesting*: Trapping light energy particles (photons) by concentrating their energy at the reaction centre (RC) through antenna part of molecular assembly, as represented by Figure 2b.
2. *Separation of charges*: As the trapped sunlight energy of the reaction centre, was used in separation of the electrical charge carriers: positive 'holes' and negative photoelectrons.
3. *Splitting of water*: As generated positive charges (holes) are used to oxidise water into hydrogen ions (protons) and oxygen and photoelectron are used to reduce water into hydrogen gas.
4. *Fuel production*: The hydrogen ions produced in step 3 was further reacted with photoelectrons to reduce water it into the hydrogen gas. And these photoelectrons are also used in reduction of CO_2 and water into the carbon-based fuels, as mentioned in the Equations 3-8.

Figure 2a showed the various photocatalytic materials with the comparative band gaps along with their conduction band and valance band positions and relative redox potentials (Nahar et al. 2017) at pH = 7 that are used in the artificial photosynthesis (Balzani et al. 1997) through water splitting and reduction of carbon dioxide in water for the production of water, methanol, methane, formaldehyde, carbon monoxide, formic acid, etc. Artificial water splitting for the above product generation is shown by Figure 2b.

2. RENEWABLE/SUSTAINABLE ENERGY PRODUCTION

Considering the day by day diminishing conventional fuels and the associated environmental pollution and their signature on our planet in the form of climate change, all needs considerable attention directed towards the finding of alternatives without dark face of pollution but in a more sustainable and recyclable ways to store and distribute energy. Some of the notable existing alternatives are shown in Figure 3: solar panels, geothermal, wind-, tidal-, hydroelectric-power plants are widely used but have the disadvantage of providing the converted energy in the form of electricity or heat, which is difficult to store and transport (Chouhan et al. 2017). Therefore, if the energy stored in a chemical form as an energy carriers, same as the fossil fuels do, will be highly advantageous. The United Nations General Assembly underlined the importance of the renewable/ sustainable energy production in current scenario by declaring the year 2012 as an International Year of Sustainable Energy. Affordable energy services with zero carbon emission are the key ingredient for the millennium development goals (UN, 2012).

Many renewable are being putting forwarded as a potential substitute for the clean energy and climate change problems such as greenhouse effect, hole in ozone layer, drought, floods, land sliding, typhoons, farming/residential land acquiring by sea/river, enhanced pollution level in cities, El Niño, La Niña, frequent earthquakes, hurricanes, and pandemic, shifting in the weather time cycle and many more adverse effects. These problems are proportionally augmented with the growth of human population, which is about to reach 10 billion with the rate and degree of modernization because both guzzle almost 500 EJ per year (≈20 TW) energy, which is mostly acquired by the burning of the carbon-based photosynthesis fuels (such as oil, coal and natural gas) (Rogner, 2004; US EIA, 2011). The second factor contributes to the prolonged adverse environmental consequences through the cost decreasing new resource discoveries and developments such as those involved in shale oil and gas extraction as well as 'fracking'. Yet, no new technology has been developed that might had long-term potential for the radical transformation of our planet towards sustainability of the artificial photosynthesis. An 'off-grid' zero-carbon energy solution (alone or together with other technologies) for all of our constructional building materials or indiscreet practical devices or structures (i.e., buildings, roads, vehicles) adds volume to the clean environment. Therefore, we advocate for the common policy to be developed for global artificial photosynthesis deployment.

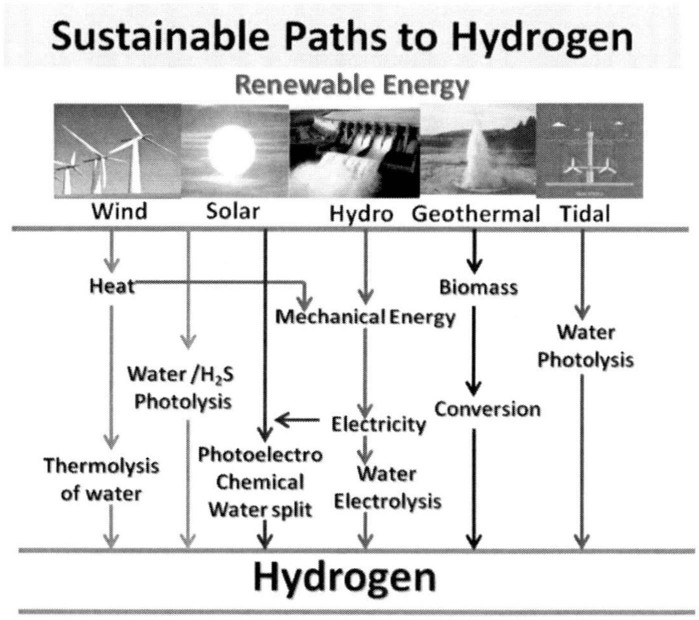

Figure 3. Various renewable energy sources using different paths to hydrogen generation. (Permission not required because it is an original figure.)

Ethanol is a particularly important fuel and fuel additive. Of course, it comes from many sources: waste, cellulose, corn, sugarcane, palm oil, sweet sorghum, saw grass, and so on, so agricultural polices throughout the world were adjusted to encourage this renewable supply. Genetic research into new, higher-alcohol-producing varieties was encouraged. Engine designs were altered to accept fuel blends in which ethanol (and other alcohols) added to fuel that represented a higher fuel efficiency. Brazil is the world's first sustainable sugarcane-based ethanol biofuels economy and became a major exporter of the ethanol fuel by exporting its half of the production to Japan from 2010. The parade of ethanol exporters growing day by day, namely are: Argentina, Australia, Central and South American countries (such as El Salvador), Malaysia, Mexico, South Africa, and Poland. India also inspired to establish the programs to encourage the ethanol production from 2004 (Posada and Cristiano, 2015).

The EU (Germany and France), with its huge agricultural production of sugar and grain, converted a major portion of its surplus food into fuels led in the production of biofuels. To boost the European biofuels industries, EU introduced the protective tariffs on imported ethanol. The U.S. and other countries carried "protectionism" and created ethanol reserves. Anti-genetic modification moves in Europe were deeply ingrained and continued the production of the crops that needed for these embryonic industries. The European countries opposing genetic modification included Austria, France, Portugal, Greece, Denmark, and Luxembourg, which lowers the crop production. Moreover, the emphasis on ethanol, creates the imbalanced in food supply that increased the hunger in world. There were brave experimental moves that attempted to use marginal lands and salty water for the production of the alcohol crops, but these added only marginal lead. Currently, it seems that the world could not have both adequate food and expanded production of alcohol grains.

The environmental backlash had been gathering momentum for years—from both sides i.e., nature and environmentalists. From the 1970s onward, forecasts of climate change and its impacts have proved to understand the ground level situation. It was observed that within last 10 years, the major areas of Tundra have been melted and released a huge amount of methane, (a gas 22 times more dangerous for the climate than CO_2). Furthermore, Nature's backlash was felt most directly in the form of increasing numbers in droughts, flooding, hurricanes, tornadoes, new diseases, fires, sandstorms, falling crop yields, and social unrest among millions of environmental refugees from floods, dying rivers and lakes. During the past 10 years East Africa experienced massive famine, which killed almost 20 million. Many fishing industries around the world are gone. The water tables have fallen dramatically in India and China over the last 20 years that leaves well dry for hundreds of miles in many locations, forcing millions to flee over to the already overcrowded cities. Furthermore, the Kashiwazaki-Kariwa Nuclear Power Plant accident of 2007 polluted the Indian ocean with radioactive waste and galvanized the brewing environmental movements with a new dynamic force all around the world

(The World Nuclear Industry Status Report, 2007). As a result pro-environment politicians of the G8 countries hammered out an agreement to create and implement the Global-Local Energy-Environment Marshall Plan (GLEEM Plan) with an Apollo-like mandate to fix the energy situation and to reduce the climate change impact on earth. Beside all discussed renewable energy sources in Table 1, hydrogen gas has good capacity of being used as fuel due to its fuel efficiency (75%) and high energy density (143 KJ/Kg) (Chouhan et al. 2017). Moreover, almost 95% of the visible materials on earth contain hydrogen. Out of which, water is the richest source of hydrogen as almost 75% of plant is water. If we use renewable energy sources (depicted in Figure 3) for generation of the hydrogen from water, than we will able to solve our energy problem at some extent. The combination of the apt photocatalyst (smart semiconducting material) and solar energy can do miracles in this case.

Table 1. Year wise evolution of the World Energy Mix and their percentage and TOE consumption

Sl. No.	Consumption	2005 consumption (mill TOE)	2020 consumption (mill TOE)	Gain/Loss (mill TOE)	Percentage Change
1.	Total (sum of the components)	11409	15544	4135	36.2
2.	Oil	3678	4300	622	16.9
3.	Natural gas	2420	3600	1180	48.8
4.	Coal	2778	3193	415	14.9
5.	Biomass and waste	793	1400	607	76.5
6.	Nuclear Fission	624	790	166	26.6
7.	Hydro	634	750	116	18.3
8.	Biomass methanol/ ethanol	370	388	18	4.8
9.	Tar sands and shales	88	350	282	287.7
10.	Processed coal	---	500	-	1
11.	Solar	11	100	89	809.1
12.	Wind	8.5	100	92	1076.5
13.	Nuclear Fusion	0	0	0	-
14.	Methane gas hydrates	0	22	22	-
15.	Geothermal	4.8	50	45	941.7
16.	Tides	0.1	1	1	900.0

Source: The Millennium Project based on 2006 energy survey.

3. Photocatalytst for Over All Water Splitting

Day by day increasing demand to satisfy the energy needs of our society along with the need of keeping our environment intact from pollution, increases the demand of an environment-amicable solution of this problem. Surrogating the energy of fossil fuels and delivering the clean and renewable energy, might be a good solution to the current energy crisis. The nature gives us a great clue in the form of photo-synthesis process. Where, the nature working in its own way to store/supply energy (carbohydrates and oxygen) by utilizing abundant free CO_2 and water in presence of sunlight in plant and various microorganisms through effective rearrangement of electrons (Barber, 2008). In this nature's energy harvesting process, the plant take advantage of their antenna system to harness light and to tunnel the excitation energy to the active reaction centers, where the charge segregation process occurs. So, we can say the photosynthesis process is the signboard that indicates the way to convert the solar energy in to the chemical energy via photo induced charge separation (Fukuzumi et al., 2008; Alstrum-Acevedo et al. 2005). To imitate the photosynthesis reactions of plant, artificial photosynthesis is proposed, where the photocatalyst replicates the role of plants to oxidize water and reduce protons or other organic compounds to produce the useful chemical fuels, under the sun light using following reactions (Equations 9-11):

Anode reaction: $2H_2O \rightarrow O_2 + 4H^+ + 4e^-$ ($\Delta E_0 = +0.82$ V; *Vs* NHE at pH = 7) (9)

Cathode reaction: $4H^+ + 2e^- \rightarrow H_2$ ($\Delta E_0 = -0.41$ V; *Vs* NHE at pH = 7) (10)

Overall reaction: $2H_2O \rightarrow O_2 + 2H_2$ ($\Delta E_0 = 1.23$ V; $\Delta G_0 = -4.92$ eV or +237.2 kJ mol) (11)

Artificial photosynthesis is divided into two parts: breaking of water to generate O_2 and H_2 and reduction of CO_2 and water for hydrocarbon and O_2 generation, under the sunlight. That can be proceed in following three ways: First, includes the man made artificial systems (photocatalyst with acceptors and donors) that can be designed to capture light through light harvesting antenna for breaking of water to generate O_2 and H_2 by capturing sunlight (Figure 4a) (Fukuzumi, 2008; Alstrum-Acevedo 2005). Second, leads to develop photoelectrochemical cells for the fabrication of the solar fuel cell similar to the dye sensitized solar cells (DSSC) for driving fuel energy by water cleavage (Figure 4b) (Tachibana et al., 2012). Third, Photocatalytic or Photoelectrochemical reduction of CO_2 and water for hydrocarbon and O_2 generation, under the sunlight (Figure 4c) by pursuing the following reaction path of Equations 12-18.

$2H_2O \rightarrow O_2 + 4H^+ + 4e^-$ ($\Delta E_0 = +0.82$ V; *Vs* NHE at pH = 7) (12)

$CO_2 + 2H^+ + 2e^- \rightarrow HCOOH + O_2$ ($\Delta E_0 = -0.61$V; *Vs* NHE at pH = 7) (13)

$CO_2 + 2H^+ + 2e^- \rightarrow CO + H_2O$ ($\Delta E_0 = -0.53$ V; *Vs* NHE at pH = 7) (14)

$$2CO_2 + 4H^+ + 4e^- \rightarrow (HCOOH)_2 + O_2\ (\Delta E_0 = -\ 0.49\ V;\ Vs\ NHE\ at\ pH = 7) \quad (15)$$

$$CO_2 + 2H^+ + 4e^- \rightarrow HCHO + O_2\ (\Delta E_0 = -\ 0.48V;\ Vs\ NHE\ at\ pH = 7) \quad (16)$$

$$CO_2 + 2H^+ + 6e^- \rightarrow CH_3OH + O_2\ (\Delta E_0 = -\ 0.38V;\ Vs\ NHE\ at\ pH = 7) \quad (17)$$

$$CO_2 + 2H^+ + 8e^- \rightarrow CH_4 + O_2\ (\Delta E_0 = -\ 0.24V;\ Vs\ NHE\ at\ pH = 7) \quad (18)$$

Figures 4a, 4b and 4c, describes the elements of the molecular assembly used for the artificial photosynthesis along with their respective electron transfer reactions. After absorption of sunlight the reaction center: chromophore (C) became sensitized (C*) as a result of the energy-transfer reactions by using antenna of multilayer array. During the electron-transfer or quenching process, the photocatalytic assembly made of the array of donor-chromophore-acceptor (D-C-A) either get oxidized: $D\text{-}C^*\text{-}A \rightarrow D\text{-}C^+\text{-}A^-$ or reduced: $D\text{-}C^*\text{-}A \rightarrow D^+\text{-}C^- \text{-}A$, which so ever reaction favors the relation $\Delta G° < 0$. Redox electron transfer, result in the products: $D\text{-}C^+\text{-}A^- \rightarrow D^+\text{-}C\text{-}A^-$ or $D^+\text{-}C^{--}\text{-}A \rightarrow D^+\text{-}C\text{-}A^-$, are driven by relation $\Delta G° < 0$. Redox couple D^+ and A^- is equivalent to the p/n junction of Si semiconductor, which takes apart the photochemically produced photoelectrons and photoholes pairs (Carabe and Gandia, 2004; Tobias et al., 2003; Schropp, 2004; Shah et al. 2003; Zeman, 2002; Torchynska and Polupan, 2002). For the photocatalytic water splitting process, the electron-transfer activation of the photocatalysts proceed by the electron transfer from A^- to a reduction catalyst cat^{red} for reduction of water, and to D^+ from a second catalyst cat^{ox}, for oxidation of water (Figure 4a). The potentials of the $D^{+/0}$ and $A^{0/-}$ couples dictate E° for the individual fuel-forming half reactions and the overall free-energy change, ΔG°, which can be exceed from the free-energy content of the excited state above the ground state, $\Delta G_{ET}°$, with $\Delta G°\ (eV)) = -F[E°(D^{+/0})\text{-}E°(A^{0/-})] \leq \Delta G_{ET}°\ (eV)$. Where, F is the Faraday constant (96,485 C/mol of electrons or 1 eV/V in SI units). Photoelectrochemical water splitting electron transfer reactions are exhibited by the Figure 4b. Multiple electron transfer and repetition of the light absorption-electron-transfer sequence demonstrated by the Figure 4c, which exhibits the required number of reductive or oxidative equivalents at cat^{ox} and cat^{red} sites to carry out the respective half-reactions. These reactions are reduction of CO_2 to HCOOH and oxidation of water to O_2.The reaction of the activated catalysts, with H_2O, H^+, CO_2, etc., gives the final energy conversion products (H_2, O_2, HCCOH, CH_3OH, CO, HCHO, $(COOH)_2$, CH_4, etc.) and returned to the cat^{ox} and cat^{red} states of the catalysts that ends the catalytic cycle. Anthropogenic CO_2 being produce in atmosphere by combustion of the fossil fuels. Therefore, CO_2 present in atmosphere as an important non-toxic, highly abundant and cheap carbon feedstock and should not be viewed as a waste material. In particular, the transformations of the carbon dioxide into methanol or formaldehyde or formic acid are important because these chemicals could be used as liquid fuels in Flexifuels or can be used as chemical for industrial purpose.

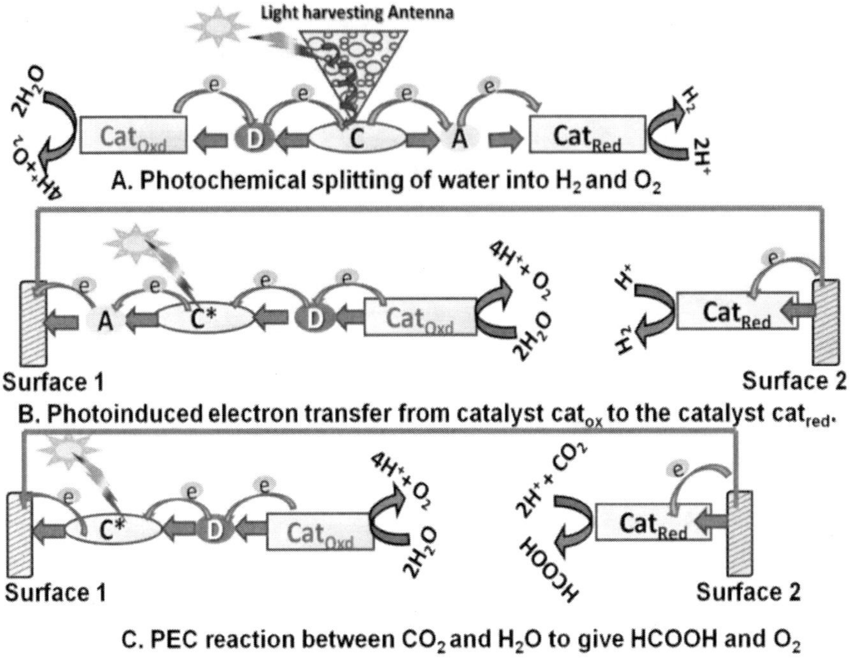

Figure 4. Artificial photosynthesis: (a) photocatalytic water splitting for H_2 and O_2 generation (b) photoelectrochemical water splitting for H_2 and O_2 generation and (c) photocatalytic reduction of CO_2 and water for hydrocarbon and O_2 generation. Figures 4a, 4b and 4c reprinted with permission from (Alstrum-Acevedo et al. 2005). Copyright (2005) *American Chemical Society.*

The decomposition of the pure water is usually found to be rather difficult than water in electrolytes because the simultaneous reduction and oxidation of water is a complex and multistep reaction process that involved at least four electrons. By using the electron donor-typed sacrificial molecules (glucose ($C_6H_{12}O_6$), methanol (CH_3OH), ethanol (CH_3CH_2OH), ethylene diamine tetra acetic acid (EDTA), cyanide (CN^-), lactic acid ($CH_3CHOHCOOH$), hydrocarbons, NaI, and formaldehyde (HCHO)) (Bamwenda et al. 1995; Lee et al. 2001; Lee et al. 2003; Gurunathan et al. 1997, Nada et al. 2005) the H_2 production rate remarkably improved, where the holes are being scavenged by these molecules as the result in suppression of the charge carrier recombination process. As O_2 is not produced than the back reaction of water production is suppressed that leads to increase the H_2 yield, subsequently and avoid the gas separation step. However, it should be noted here that the yield of the H_2 formation will eventually be reduced due to the recombination of holes and electrons (Ni et al. 2007) and competing reduction reactions by the oxidation of the sacrificial reagents and their respective products formation. The typical mechanism of the water splitting process that proceeds in presence of the hole-scavenger is as shown by the Equations 19 to 21 and Figure 5a.

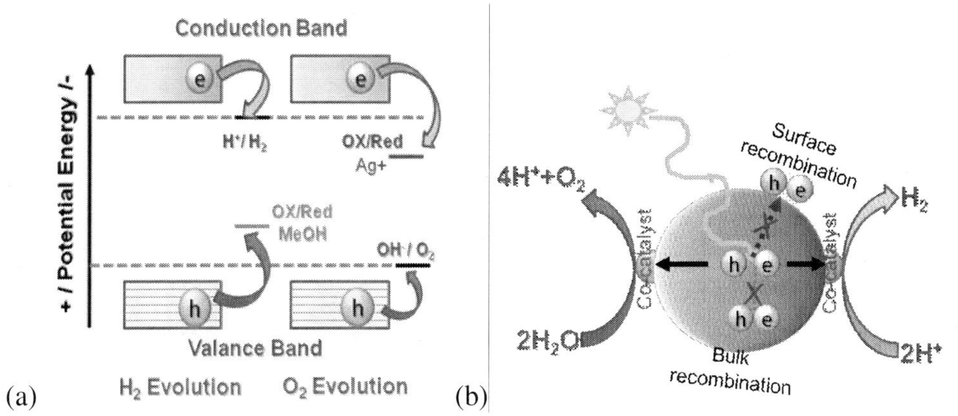

Figure 5. (a) Schematics of the H_2 and O_2 evolution reactions of water splitting in presence of the hole and electron sacrificial reagents and (b) charge separation at the active sites for hole and electron accumulation for water reduction and oxidation, which can generate hydrogen and oxygen, respectively. (Permissions not required because both are an original figures.)

$$\text{Photocatalyst} + h\nu \rightarrow e^-(CB) + h^+(VB) \tag{19}$$

$$2H_2O + 2e^-(CB) \rightarrow H_2 + 2OH^- \tag{20}$$

$$2I^- + 2h^+(VB) \rightarrow I_2 \tag{21}$$

Similarly, the electron scavengers ($AgNO_3$, etc.) used to produce oxygen by utilizing electrons for reduction of other species and sacrificial electrolytes itself as shown by the Equations 22 to 25 and Figure 5a. However, no O_2 is formed with electron acceptors such as carbon tetrachloride or tetranitromethane are used even though their irreversible one-electron reduction is readily observed.

$$AgNO_3 + e^- \rightarrow Ag + NO_3^- \ (CB) \tag{22}$$

$$2H_2O + 4h^+ \rightarrow O_2 + 4H^+ \ (VB) \tag{23}$$

$$H^+ + NO_3^- \rightarrow HNO_3 \tag{24}$$

$$CCl_4 + 2H_2O + 4h^+ \rightarrow CCl_4 + O_2 + 4H^+ \ (VB) \tag{25}$$

While making the choice of sacrificial electron donors is comparatively easy due to the large number of the donor scavengers for the photocatalytic formation of molecular H_2 as a result of water splitting but for the sacrificial photocatalytic oxidation of water the choices are rather limited for assortment of the electron acceptors (Schneider and Bahnemann, 2013).

After sacrificial reagents one has to be select proper cocatalyst to suppress the recombination of the photo electrons and photoholes as exhibited by the Figure 5b. Few of the well-known cocatalysts are: Pt, Rh, Au, NiO, Ru_2O, etc, for hydrogen evolution and Ir_2O for oxygen evolution reactions.

Following are the important aspects that should be taken care while examination of the overall photocatalytic water splitting:

(a) All reaction products including the intermediates should be properly acknowledged and if it is not possible then alternative analysis should be given.
(b) While evaluating the photocatalytic water splitting efficiency of the synthesized material, comparison with at least two different reference materials should be made.
(c) During the estimation of the quantum efficiency/yield certain substances are not recommended as a standard compounds that tends to produce the unstable intermediates and donate the second electron to conduction band (doubling effect exhibited). These substances are: alcohols (methanol or ethanol), formaldehyde, acids (formic, oxalic, citric, succinic, phthalic), As^{3+}, Sn^{2+}, Hg^{2+} (Finklca, 1988).
(d) Usually, the photocatalytic reactions are governed by the photo flux characters (light power, photocurrent density, reactor geometry, irradiance intensity). Therefore, they should be carefully considered as standard (1Sun = AM100G) light exposure during the analysis while the proposing mechanism.
(e) Most of the water splitting process exhibits the pseudo first order kinetics and follows Hinselwood mechanism/relationship. Therefore, the absorption equilibrium should be maintained and amount of the adsorbed substrate should be proportional to the concentration of the solution during the photocatalysis (Chouhan et al. 2016).
(f) High photocatalytic activities of the new materials are often attributed to the large specific area or novel structures (shape, size, crystalline, phase). Researcher should provide analytical evidences for connecting the photocatalytic activities and the physicochemical properties (Ohtani, 2008).
(g) Beside the estimation of the Eg (band gap) using diffuse reflectance or Tauc/ Kubelka-Munk plots, the accurate CB and VB positions should be calculated using electrochemical analysis (electrochemical impedance) or surface analysis (e.g., photoelectron spectroscopy) during the photocatalytic activity measurement under the visible light especially for the OH*-radical mediated oxidations.
(h) While estimating the photocatalytic activity of the binary photocatalysts or composites (semiconductors A and B), the synergic effect dominates means the overall photocatalytic activities are much higher than the activity of the each component, either A or B.

(i) It is suggested to avoid the use of dye sensitizer as a role model substance due to following reasons; First, they absorb the wide range of the solar spectrum in visible region that made complicate the measurements of the photocatalytic effectiveness because the dye may display the spectral overlap with the semiconductor that leads to the biased comparison (Kudo and Miseki, 2009). Second, typical dye concentration of the dye is quite low than those of the semiconductors because of the higher absorption coefficient. Third, complicated molecular structure of the dye and poorly known intermediates makes their role difficult to be characterized (Bae et al. 2014). Fourthly, commercially available dyes are not 100% pure and accurate information about the impurities are not known. Color content of the dye may not represent real composition of the dye. Fifth, some colorant have complicate redox reactions and their behavior is quite sensitive to the chemical environment (pH, dissolve oxygen, etc.) for example methylene blue decolorized by the oxidation and reduction process when the reduced dye exposed to oxygen the colored of the dye reappears gradually. Absorbance value measured in-situ under the well-controlled conditions.

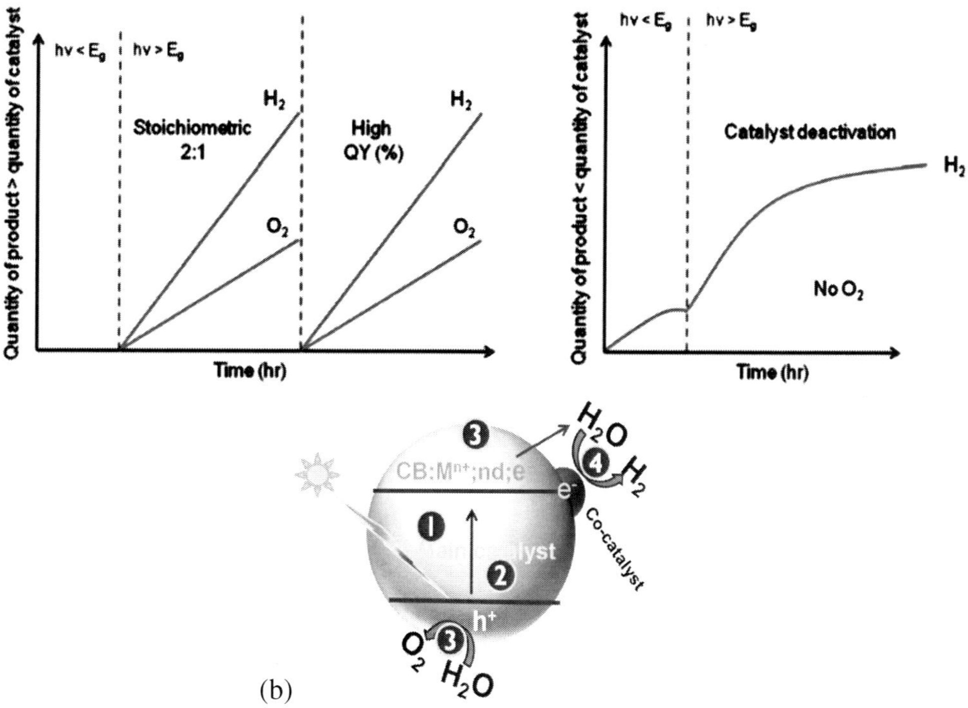

(a)

(b)

Figure 6. (a) Comparative typical reliable (left) and unreliable (right) data for overall water splitting and (b) Mechanism of the water splitting process. Figure 6a reprinted with permission from (Bae et al. 2014). Copyright (2014) Elsevier (Permission yet to be taken as it is paid) and Figure 6b is created by authors.

(j) The stoichiometric production of H_2 and O_2 should follow the ratio of 2:1, respectively even in the absence of a sacrificial reagent.
(k) The evolution of H_2 and O_2 should be directly proportional to time and should increase with an increasing irradiation time (Figure 6a). In overall water cleavage, the production of H_2/O_2 should be prominently higher than the amount of catalyst used. If not so then it is difficult to judge either the reaction is photocatalytic or not.

Adsorption, surface structure, interface, charge diffusion, charge trapping, particle size, surface electronic states, dark reactions, and photoreaction, are the most important driving factors that can control the whole water splitting process, shown in Figure 6b.

In reliable data steady stoichiometric evolution of both H_2 and O_2 exhibited with no activity prior to the illumination of light and no significant deactivation in the catalyst activity will be found within the increasing time frame. Where, the unreliable data show activity which could not be attributed to the photocatalytic activity such as H_2 evolution under no illumination and lack of O_2 evolution.

4. PHOTOCATALYTST FOR HYDROGEN PRODUCTION

Hydrogen is a versatile fuel and is found in 90% of the Universe. It can be made from any source, used for any service and ready to be stored in large amounts.1 Kg of H_2 (2.2 pounds) contains the same amount of energy as 1 U.S. gallon (6.2 pounds) of the gasoline. Most of the hydrogen (83%) produced at refineries and utilized at site is known a Captive Hydrogen and 17% hydrogen available to sale outside the production site, are known as Merchant Hydrogen. There are eight main ways to generate hydrogen through water splitting e.g., (1) thermochemical water splitting (2) biophotolysis (3) photocatalytic (4) photoelectrochemical (5) magnetolysis, (6) mechanolysis (7) radiolysis and (8) plasmolysis (Chouhan et al. 2017). Where, the thermochemical method is the simplest one but it requires the huge solar concentrators that increases its cost and make it less ~~un~~favourable. Moreover, biophotolyses are of two types: water biophotolysis and organic biophotolysis. Organic biophotolysis depends on the type of the microorganism used and reaction mechanisms. With no CO_2 emissions, water biophotolysis is the cleaner approach than the organic biophotolysis but produce low hydrogen, possess toxic effects of enzymes, and tough scaling up process. Furthermore, the photoelectrochemical water splitting needs electricity that used conventional sources for electricity generation. Rest of the methods such as: magnetolysis, mechanolysis, radiolysis and plasmolysis, needs sophisticated instrumentation.

Lastly, the photocatalytic water splitting possesses several advantages over the above discussed method:

(1) low cost (capable of reducing the photovoltaic arrays);
(2) relative high solar-to-H_2 conversion efficiency;
(3) capable for separating H_2 and O_2 streams; and
(4) adjustable reactor size can be applicable for the small scale usage.

The oxides are used as a main photocatalytic material due to their stability and non-toxicity, favorable band gap and VB/CB positions. Whereas many III–V nitrides and phosphides, and II–VI chalcogenides have shown assurance for good activity but their implementation is limited due to its poor stability under light exposure and oxidative conditions. There are two most popular approaches to select the photocatalyst for water splitting. First, class includes the wide-band gap oxide semiconductor for water splitting (i.e., anatase TiO_2) since the discovery of the water splitting process discovered by the Fujishima and Honda. The alternative wide-band gap oxides used for the water splitting are ZnO and SnO_2. Nevertheless, the biggest disadvantage of TiO_2 is its inability to harvest the visible light, which accounts for a major portion (45% of sunlight radiation) of sunlight. To overcome this shortcoming, several techniques, such as metal doping, ion doping and dye/quantum dot sensitization, have been studied extensively. Despite the successful development of the several visible-light-driven photocatalysts, low H_2 or O_2 production yield can be achieved, which attributed to the intrinsic band gap limitation of the photocatalyst.

Photocatalytic reduction of the water into molecular hydrogen gas is a two electron reduction process (Equation 10), and all semiconductor photocatalysts owning the reductive potential more than the reduction potential $E_{(H2O/H2)}$ = -0.42 V or 0.00eV vs. NHE at pH = 7 are able to generate H_2. However, these reactions are kinetically not favored and carried on too slowly in the absence of a suitable catalyst. Transition metal complexes can store electrons and form products through various multiple redox states. There are suitable numbers of candidates to catalyze these types of reaction. Formation of the high energy redox intermediates via the coupled oxide reduction processes. Reduction of water to hydrogen and oxidation to O_2 occur in artificial water splitting systems that can be considered into two parts: First, a photochemical or electrochemical component that required the oxidizing or reducing equivalents to be generated. Second stage, in which the suitable redox catalyst, assist to the formation of molecular gases products. Most of the efforts till date still revolved around this second component to identify the suitable redox catalyst. Noble metals such as Pt (worked as cathode) are known as good catalysts for H_2 evolution but not for O_2. And a metal oxide such as RuO_2 or IrO_2 used as an anode (Ftoy and Critchley, 2005; Kalyanasundaram et al. 1978; Graetzel, 1983; Borgarello et al. 1981; Kalyanasundaram et al. 1981; Dimitrijevic, 1984). Anode materials show quite low overvoltage for water oxidation to molecular O_2. In photochemical / electrochemical studies of the H_2 evolution, a popular one-electron redox reagent like Methyl viologen (4,4'-dimethyl-bipyridinium chloride, MV^{2+}) used as a key intermediate reagent. MV^{2+} is highly soluble in water and has slightly more negative redox potential than that of normal potential of hydrogen (E^0 =-0.42 V). In its oxidized form (MV^{2+}), the reagent is colorless but it turns up into the deep blue colored reduced form of MV^{2+} i.e., MV^+. Furthermore, in the presence of the suitable redox catalysts the reduced form further oxidized to evolve H_2 gas from water, as followed by Equations 26 and 27:

$$MV^{2+} + e^{-} \rightarrow MV^{+} \quad (26)$$

$$2MV^{+} + 2H^{+} \rightarrow 2MV^{2+} + H_2 \quad (27)$$

Photolysis of a $[Ru(bpy)_3]^{/2+}$complex take place in presence of a redox catalyst (potassium hexa- niobate nanoscrolls (NS-$K_4Nb_6O_{17}$)), e-acceptor (MV^{2+}) and sacrificial donor like EDTA disodium salt, under visible light exposure that leads to the sustainable evolution of H_2 gas from water, as represented by the Figure 7a (Hoertz et al. 2007). In biomimetic systems enzymes such as hydrogenases was used for photolysis of the neutral aqueous solutions. There are two different approaches for hydrogen production from cleavage of water. First, use of photocatalytic material with CB position beyond the potential of reduction of water (0.00eV at pH = 7.00) and secondly, use of hole eater scavenger electrolyte.

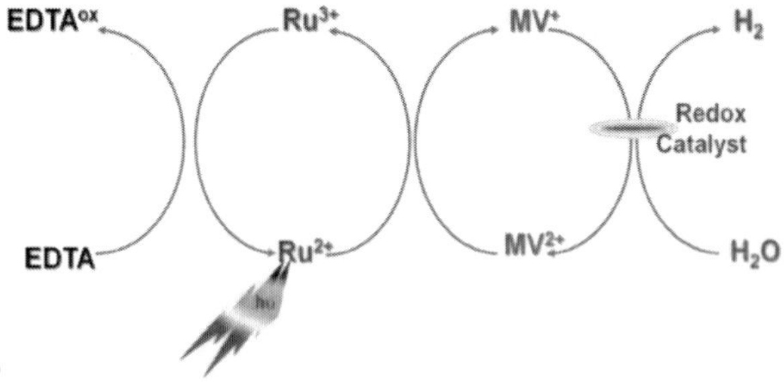

(a)

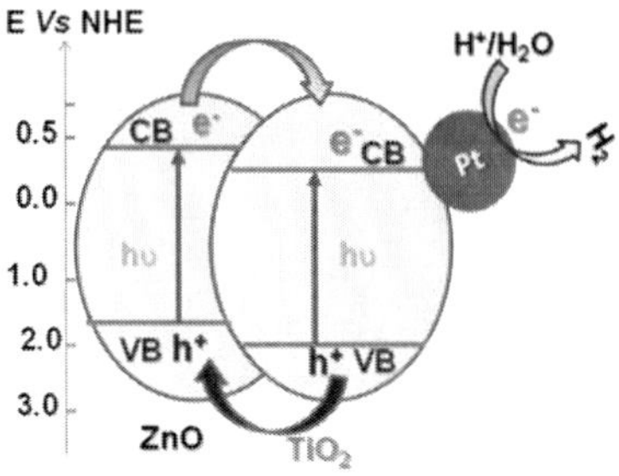

(b)

Figure 7. (a) Coupled electron transfer processes occurred during the photolysis of $[Ru(bpy)_3]^{/2+}$ complex in presence of a redox catalyst, e-acceptor (MV^{2+}) and sacrificial donor (EDTA disodium salt) (Hoertz et al. 2007) and (b) Pt-doped TiO_2–ZnO system used for hydrogen generation under visible light exposure. Figure 7a reprinted with permission from (Xie et al. 2017) copyright (2017) *American Chemical Society* and Figure 7b is open access article so permission not taken.

Pt-doped TiO_2–ZnO is another good example of the photocatalytic material, which can liberate hydrogen of 203 μmol$h^{-1}g^{-1}$ from TiO_2–ZnO (Ti/Zn = 10) and 2150 μmol$h^{-1}g^{-1}$ from 0.5 wt% Pt/TiO_2–ZnO (Ti/Zn = 10) under 400 W mercury arc lamp with a cutoff filter which, filtered out all the wavelengths under 400 nm as the visible light source (Figure7b) (Xie et al. 2017). Moreover, 0.1 wt% Pt/TiO_2–ZnO sample can show a satisfying long-term stability with 88% and 77% hydrogen productivity even after 7 and 14 days reactions, respectively (Xie et al. 2017).

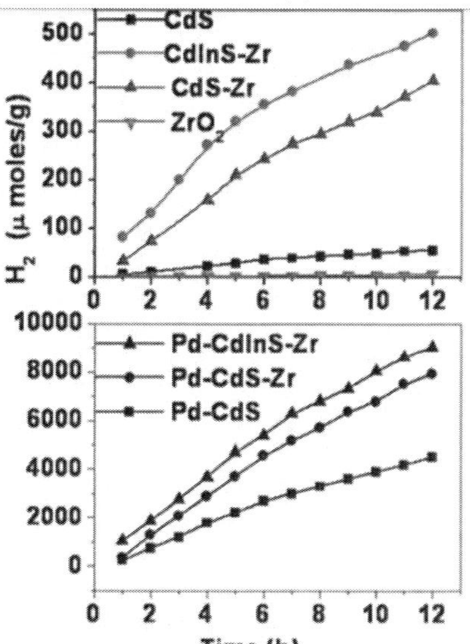

(c)

Figure 8. (a) SEM images of TiO_2 dispersed on Al_2O_3, CeO_2 ZrO_2, and zeolite and (b) their comparative hydrogen generation rate with and (c) without Pd loading- TiO_2/ZrO_2, TiO_2/zeolite and CdS/ZrO_2 systems (Sasikala and Bharadwaj, 2012). Figure 8a, 8b and 8c is open access article so permission not taken.

Dholam et al. (Dholam et al. 2009) synthesized Cr- or Fe-ion-doped TiO_2 thin films by radio-frequency magnetron sputtering and a sol–gel method and used to generate hydrogen by photocatalytic water-splitting under visible light irradiation. It was found that the Fe-doped TiO_2 (15.5 μmol/h) produced more hydrogen than for Cr-doped TiO_2 (5.3 μmol/h) due to the ability of Fe ions to trap both electrons and holes but Cr can trap one type of carrier (Dholam et al. 2009).

Figure 8 exhibits an enhanced photocatalytic activity for TiO_2 - dispersed on zeolite, ZrO_2, Al_2O_3 and CeO_2 surface. And due to the high degree dispersion of TiO_2 on zeolite surface, the number of active sites enhanced with the extent of dispersion. Na–Y zeolite shown the maximum Lewis acidity go after by ZrO_2, Al_2O_3 and CeO_2. Optical absorption property of TiO_2 dispersed on different surface of show that the zeolite and CeO_2, exhibits better photocatalytic activity than that of unsupported TiO_2. Photocatalytic studies for hydrogen generation from water showed the enhanced activity for CdS dispersed on ZrO_2 and furthermore, the indium doping for CdS exhibits the highest activity further. Fluorescence lifetime studies indicate that the In- supported CdS, have higher lifetime for the charge carriers than the unsupported CdS (Sasikala and Bharadwaj, 2012; Sasikala et al. 2010; Sasikala et al. 2011).

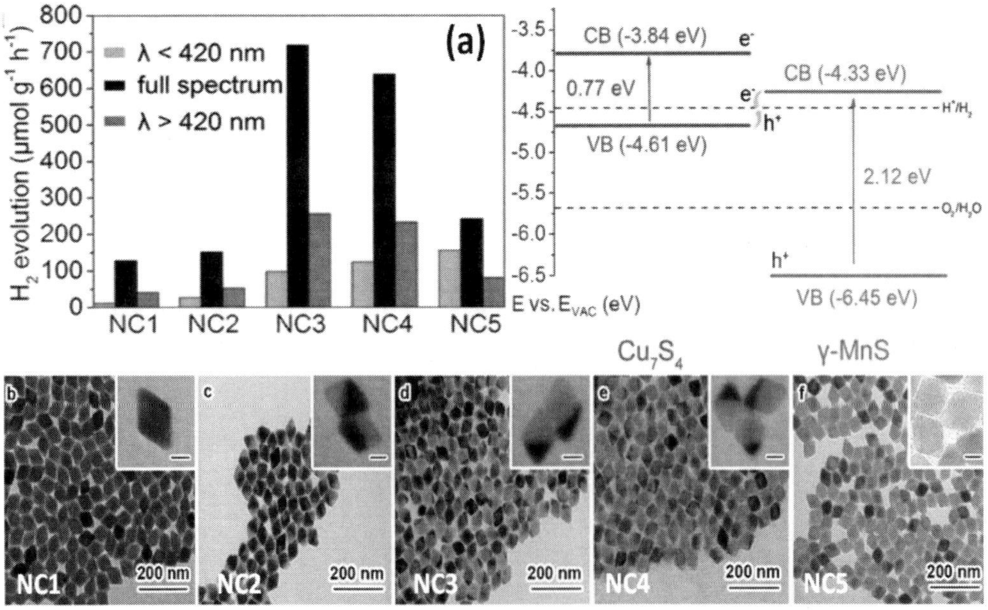

Figure 9. (a) Photoctalytic hydrogen generation by splitting of water using catalyst: (b) Cu_7S_4 nanocrystals (NC1), (c) γ-MnS/Cu_7S_4 sample after cation exchange for 0.5 h (NC2), (d) γ-MnS/Cu_7S_4 sample after cation exchange for 1 h (NC3), (e) γ-MnS/Cu_7S_4 sample after cation exchange for 1.5 h (NC4), and (f) the γ-MnS sample after nearly complete cation exchange (NC5). Permission yet to be taken as it is paid (Yuan et al. 2017).

The particular type of Janus- structures of compound γ-MnS/ Cu_7S_4 i.e., NC1, NC2, NC3, NC4 and NC5 were synthesized at different time interval (0, 0.5h (Mn/Cu = 0.6), 1h (Mn/Cu = 3), 1.5h ((Mn/Cu = 6) and 1.5h (Mn/Cu = Mn/Cu = 256) displayed dramatically enhanced photocatalytic hydrogen production rates i.e., 128, 151, 718, 639, 242 mmol g^{-1} h^{-1} under full-spectrum irradiation in aqueous electrolyte solution 0.35 M Na_2S and 0.25 M Na_2SO_3 (Figure 9) (Yuan et al. 2017).

Usually, oxides (ZnO, TiO_2 SiO_2, SnO_2, pervoskites, etc.) oxynitrides, oxysulfides, chalchoginides, used for hydrogen production photocatalyst but oxides has large band gap so able to harvest small portion of light UV portion (5% of whole light). Therefore, loading of noble metals, anion- doping, metal-doping, dye/quantum dots, were done to improve their band gap and water splitting capacity (Chouhan et al. 2017).

5. PHOTOCATALYTST FOR OXYGEN PRODUCTION

The ultimate goal of the artificial photosynthesis is to develop an apt material that can convert solar energy into the chemical bonds without any assistant/mode of energy/chemicals. Water splitting into H_2 and O_2 is accompanied by a large positive change in the Gibbs free energy i.e., it is an uphill reaction and on the other hand, degradation reactions such as the photo-oxidation of organic compounds using oxygen molecules are usually downhill reactions (Chouhan et al. 2017). The major obstacle in the high performance for solar-to-fuel conversion is the low photoanodic performance (oxidation) for driving an overall water splitting reaction under the sunlight. Because this reaction requires the four or more electrons, which made this reaction: a thermodynamically tough reaction as seen in following Equations 28-31.

$$H_2O \rightarrow HO^* + H^+ + e^- \; E^o = 2.39 \text{ V (vs. NHE at pH 7)} \quad (28)$$

$$2H_2O \rightarrow HOOH + 2H^+ + 2e^- \; E^o = 1.37 \text{ V (vs. NHE at pH 7)} \quad (29)$$

$$2H_2O \rightarrow HOO^* + 3H^+ + 3e^- \; E^o = 1.26 \text{ V (vs. NHE at pH 7)} \quad (30)$$

$$2H_2O \rightarrow O_2 + 4H^+ + 4e^- \; E^o = 0.82 \text{ V (vs. NHE at pH 7)} \quad (31)$$

Out of above four equations, the Equation 32 is the lowest energy path that involves the removal of $4H^+$ and $4e^-$ along with the formation of an O–O bond, which reflect the mechanistic complexity of the reaction. There are two different type of photocatalyst are in trend: homogeneous and heterogeneous catalysis. The capacity of the homogeneous catalyst increases with the (i) performance based rationally designer OER catalysts by ligand modifications, choice of the transition metal with its oxidation state and geometry, space interactions, electronic coupling, and active site hindering; (ii) increase the solubility of catalyst by adding additional functionalities; and (iii) enable the anchoring capacity of the catalyst onto electrodes.

Beside above, the two different types of the heterogeneous molecular species are the basic driving forces for oxygen production from heterogeneous photocatalytic water cleavage. First, the main photocatalytic material with the VB position beyond the oxidation potential of water (1.23 eV at pH = 7.00) and second, electron acceptor - scavenger electrolyte. We had already discussed scavengers in section 3. Most of the main photocatalytic units are the oxide semiconductors such as WO_3, Fe_2O_3, TiO_2, ZnO, SnO_2, and CeO_2, etc. Usually, the efficiency of the oxide-photocatalytic water splitting is usually found lower than their other molecular counter parts because of the following reasons:

(1) *Recombination:* of the photo-generated charge carriers i.e., electron/hole pairs. Because when CB electrons recombine with VB holes they release energy in the form of heat or photons;
(2) *Fast backward reaction:* Decomposition of water into hydrogen and oxygen is an energy increasing process, thus backward reaction (recombination of hydrogen and oxygen into water) easily proceeds.
(3) *Inability to utilize visible light:* most of the oxides are wide band gap semiconductors (TiO_2, ZnO ~Eg = 3.2 eV and only UV light can be utilized for hydrogen production. Since the UV light only accounts for about 4% of the solar radiation energy while the visible light contributes about 50%, the inability to utilize visible light limits the efficiency of solar photocatalytic hydrogen production.

These limitations can be overcome by the modification of main semiconductor by addition of the noble metal loading, metal/non-metal loading, nano-structuring, sensitization (dye/quantum dots), composite formation, solid solutions (Chouhan et al. 2017).

There are following two different mechanisms, which can be operated in the oxygen gas formation during homogeneous catalysis as demonstrated in Figure 10:

(i) *Water nucleophilic attack (WNA)* where a water molecule nucleophilically attacks the oxo group of the electrophilic M–O moiety (Figure10a left). The O–O molecule and the reduced metal centre are the result of the interaction between the HOMO of the water molecule and the LUMO of the metal–oxo (M–O) complex that promote cleavage of the M–O bond (Bofill et al. 2013). Most mononuclear and some polynuclear OER catalysts (tetranuclear polyoxometalate) are reported to oxidise water through this mechanism (Derek et al. 2013).

(ii) *Interaction between two M–O entities (I2M):* As the name indicates, this pathway consists of the interaction between the two metal–oxo moieties. I2M can be of two types: either a radical coupling or a reductive elimination. The radical coupling can take place both in an intra-and in an inter-molecular way. The reductive elimination depends upon the oxidation state of the metal centre and on the number of oxygen atoms attached (Figure 10a right) (Berardi et al. 2014). The binuclear complexes undergo the water oxidation through this mechanism by the intra-molecular interaction of the two RuQ-O moieties. Where, the highly active mononuclear ruthenium complex promotes the oxygen–oxygen bond formation by an inter-molecular interaction between two complexes.

(a)

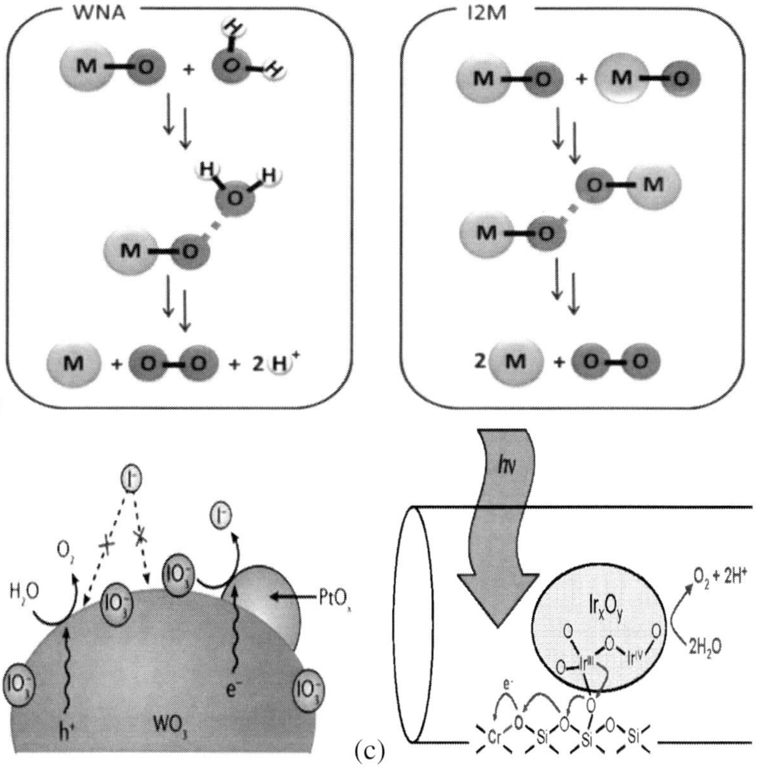

Figure 10. (a) Proposed mechanisms for O–O bond forming using homogeneous catalysis: Left, high-valent transition metal–oxo complexes with water [water nucleophilic attack; WNA] and right, O–O bond produced by transition metal complexes interaction between two M–O entities (I2MO), (Berardi et al. 2014) (b) schematic illustration of the heterogeneous photocatalytic water oxidation reactions over PtO_x/WO_3 in the presence of IO^{3-} and I^- ions and (c) nano-sized crystals of Co_3O_4 impregnated on mesoporous silica used as oxygen evolution photocatalytic water splitting. Permission not taken for 10a and Figure 10b reprinted with permission from (Abe, 2011) copyright (2011) *American Chemical Society.*

Figure 10b represented the water splitting mechanism of the particulate system i.e., PtO_x (co-catalyst) loaded WO_3 in photooxidation of water in presence of IO^{3-} and I^- ions. Where, the holes oxidize water to O_2 and electrons at PtO_x surface reduce IO^{3-} ions into the I^- ions (Abe, 2011). Frei et al. (Nakamura and Frei, 2006) recently demonstrated that nano-sized crystals of Co_3O_4 impregnated on mesoporous silica (Figure 10c) work efficiently as an oxygen-evolving catalysts.

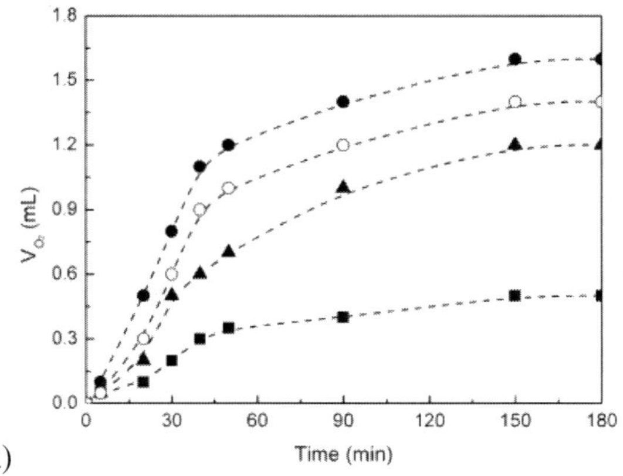

(a)

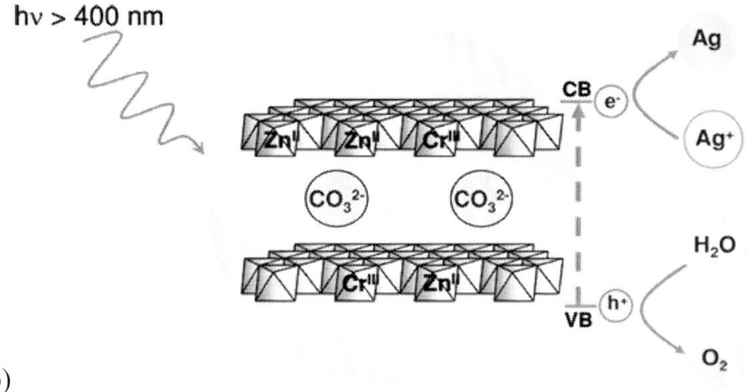

(b)

Figure 11 (a) Double layered hydroxides for oxygen generation through photocatalytic water splitting and (b) their water splitting mechanism in $AgNO_3$. Figure 11a and 11b reprinted with permission from (Silva et al. 2009) copyright (2009) *American Chemical Society.*

Typical oxygen evolution can be achieved by using either bulk electrodes/ heterogeneous or homogeneous redox catalytic system. It was mentioned earlier that metal oxides such RuO_2 are known as the best electrocatalytic material for use as anodes for water oxidation. Finely divided particles of these materials have been used to test oxygen evolution using electrochemical oxidation reactions or using strong chemical oxidant reagents such as: Ce(IV) or $Ru(bpy)_3/^{3+}$ produced via light-induced electron-transfer reactions. Mallouk et al. (Hoertz et al. 2007) had shown that nanocrystals of catalytic metal oxides such as Ir-oxide or Nb-oxide work efficiently as catalysts for water oxidation. Nickel–bismuth (Ni–Bi) electrocatalyst used as the oxygen evolution reaction (OER) catalyst film in a potassium borate aqueous solution with Ag/AgCl reference electrode and Pt cathode, which is related with the formation of nanoscale order domain of edge-sharing NiO_6 octahedra in the Ni–Bi electrocatalyst (Yoshida et al. 2015).

Silva et al. tested a series of Zn/Ti, Zn/Ce, and Zn/Cr layered double hydroxides (LDH) at different Zn/metal atomic ratio (from 4:2 to 4:0.25) for oxygen generation through photocatalytic water splitting under visible light irradiation (Figures 11a and 11b). (Zn/Cr)LDH with an atomic ratio of 4:2 found to be the most active material at λ_{max} = 410 and 570 nm with apparent quantum yields for oxygen generation (Φ apparent = 4 × mol oxygen/mol incident photons) are of 60.9% and 12.2%, respectively (Silva et al. 2009). This activity is 1.6 times higher than that of WO_3 under the same conditions.

Figures 12a and 12b illustrated that when 1 wt % gold-supported ceria nanoparticles irradiated with visible light (λ > 400 nm) they generate oxygen by breaking of water (10.5 μmol.h^{-1}) more efficiently than the standard WO_3 (1.7 μmol.h^{-1}) even under UV irradiation (9.5 μmol.h^{-1}) (Primo et al. 2012).

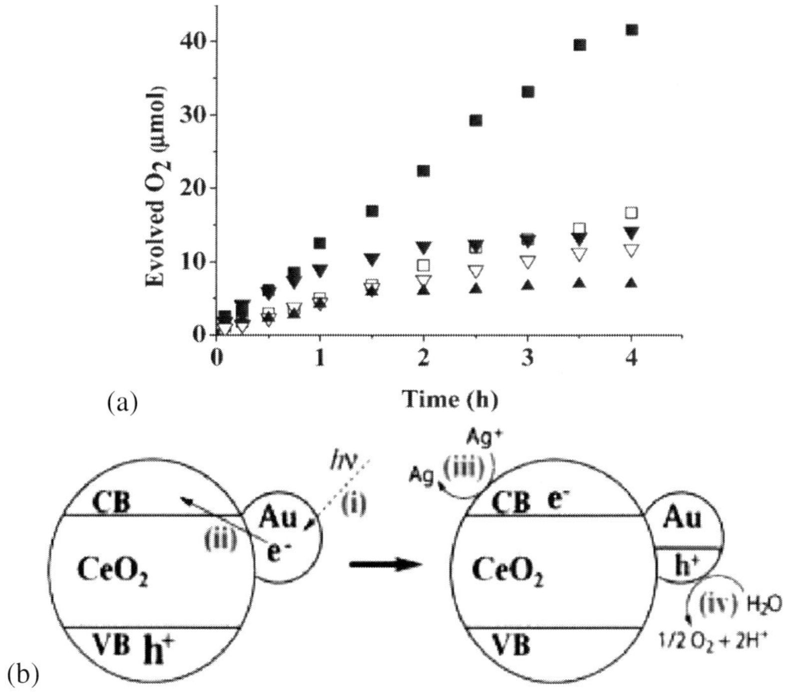

Figure 12. (a) Oxygen evolved upon visible light (λ > 400 nm) illumination of an aqueous $AgNO_3$ suspension containing the photocatalyst. ■ Au(1.0 wt %)/CeO_2(A); □ Au(1.0 wt %)/CeO_2(B); ▼ Au(3.0 wt %)/CeO_2(A); ▽ Au(3.0 wt %)/CeO_2(B); ▲ WO_3 i) Photon absorption; ii) electron injection from Au to ceria conduction band; iii) electron quenching by Ag^+; iv) water oxidation by h^+.
Figure 12a and 12b reprinted with permission from (Primo et al. 2012) copyright (2012) *American Chemical Society.*

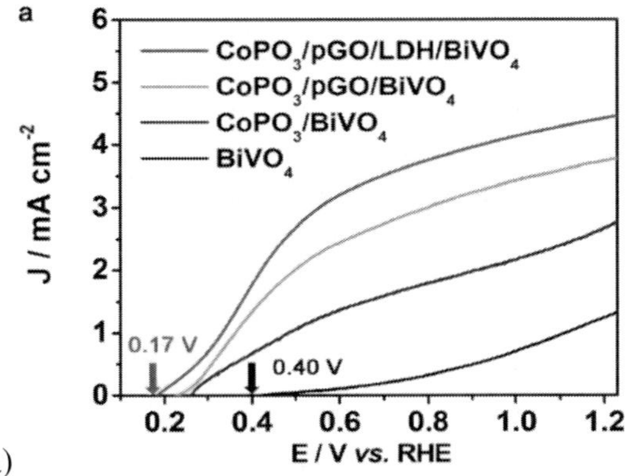

(a)

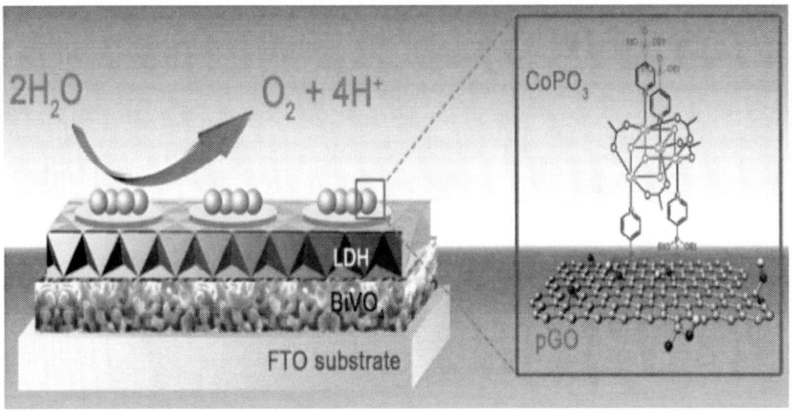

(b)

Figure 13. Photoelectrocatalytic water oxidation power of the studied systems for artificial photosynthesis system as: (a) comparative current–potential plots of $BiVO_4$, $CoPO_3/BiVO_4$, $CoPO_3/pGO/BiVO_4$, and $CoPO_3/pGO/LDH/$ $BiVO_4$ photoanodes and (b) schematic representation of integrated $BiVO_4$ photoanodic system. Figures 13a and 13b reprinted with permission from (Ye et al. 2018) copyright (2018) *American Chemical Society*.

Ye et al. developed a photoelectrocatalytic system (Figures 13a and 13b) which can mimic the functions of photosystem II (PS-II) with $BiVO_4$ semiconductor as a light harvester that guarded by a hole storage layer of layered double hydroxide (NiFe LDH), a partially oxidized graphene (pGO) for layer charge transfer that act as a biomimetic tyrosine and molecular Co cubane as oxygen evolution complex. The integrated system exhibited the low onset potential (0.17 V) with a high photocurrent (4.45 mA cm^{-2}) at 1.23 V versus reversible hydrogen potential (RHE) indicating a greater driving force for water oxidation with more efficient charge separation in the photoanode (Ye et al. 2018).

Future Prospective

Artificial photosynthesis can be able to provide an efficient way for producing solar fuels and holds the great promise for future of green fuels. More and more talented researchers are associated with the challenge of the search of the robust catalytic systems, which can mimic the four general steps of the photosynthesis, efficiently for the water oxidation and fuel production. We are looking forward with the great expectation to the molecular crystallography and sophisticated spectroscopic tools, which can reveal the secrets of the precise mechanism of the water splitting reaction within next 10 years. Another approach is to develop the ability to utilise the whole the portion of the sunlight spectrum by the proper combination of the semiconductor materials. For an example a combination of $BaTaO_2N$ and WO_3 is restricted by the longest working wavelength of the WO_3 (450 nm), in spite of the fact that $BaTaO_2N$ take in the wavelengths up to 660 nm. But the recently reported system $ZnRh_2O_4$–$Bi_4V_2O_{11}$ composite bridged with Ag as an electron mediator, (Kobayashi et al. 2005) and $LaMg_{1/3}Ta_{2/3}O_2N$-based photocatalyst (Pan et al. 2015) could be operated up to the wavelength~740 nm in a two-step water oxidation photocatalysis. The maximum solar energy conversion efficiency of the water-splitting reaction could be achievable with photovoltaic devices (Hanna and Nozik, 2006; Shockley and Queisser, 1961). Study indicating that the theoretical efficiencies for single-junction devices could approach 31% under 1 Sun (1 kW m^{-2}) at AM1.5G illumination (Walter et al. 2010). Up-gradation in the photocatalytic efficiency to the required level (STH > 10%; AQY = 40–60% at 600–700 nm) should be made by using narrow-band gap photocatalyst that consist of inexpensive and abundant elements, which is a real challenge at present level of knowledge and technology. It would also be a wise step to examine the design of solar hydrogen plants at the bench scale, including photocatalytic reactor's category and the safe and sound severance of the explosive mixture of H_2 and O_2. Field examinations of the durability (thermal and photo stability) of the photocatalytic material and related equipment's are also a necessary requirement for the life span of the decades in future practical applications. Therefore, the particulate photocatalytic organisms would be advantageous over the PEC or PV-E systems due to their greater scalability and operability. These advancements can help to provide the remarkable 'blueprint' of the next higher generation of catalysts that can replicate the 'smart matrix (such as: enzyme)' of the plants and will truly copy the catalytic power of the natural system. Just like the plant biology that uses self-assembling and self-repairing process to produce the long-term stability in photosynthesis, can strengthen the artificial photosynthesis for being scalable. In looking to the present energy status and energy scenario, we the researcher needs to concentrate the focus on this emerging field. That can bring research in this area at a fascinating stage. We know where to go but, what we do not yet know, is how to get there. By the disciplined approached and proper use of the knowledge of the interdisciplinary approach (material, biology, physics and chemistry) involved in biological process, we can achieved our target of sustainable production of carbon-neutral fuels without hurting mother nature. For this, we should give rise to the technological advancement to save mankind at the stage, when oil and gas run out.

Conclusion

Day by day soaring prices of gasoline and LPG along with increasing number of natural disaster and environmental issues compels us to act sincerely on the affordable clean energy. Whenever, we are in fix, we look forward to nature for solution, here also nature provide us the great hint in the form of photosynthesis process occurred in plants. Therefore, in this chapter, we discussed artificial photosynthesis process for generation of O_2 and H_2 as a clean fuel by cleavage of water using photocatalyst under the sunlight as plants do. Renewable energy sources are great hope in current energy tight situation. A form of artificial photosynthesis i.e., photocatalytic water splitting can produce clean energy (O_2 and H_2) without much disturbing our environment. Hence, overall water splitting and restricted water splitting either for only hydrogen or oxygen production, ~~is~~ are discussed in the chapter with their mechanism and futuristic prospective.

References

Abe, Ryu. 2011. "Development of a new system for photocatalytic water splitting into H_2 and O_2 under visible light irradiation." *Bull. Chem. Soc. Jpn.* 84(1): 1000–1030. DOI: /10.1246/bcsj.20110132.

Alstrum-Acevedo, James H. Brennaman, Kyle M. and Thomas J. Meyer. 2005. "Chemical approaches to artificial photosynthesis. II." *J. Inorg. Chem.* 44: 6802-6827. DOI: 10.1021/ic050904r.

Bae, Sugyeong, Sujeong Kim, Seockheon Lee and Choi Wonyong. 2014. "Dye decolorization test for the activity assessment of visible light photocatalysts: Realities and limitations." *Catalyst Today* 224:21-28. DOI: 10.1016/j.cattod.2013.12.019.

Balzani, Vincenzo, Credi Alberto and Margherita Venturi. 1997. "Photoprocesses" *Curr. Opin. Chem. Biol.* 1(4): 506–513. DOI: 10.1016/S1367-5931(97)80045-2.

Bamwenda, Gratian R., Tsubota Susumu, Nakamura Toshiko and Masatake Haruta. 1995. "Photoassisted hydrogen production from a water-ethanol solution: a comparison of activities of Au-TiO_2 and Pt-TiO_2." *J. Photochem. Photobiol. A* 89: 177−189. DOI:10.1016/1010-6030(95)04039-I.

Barber, James. 2008. "Crystal Structure of the Oxygen-Evolving Complex of Photosystem II" *Inorg. Chem.* 47(6): 1700–1710. DOI: 10.1021/ic701835r.

Berardi Serena et al. 2014. "Molecular artificial photosynthesis" *Chem. Soc. Rev. 43:* 7501-7519. DOI: 10.1039/C3CS60405E.

Bofill, R. et al. 2013. *"Comprehensive Inorganic Chemistry- II"* eds.: J. Reedijk and K. Poeppelmeier, 2nd edn Elsevier, Amsterdam, pp. 505–523.

Borgarello Enrico, Kalyanasundaram Kuppuswamy, Okuno Yohmei and Michael Gra¨tzel. 1981. "Visible Light-induced Oxygen Generation and Cyclic Water Cleavage Sensitized by Porphyrins." *Helv Chimica Acta*, 64(6):1937-1942. DOI: 10.1002/hlca.19810640626.

Carabe, J. and J. J. Gandia. 2004. "Thin-film-silicon solar cells." *Opto-Electronics Review* 12(1): 1-6. (http://www.wat.edu.pl/review/optor/12(1)1.pdf).

Castillo, Mauricio. 2012. "The Omega Point and Beyond: The Singularity Event." *American Journal of Neuroradiology* 33 (3): 393–395. DOI: 10.3174/ajnr.A2664.

Chouhan Neelu, Liu Ru-Shi and Jiujun Zhang. 2017. "Photochemical Water Splitting: Materials and Applications"; *Series Name: Electrochemical Energy Storage and Conversion Series* Editor: Jiujun Zhang, CRC Press, Taylor & Francis Group, Florida-USA, pp: 358. ISBN 9781482237597 - CAT# K23177

Chouhan Neelu, Ameta Rakshit, Meena Rajesh Kumar, Mandawat Niranjan and Rahul Ghildiyal. 2016. "Visible light harvesting Pt/CdS/Co-doped ZnO nanorods molecular device for hydrogen generation" *International Journal of Hydrogen Energy* 41(4): 2298-2306. DOI:10.1016/j.ijhydene.2015.11.019

Derek J., Ryan D. Palmer and P. Berlinguette Curtis. 2013. "Homogeneous water oxidation catalysts containing a single metal site." *Chem. Comm.* 49: 218–227. DOI: 10.1039/C2CC35632E.

Dholam, R., Patel, N., Adami M. and A. Miotello. 2009. "Hydrogen production by photocatalytic water-splitting using Cr- or Fe-doped TiO_2 composite thin films photocatalyst." *International Journal of Hydrogen Energy* 34(14): 5337-5346. DOI: 10.1016/j.ijhydene.2009.05.011.

Dimitrijevic, Nada M., Shuben Li and Michael Graetzel. 1984. "Visible light-induced oxygen evolution in aqueous cadmium sulfide suspensions." *Journal of the American Chemical Society* 106(22):6565. DOI: 10.1021/ja00334a018.

Finklca. 1988. "Semiconductor electrodes" Elsevier, New York.

Ftoy, Collings Anthony and Critchley Christa (Eds).2005. "*Artificial Photosynthesis: From Basic Biology to Industrial Applications.*" Weinheim: Wiley-VCH.

Fukuzumi, Shunichi. 2008. "Bioinspired Energy Conversion Systems for Hydrogen Production and Storage." *Eur. J. Inorg. Chem.* 2008(9): 1351-1362. DOI: 10.1002/ejic.200701369.

Graetzel M. (Ed). 1983. *"Energy Resources through Photochemistry and Catalysis."* Academic Press, New York.

Gurunathan, K. Maruthamuthu, P. and M. V. C. Sastri. 1997. "Photocatalytic hydrogen production by dye-sensitized Pt/SnO_2 and Pt/ SnO_2/RuO_2 in aqueous methyl viologen solution." *Int. J. Hydrogen Energy* 22: 57−62. DOI: 10.1016/S0360-3199(96) 00075-4.

Hanna, Mark C. and Arthur J. Nozik 2006. "Solar conversion efficiency of photovoltaic and photoelectrolysis cells with carrier multiplication absorbers." *J. Appl. Phys.* 100(7): 074510. DOI:10.1063/1.2356795.

Hoertz P. G., Kim Y., Youngblood W.J. and T. E. Mallouk. 2007. "Bidentate Dicarboxylate Capping Groups and Photosensitizers Control the Size of IrO2 Nanoparticle Catalysts for Water Oxidation" *J Phys Chem B*, 111:6845-6856. DOI: 10.1021/jp070735r.

Kalyanasundaram K. and M. Graetzel 2010. "Artificial photosynthesis: biomimetic approaches to solar energy conversion and storage" *Current Opinion in Biotechnology* 21:298–310. DOI: 10.1016/j.copbio.2010.03.021.

Kalyanasundaram, Kuppuswamy, Borgarello Enrico, Duonghong Dung and Michael Grätzel. 1981. Cleavage of Water by Visible-Light Irradiation of Colloidal CdS Solutions; Inhibition of Photocorrosion by RuO_2 *Angew. Chem Int. Ed.* 1980, 20:987-988. DOI: 10.1002/anie.198109871.

Kalyanasundaram, Kuppuswamy, Kiwi John, and Michael Grätzel. 1978. "Hydrogen Evolution from Water by Visible Light, a Homogeneous Three Component Test System for Redox Catalysis." *Helvetica* 61(7): 2720-2730. DOI: 10.1002/hlca.19780610740.

Kardashev, Nikolai. 1984. "On the Inevitability and the Possible Structures of Supercivilizations", The search for extraterrestrial life: Recent developments; *Proceedings of the Symposium,* Boston, MA, June 18–21, (A86-38126 17-88).

Kobayashi, Ryoya, Toshihiro Takashima, Satoshi Tanigawa, Shugo Takeuchi, Bunsho Ohtani, and Hiroshi Irie. 2016. "A heterojunction photocatalyst composed of zinc rhodium oxide, single crystal-derived bismuth vanadium oxide, and silver for overall pure-water splitting under visible light up to 740 nm." *Physical Chemistry Chemical Physics* 18 (40): 27754-27760. DOI: 10.1039/C6CP02903E.

Kudo A, Miseki Y. 2009. "Heterogeneous photocatalyst materials for water splitting." *Chem. Soc. Rev.* 38:253–278. 10.1039/B800489G.

Lee, Sang Gi, Lee Sangwha, and Ho-In Lee. 2001. "Photocatalytic production of hydrogen from aqueous solution containing CN- as a hole scavenger." *Appl. Catal. A,* 207: 173−181. 10.1016/S0926-860X(00)00671-2.

Lewis, Nathan S. 2014. *"Artificial Photosynthesis: Direct Production of Fuels from Sunlight"* Joint Center for Artificial Photosynthesis, California Institute of Technology Pasadena, CA 91125.

Li, Yuexiang, Lu, Gongxuan, Li, Shuben. 2003. "Photocatalytic production of hydrogen in single component and mixture systems of electron donors and monitoring adsorption of donors by in situ infrared spectroscopy." *Chemosphere* 52: 843−850. DOI: 10.1016/S0045-6535(03)00297-2.

Morikawa, Motoharu et al. 2014. "Photocatalytic conversion of carbon dioxide into methanol in reverse fuel cells with tungsten oxide and layered double hydroxide photocatalysts for solar fuel generation." *Catal. Sci. Technol.* 4: 1644-1651. DOI:10.1039/C3CY00959A.

Nada, A. A., Barakat, M. H., Hamed, H. A., Mohamed, N. R. and T. N. Veziroglu. 2005. "Studies on the photocatalytic hydrogen production using suspended modified TiO_2 photocatalysts." *Int. J. Hydrogen Energy* 30: 687–691. 10.1016/j.ijhydene. 2004.06.007.

Nahar, Samsun et al. 2017. "Advances in Photocatalytic CO_2 Reduction with Water: A Review." *Materials* 10(6): 629. DOI: 10.3390/ma10060629 (open access).

Nakamura Ruyhei and Heinz Frei. 2006."_Visible Light-Driven Water Oxidation by Ir Oxide Clusters Coupled to Single Cr Centers in Mesoporous Silica." J. *Am Chem Soc*. 128:10668-10669. DOI: 10.1021/ja0625632.

Ni, Meng, Michael KH Leung, Dennis YC Leung and K. Sumathy. 2007. "A review and recent developments in photocatalytic water-splitting using TiO_2 for hydrogen production." *Renewable and Sustainable Energy Reviews.* 11: 401–425. DOI: 10.1016/j.rser.2005.01.009.

Ohtani, Bunsho. 2008. "Preparing articles on photocatalysis—beyond the illusions, misconceptions, and speculation." *Chem. Lett.* 37(3):217–229.DOI: 10.1246/cl. 2008.216.

Pan, Chengsi, Tsuyoshi Takata, Mamiko Nakabayashi, Takao Matsumoto, Naoya Shibata, Yuichi Ikuhara, and Kazunari Domen. 2015. "A complex perovskite-type oxynitride: the first photocatalyst for water splitting operable at up to 600 nm." *Angewandte Chemie International Edition* 54(10): 2955-2959. https://doi.org/ 10.1002/anie.201410961.

Posada, Francisco and Façanha Cristiano. 2015. "Brazil Passenger Vehicle Market Statistics: International comparative assessment of technology adoption and energy consumption." *International Council on Clean Transportation (ICCT). Retrieved 2015-11-24. pp. 3 and 14.).*

Primo, Ana Marino, Tiziana Corma, Avelino Molinari Raffaele and Hermenegildo García. 2012. Efficient Visible-Light Photocatalytic Water Splitting by Minute Amounts of Gold Supported on Nanoparticulate CeO2 Obtained by a Biopolymer Templating Method. *J. Am. Chem. Soc.* 134 (3): 1892–1892. DOI: 10.1021/ ja211206v.

Rogner Hans-Holger. 2004. *United Nations Development World Energy Assessment*, United Nations, Geneva, pp. 162.

Sasikala, R., Shirole A. R., Sudarsan V., Kamble V. S., Sudakar C., Naik R., Rao R. and S. R. Bharadwaj. 2010. "Role of support on the photocatalytic activity of titanium oxide." *Applied Catalysis A: General* 390: 245-252. DOI: 10.1016/j. apcata.2010.10.016.

Sasikala, R. and S. R. Bharadwaj. 2012. "Photocatalytic Hydrogen Generation from Water using Solar Radiation." *BARC Newsletter* 325: 10-15.

Sasikala, Rajamma, Shirole Archana Ramchandra, Sudarsan Vasanthakumaran, Girija Kalpathy Ganapathy, Rao Rekha, Sudakar Chandran and Shyamala Ramkumar Bharadwaj. 2011. "Improved photocatalytic activity of indium doped cadmium sulfide dispersed on zirconia." *J. Mater. Chem.*, 21 (14): 16566 – 16573. DOI: 10.1039/C1JM12531A.

Schneider, Jenny and Detlef W. Bahnemann. 2013. "Undesired Role of Sacrificial Reagents in Photocatalysis", *J. Phys. Chem. Letters* 4(2013): 3479-3483. DOI: 10.1021/jz4018199.

Schropp, Ruud EI. 2004. "Present status of micro-and polycrystalline silicon solar cells made by hot-wire chemical vapor deposition." *Thin Solid Films* 451: 455-465. DOI: 10.1016/j.tsf.2003.10.126.

Shah, A. V., Meier Johannes, Vallat-Sauvain Evelyne, Wyrsch Nicolas, Kroll U., Droz C. and U. Graf. 2003. "Material and solar cell research in microcrystalline silicon." *Sol. Energy Mater. Sol. Cells,* 78: 469-491. DOI: 10.1016/S0927-0248(02)00448-8.

Shockley, William and Queisser Hans J. 1961. "Detailed balance limit of efficiency of p–n junction solar cells." *J. Appl. Phys.* 32: 510–519. DOI: 10.1063/1.1736034.

Silva, *Cláudia Gomes*, Bouizi, *Younès*, Fornés, *Vicente*, and García *Hermenegildo*. 2009. Layered Double Hydroxides as Highly Efficient Photocatalysts for Visible Light Oxygen Generation from Water. *Journal of the American Chemical Society*, 131(38): 13833–13839. DOI: 10.1021/ja905467v.

Tachibana, Yasuhiro, Vayssieres, Lionel and James R. Durrant. 2012. "Artificial photosynthesis for solar water-splitting." *Nat. Photonics*, 6(8): 511-518. DOI: 10.1038/nphoton.2012.175.

The Millennium Project based on 2006 energy survey report.

The World Nuclear Industry Status Report 2007" Archived 2008-06-25 at *the Wayback Machine*. pp. 23.

Tobias, I., Del Canizo C. and J. Alonso 2003. Crystalline silicon solar cells and modules." *In Handbook of PhotoVoltaic Science and Engineering*. Tobias, I., Del Canizo, C., Alonso, J., Eds.; John Wiley & Sons Ltd.: Chichester, U.K.

Torchynska, T. V. and G. P. Polupan. 2002. "III-V material solar cells for space application." *Semicond. Phys., Quantum Electron. Optoelectron.* 5: 63.

UN. 2012. "United Nations Global Initiative on Sustainable Energy for All." http:// sustainableenergyforall.org/objectives, accessed November 2012.).

US EIA. 2011. United States Energy Information Administration, *International Energy Outlook 2011.*

Walter, M. G et al. 2010. "Solar water splitting cells." *Chem. Rev.* 110(11): 6446–6473. DOI: 10.1021/cr1002326.

Xie, Meng-Yu, Su Kang-Yang, Peng Xin-Yuan, Wu Ren-Jang, Chavali Murthy, and Wei-Chen Chang. 2017. "Hydrogen production by photocatalytic water-splitting on Pt-doped TiO_2–ZnO under visible light." *Journal of the Taiwan Institute of Chemical Engineers* 70: 161-167. DOI: 10.1016/j.jtice.2016.10.034.

Ye, Sheng, Ding Chunmei, Chen Ruotian, Fan Fengtao, Fu Ping, Yin Heng, Wang Xiuli, Wang Zhiliang, Du Pingwu, and Can Li. 2018. "Mimicking the Key Functions of Photosystem II in Artificial Photosynthesis for Photoelectrocatalytic Water Splitting." *J. Am. Chem. Soc.* 140(9): 3250-3256. DOI: 10.1021/jacs.7b10662.

Yoshida, Masaaki et al. 2015. "Direct Observation of Active Nickel Oxide Cluster in Nickel−Borate Electrocatalyst for Water Oxidation by In Situ O K- Edge X-ray Absorption Spectroscopy." *J. Phys. Chem. C* 119:19279−19286. DOI: 10.1021/ja2011498.

Yuan, Qichen et al. 2017. "Noble-Metal-Free Janus-like Structures by Cation Exchange for Z-Scheme Photocatalytic Water Splitting under Broadband Light Irradiation." *Angew. Chem. Int. Ed.* 56: 4206 –4210. DOI: 10.1002/ange.201700150.

Zeman, Miro. 2002. "New trends in thin-film silicon solar cell technology." In *Advanced Semiconductor Devices and Microsystems, The Fourth International Conference ~~on~~*, pp. 353-362. Zeman, M. Ed.; Institute of Electrical and Electronics Engineers: New York.

In: Photocatalysis
Editors: P. Singh, M. M. Abdullah, M. Ahmad et al.
ISBN: 978-1-53616-044-4

Chapter 4

BIOPOLYMER BASED PHOTOCATALYSTS AND THEIR APPLICATIONS

***Marzieh Badiei*[1,*], *Nilofar Asim*[2], *Masita Mohammad*[2], *Md Akhtaruzzaman*[2], *Nowshad Amin*[3] and *M. A. Alghoul*[4]**

[1]University of Applied Science & Technology, Mashhad, Khorasan Razavi, Iran
[2]Solar Energy Research Institute, Universiti Kebangsaan Malaysia, Malaysia
[3]Institute of Sustainable Energy, Universiti Tenaga Nasional, Malaysia
[4]Center of Research Excellence in Renewable Energy Research Institute, King Fahd University of Petroleum & Minerals, Dhahran, Saudi Arabia

ABSTRACT

Photocatalysis is an innovative practical technique which has been widely used to overcome environmental challenges. Photocatalysts are chemically activated upon exposure to ultraviolet irradiation that causes degradation and even demineralization of pollutants. So far, studies have been focused on titanium dioxide (TiO_2) as semiconductor photocatalyst. Experimental researches have been directed toward the development of TiO_2-based photocatalysts with proper choice of immobilization support to prepare bifunctional adsorbent-photocatalyst composites with industrial photo-oxidation performance. The natural biopolymers have a special affinity for metal and metal oxides. They are new biomatrix for semiconductor photocatalysts that are appropriate for wastewater treatment. This review consists of three main sections. Initially, the necessity of immobilization of photocatalytic nanoparticles is explained. Then, the main structural characteristics of chitosan, alginate and cellulose as adsorbent and templating support are

* Corresponding Author's E-mail: gbadiei317@gmail.com.

explained. The third section of this chapter summarizes the potential applications of biopolymer-based photocatalysts composites.

Keywords: photocatalysis, photocatalyst, TiO_2, biopolymer, chitosan, alginate, cellulose

1. INTRODUCTION

Photocatalysis technique is generally illustrated as an oxidative activity of metal oxide semiconductors under the UV/visible irradiation to accelerate the reaction (Herrmann, 1999; Golabiewska et al. 2018). Photocatalysis process is an essential engineering method in environmental remediation and wastewater treatment for removal of pollutants (Fabiyi and Skelton, 2000; Zainal et al. 2005; Alinsafi et al. 2007). Development of photocatalyst systems for industrial applications is of great importance. Photocatalysts in current scientific researches are mainly consisted of titanium dioxide (TiO_2), Zinc oxide (ZnO), tungsten trioxide (WO_3), etc., however, it is quite clear that they are not the ideal catalysts for this purpose. According to reports (Hashimoto and Fujishima, 2005; Magalhaes et al. 2017), titanium dioxide (TiO_2) as the low-cost reactive photocatalyst, is attracted to scientists more than other metal oxides for the environmental treatment process (Zaleska-Medynska A., 2018).

In recent years, semiconductors were found responsive catalysts for the elimination of contaminants in the industrial and environmental treatment process (Pozzo et al., 1997). The mechanism of photcatalysis is based on advanced oxidation processes (AOPs). In this process, the reactive oxygen species are able to degrade the pollutants in wastewater (Stasinakis, 2008). Activation of catalyst takes place when a photon of ultraviolet band gap energy is absorbed to produce electron donor/electron acceptor as oxidation and reduction positions. Organic molecules are then oxidized to carbon dioxide (CO_2) and water whereas inorganic particles are transformed into their ionic species. photocatalyst nanoparticles with the large surface area are generally utilized in the suspension system in the conventional treatment process (Haarstrick et al. 1996; Wang et al. 2013). The process, however, is time consuming since colloidal catalyst particles should be separated and recycled from the liquid. Metal oxide nanoparticles tend to aggregate in water solutions, so, their effective surface area as well as their catalytic activity is decreased (Zhang et al. 2008). In parallel, polluting materials mostly organic compounds with polar surfaces have low adsorption ability to catalyst surface (Leopore et al. 1996; Torimoto et al. 1997). Another reason for decreasing the catalytic efficiency is the absorption of irradiated sunlight by particles mainly catalysts particles as well as dissolved organic compounds that gradually inhibit light penetration in the slurry. Having homogenous and well separate nanosized TiO_2, the photocatalyst can be immobilized on an appropriate porous solid materials such as silica (Li et al. 2010), glass (Khataee et al.,

2009), carbon nanotube (Woan et al. 2009) and polymers as stabilizing or protective agents to provide adsorbent-photocatalyst composite. The synergy between large surface area and high absorption capacity provides composites of higher activity and selectivity for generation of desired products. In this composite, adsorbent has acted as a support for immobilization and dispersion of the active phase while increases the surface area of the final solid. The adsorbed reactants on adsorbent site readily transfer to active sites of semiconductor through the interface between support and catalyst nanoparticles. The supported nanoparticle catalysts as the active phase are easily dispersed into the solution that consequently reduces the wash out of catalyst as an important issue. Providing an appropriate porous network likewise assist the particular target species to migrate to the active site of the catalyst. The valuable properties of polymers including high porous structure, surface area as well as mechanical strength have made them ideal as support for metal oxide nanoperticles. The first composite of TiO_2 nanoparticles immobilized on the surface of synthetic polythene (PE) film was used for photodegradation of phenol (Tennakone et al. 1995). The process was reported as an effective and cheap method. Nevertheless, synthetic polymers release toxic chemical by-products such as antioxidants and plasticizers when exposed directly to oxidizing radical species generated during the photocatalytic process by sunlight. The catalysts are not considered environmentally-friendly and the process is not sustainable.

Recently, natural biopolymers have been utilized as templates for TiO_2 nanoparticles (Crini, 2005; Kim et al. 2010) due to distinctive characteristics mainly biodegradability, high activity, biocompatibility, non-toxicity and adsorption properties. There are numerous reports about preparation methods and photocatalytic activity of TiO_2 – biopolymer composites for the elimination of contaminants and wastewater treatment (Mahmoodi et al. 2007; Ajmal et al. 2014). The interaction of metal ions with reactive functional groups of polymers is sufficiently strong that make polymers ideal matrix to entrap the photocatalyst nanoparticle. Entrapment of nanoparticles within the polymeric matrices has been applied for immobilization of TiO_2. Natural biopolymers from renewable resources are a suitable matrix for embedment of certain particles (Sarrouh et al., 2007). Different techniques have been developed to immobilize and stabilize the TiO_2 nanopowders on the support. Deposition of the nanoparticles on the solid substrates such as glass, synthetic polymers, biopolymers, thin films, etc. is a common method.

This chapter reports the overview of the main findings of renewable natural biopolymers such as chitosan, alginate, and cellulose for immobilization of nanoparticles. In addition, the methods to modify the physico-chemical properties of biopolymers have also been reported in this chapter, which affect the productivity of supported photocatalyst.

2. Natural Polymers for Photocatalytic Materials

Natural biopolymers from renewable resources have received increased attention as template for photocatalyst nanoparticles. The unique properties of these abundant biopolymers specifically high absorption capacity enhance their potential as green support in preparation of organic-inorganic nanoparticle photocatalysts. In the following sections, structural characteristics of chitosan, alginate and cellulose biopolymers as well as biopolymer based composites have been discussed.

2.1. Chitosan

Chitosan $(C_6H_{11}NO_4)_n$ is a carbohydrate biopolymer made up D-glucosamine and N-acetyl-D-glucosamine monomers. This can be derived by partial deacetylation of natural biopolymer chitin through hydration or enzymatic hydrolysis. Chitin that is obtained from the shells of crustacean as well as fungi cell walls is a polymer composed of N-acetyl-2-amino-2-deoxy-D-glucopyranose units (Zargar et al. 2015).

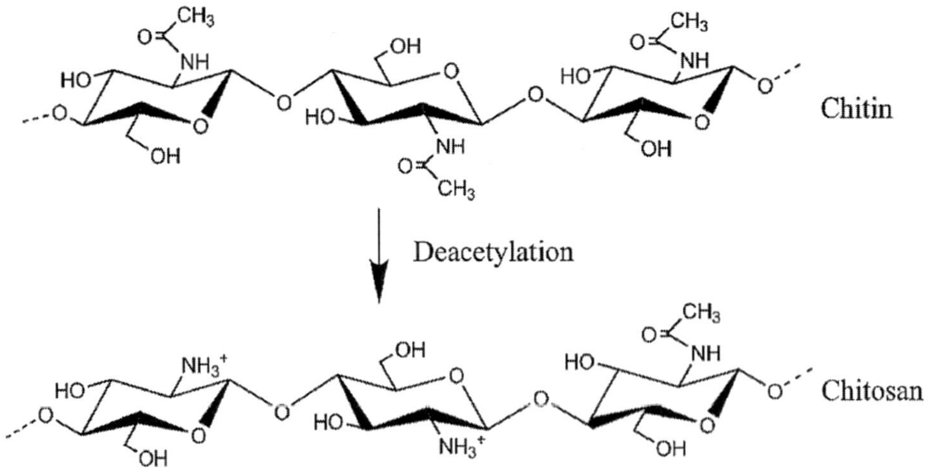

Figure 1. Molecular structure of chitosan. Reprinted from *Carbohydrate Polymers*, Vol 199, (Hamedi et al., 2018, Chitosan based hydrogels and their applications for drug delivery in wound dressings: A review, Pages 445-460, Copyright (2018), with permission from Elsevier.

Diverse applications of chitosan, mainly depends on the chemical structure as well as molecular size of this polymer (Guibal, 2004). The functional groups including amino group as well as primary and secondary hydroxyl groups at the C-2, C-3 and C-6 are responsible of high reactivity of chitosan. The amino groups are the key factors that cause intra- and inter-molecular hydrogen bonding. Chitosan with high degree of deacetylation has polycationic nature that could toughly bind to negatively charged molecules and chelates metal ions that result in modification of physiochemical characteristics of final compounds (Majeti and Kumar, 2000). The hydrogen bonds between chitosan molecules have made it insoluble in water. Protonation of amino groups is consequent of dissolution that leads to generation of positively charged chitosan ($RNH3^+$). The resultant product is readily soluble in water and organic acids while partially dissolve in some inorganic acids (Hamdine et al. 2005; Aranaz et al. 2009). Chitosan is biocompatible and biodegradable with low toxicity and antimicrobial property (Hamalainen et al. 1972).

Functional groups available on chitosan and its derivatives has made them efficient adsorbent for cationic and ionic dyes (Vakili et al. 2014) and heavy metals (Chen et al. 2008) as persistent contaminants in water and wastewater. The effectiveness of chitosan as adsorbent of acid dyes has been widely considered and is attributed to protonated amino groups of chitosan in acidic conditions. Increasing the degree of deacetylation of chitosan has normally increased the protonated amine groups that lead to increase in adsorption of anionic dyes. Although dye molecules are reasonably adsorbed on hydroxyl groups, amine groups are main active functional groups of chitosan in adsorption of dyestuffs via electrostatic attractive interaction. The studies (Azlan et al. 2009, Ngah et al. 2010, Naseeruteen et al. 2018) revealed that adsorption of acid dyes such as Acid Red 37 (AR37) onto chitosan-based adsorbents was a physical adsorption. Naseeruteen et al. (2018) studied removal of Malachite Green (MG) as cationic dye using chitosan hydrogel. Adsorption of basic dyes, however, involves hydrogen and covalent bonding with hydroxyl groups on chitosan biopolymer.

According to reports (Li et al. 2008; Walalawela, 2014), chitosan revealed as suitable biomatrix for inorganic nanoparticles such as CuO, ZnS, CdS, and ZnO. The photo degradation process of dyestuffs comprises complete mineraliztion of organic molecules that does not introduce other pollutants to environment (Ngah et al. 2011). Among the metal oxides, TiO_2 is introduced as the potential photocatalyst where complete degradation of organic pollutants achieved. Many kinds of chitosan-supported metal oxides were studied as photocatalyst such as ZnO (Salehi et al. 2010) and CuO (Chen et al. 2008), however, TiO_2 –chitosan composites are the most ever studied one in the articles. In addition, Incorporation of TiO_2 nanoparticle with chitosan as stabilizing support can effectively prevent agglomeration of nanopowders and overcome the problem of recovery of nanoparticles (Li et al. 2008).

The first titanium/chitosan composite catalysts were prepared by Zainel and his coworkers (Zainal et al., 2009) to treat organic dye contaminants under visible light illumination. Following studies also confirmed the enhanced photocatalytic performance due to higher adsorption capacity of chitosan/TiO_2 or ZnO composites. Following that (Nawi et al., 2011) simply immobilized TiO_2-chitosan onto surface of glass as support to remove the phenol contaminants. They found that porous structure improved the photocatalytic reaction due to increase in diffusion of contaminants and increase in light penetration on the surface of catalysts. Hence, they used the similar systems to remove reactive red 4 dye (RR4) in their later studies (Nawi and Sheilatina, 2012). TiO_2-Chitosan materials were also prepared which were able to adsorb both acidic eg. Benzopurpurin (BP) and basic e.g., Methylene Blue (MB) dyes. The UV light irradiation is responsible for photodegradation of both dyes with TiO_2 (Zubeita et al., 2008).

In spite of all advantages, chitosan suffers from serious drawbacks including low porosity and low surface areas (Zhang et al. 2016) which may lower the ability of chitosan in adsorption of pollutants. Therefore, modified chitosan composites were prepared to overcome these drawbacks. Through physical and chemical modification materials with desired properties were produced to overcome the main limitations. Physical methods usually mean converting chitosan powder into gel beads, membranes and film or converting to nanoparticles to improve the mechanical properties of chitosan chains and increase its diffusion property (Acharyulu et al. 2013). Regarding this issue (Cunff et al., 2015) showed that the immobilization of thin film photocatalyst significantly reduced some problems related to heterogeneous photocatalyst. They prepared thin and stable layer of the TiO_2/chitosan composite photocatalyst for degradation of organic dye. Chitosan/TiO_2 nanocomposite membrane gel also exhibited higher ability in separation of ethanol–water mixture as model system (Nasikhudin et al. 2018).

Besides, chemical modification improves the flexibility and chemical stability of chitosan and reduces its weakness in acidic conditions. The main chemical methods are grafting (Abdelwahab and Helaly, 2017), cross-linking (Mahmoud et al., 2018) and impregnation (Essawy, 2017). First of all, they are performed on the functional groups of chitosan and mostly on primary amino groups. However, crosslinking reactions lead to reduction of free amine groups on chitosan and lowers reactivity of chitosan biopolymer toward metal ions. The functional groups of cross-linking agents compensate this problem. Cross-linking reactions improve the mechanical strength of chitosan and increase its stability in acidic solutions (Chiou et al., 2004) As an example, photo-oxidation of cross-linked Chitosan-epichlorohydrin (CS-ECH) performed by means of immobilized photocatalyst. In this experiment formation of carbonyl groups and elimination of amine groups in oxidized CS-ECH film were detected. The new photocatalysts system revealed fast photodegradation of phenol compared to both TiO_2/CS-ECH and TiO_2/oxidized-CS catalyst composites (Jawad and Nawi, 2012).

Increasing the number of reactive groups on surface of chitosan expand its applications which is obtained through grafting of monomers and polymers to the functional groups (Abdelwahab and Helaly, 2017). For example, grafting of carboxymethyl chitosan (CMC) with polyacrylamide improved the stability, adsorption capacity as well as photocatalytic activity. The TMPAM-g-CMC photocatalyst composite employed for removing Congo Red (CR) dye under irradiation with visible light. Stability of catalyst was confirmed by its recyclability.

Impregnation of special functional groups into chitosan has improved its adsorption ability. The aim of all of these researches is to indicate how the adsorption characteristic, selectivity and reusability of chitosan are improved through physical and chemical modification processes.

In Majority of studies photoactivity of TiO_2 nanoparticles is entirely motivated under UV irradiation. However, photocatalysis reaction under visible light has attracted attentions. There have been many efforts to get this done through deposition of TiO_2 surface with noble metals (Asahi et al. 2001; Pathakoti et al. 2013, Zhang et al. 2017). Furthermore, coating and doping of metal and non-metal ions on the TiO_2 surface enhance the photocatalytic activity in the presence of visible light. Metal ion dopants such as Fe, Co, Ni, Mn or V were used in the studies (Chiou et al. 2004, Song et al. 2017). In fact, non-metal doping agents have been more effective in generating highly efficient photoactive catalysts in visible region of light (Su et al. 2008; Devi and Kavitha, 2013; Marschall and Wang, 2014). However, some weakness hinders their further development. While C doped catalysts usually need a high temperature (Dong et al. 2011; He et al. 2017), F doped ones are known as highly toxic to human and environment (Zhu et al. 2012). Then, N doping has been introduced as a suitable technique. According to (Shao et al., 2015), the TiO_2-chitosan based composites with N-doping (Cs/TiO_2) has been a better photocatalyst rather than $NTiO_2$ in the presence of visible light. Additionally, coupling of two semiconductors is a novel approach. For instance, the synthesized TiO_2/ZnO/chitosan nanocomposite thin films have demonstrated acceptable photoactivity for eliminationof methyl orange (Zhu et al. 2012). According to TEM images, the hybrid TiO_2/ZnO was dispersed and uniformed in the chitosan films.

The valuable characteristics of chitosan have promoted its application in many industrial areas; however, better understanding of this biopolymer is yet needed.

2.2. Alginate

Alginate is an abundant polysaccharide and a biodegradable and water soluble natural biopolymer. Alginate as a copolymer is composed of two different monomers including α–L guluronic acid (G-blocks) and β– D–mannuronic acid (M-blocks) with homopolymeric regions of G-blocks and homopolymeric region of M-blocks that are alternatively connected along the polymer chain. The G blocks contribute to intermolecular crosslinking with divalent Ca^{+2} that means increasing the G blocks controls the mechanical strength of hydrogel (Aslani and Kennedy, 1996; Yang et al. 2011). Alginates are provided from variety of sources and usually show different chemical structures meaning different concentration of G-blocks and M-blocks in a chain resulted in higher and lower rigidity (Hay et al. 2010). Figure 2 demonstrates the molecular structure of alginate chain.

Alginate has free hydroxyl and carboxylic groups evenly distributed in the polymer chain and are responsible of high reactivity of this polymer. Carboxylate groups, however, exist as the most reactive functional group that are protonated and form hydrogen bonds (Pawar and Edgar, 2012). As a result, alginate is able to form stable and water insoluble hydrogels with high surface areas. The spherical porous hydrogel network is produced through intermolecular crosslinking of poly guluronic acid blocks (G-blocks) with divalent cations (Sikorski et al. 2007). Varieties of divalent cations such as Pb^{+2}, Cu $^{+2}$, Cd^{+2}, Ca^{2+} and Ba^{2+} are able to chelate with alginate, however, Ca^{+2} mostly used in gel formation (Morch et al. 2006). The MG blocks provide weaker gel structures (Donati et al. 2005). Therefore, higher G block content in alginate chain is responsible for stronger gel networks. The final network structure has an appearance in common with egg-box (Grant et al. 1973). Figure 3 depicted the egg-box structure.

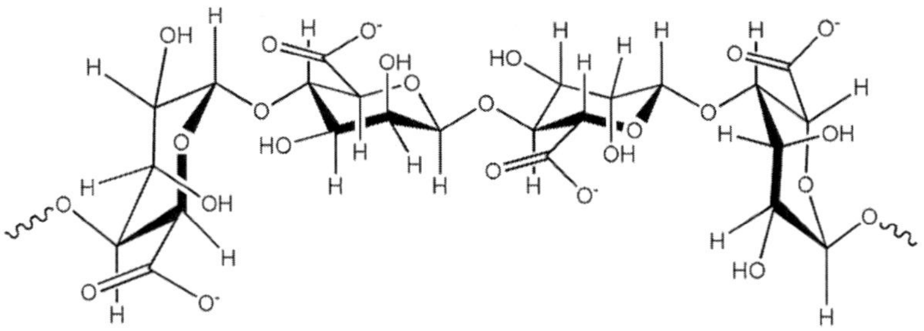

G **M** **M** **G**

Figure 2. Molecular structure of alginate, G: guluronic acid, M: mannuronic acid. Reprinted from *Journal of Hazardous Materials*, Vol 258, (Vipin et al. 2013, "Prussian blue caged in alginate/calcium beads as adsorbents for removal of cesium ions from contaminated water", Pages 93-101, Copyright (2018), with permission from Elsevier.

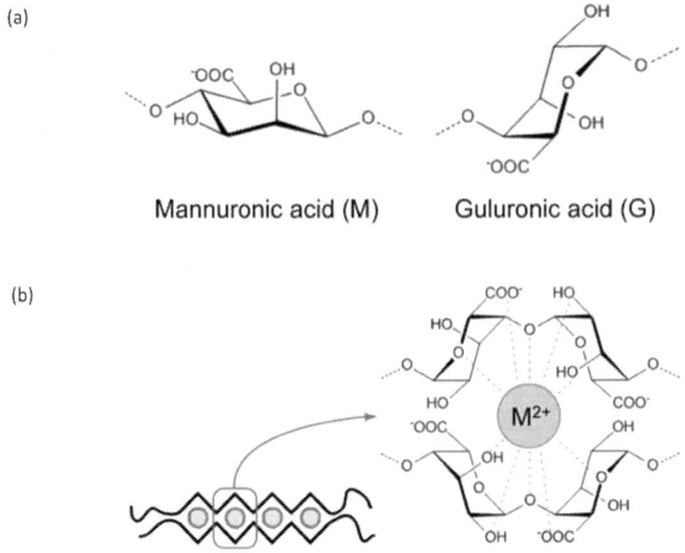

Figure 3. (a) Molecular structure of monomers in alginate: mannuronic acid (left) and guluronic acid (right). (b) "egg-box model". Reprinted from *Carbohydrate Polymers*, Vol 170, (Simo et al. 2017), "Research progress in coating techniques of alginate gel polymer forcell encapsulation", Pages 1-14, Copyright (2018), with permission from Elsevier.

In this process negative charges of the glucoronic units and mannuronic units of alginate chain at pH >5 can attract the opposite electric charge of metal cations. Enhanced hydrophobicity and further rigidity of alginate films is obtained from crosslinking with metal cations specifically calcium ions.

Calcium alginate (CaAlg) hydrogel is biocompatible and has mechanical stability with low solubility in water and ability to absorb metal ions that make it suitable biopolymer employed in industrial environmental applications. In fact, carboxylic groups of alginate beads facilitate adsorption and then elimination of contaminants. The polymeric CaAlg with controlled porosity appeared to be a suitable adsorbent to remove divalent ions of Pb^{2+} and Cd^{2+} from wastewater (Nita et al. 2007).

Recently, several studies described the use of alginate biopolymer as templating agent for preparation of metal oxide photocatalysts (Kim et al., 2010; Zhao et al. 2014). In several studies (Harikumar et al. 2013; Gopalakannan and Viswanathan, 2015; Kanakaraju et al. 2017) Calcium alginate was introduced as an efficient scaffold for immobilizing metal oxide nanoparticles and was used for development of eco-friendly immobilization system for wastewater decontamination. Significant compatibility of alginate biopolymer with TiO_2 ensures easy dispersion of nanoparticles onto CaAlg porous fibers (Gjipalaj and Alessandri, 2017). In practice the photocatalytic systems are usually expected to be reusable that is resulted from good compatibility between metal oxide and the stabilizing support to prevent the release of metal oxide nanoparticles into the solution. The beads have properties like high transparency, light penetration and ability to extend their porosity which made them possible for potential applications as stabilizing support for photocatalyst Nanoparticles (NPs). For example, TiO_2 nanoparticles immobilized in Ca–Alg beads were successful in photodegradation of methylene blue (Albarelli et al. 2009) as organic dye and removal of 86.7% of Cu(II) as heavy metals (Kanakaraju et al. 2017). The small size of the resulting NPs and their large surface area leads to higher photocatalytically potential. Following this strategy, separation and recycling procedure of photocatalyst powders from solution is also eliminated. Several reports have emphasized on priority of embedding/encapsulation of the metal oxide in alginate materials as support matrix compare to deposition of nanoparticles on external surface of adsorbent. Encapsulation provides more effective and more stable photocatalytic materials. To avoid fast degradation of photocatalyst, encapsulated particles in hydrogel should not have chemical interaction with components of hydrogel(Gjipalaj and Alessandri, 2017). According to results, the photodegradation mostly happens on the surface layer of composite film film while its inner part may not change after UV irradiation (Zhao et al. 2014). However, uniform dispensability of photocatalyst-immobilized in CaAlg beads or CaAlg fibers in reaction solution still remained a challenge for researchers. In addition, results revealed that supported TiO_2 often has lower photoactivity since the active surface area of metal oxide nanoparticles was reduced after immobilization. For example, most of the TiO_2 nanoparticles are embedded in a large microsphere and do not show their photocatalytic character.

The promising features of resulting nanoparticles are small size and large surface area. Bezbaruah remarked that open lattice structure of the calcium alginate (Bezbaruah et al. 2011), allows the interaction of photocatalysts and contaminants to facilitate fast diffusion of metal ions. Besides, biocompatibility of alginate and chitosan in the form of hydrogel beads has been effective in their application as immobilization support (Gjipalaj and Alessandri, 2017). Further, hydrogel beads with bigger size would make their recovery easier through simple sedimentation. TiO_2-gel beads presented good stability in water after several uses under UV irradiation (Albarelli et al. 2009). In one study (Molly et al. 2016), crosslinked composite of Ba/Alg/CMC/TiO_2 hydrogels was synthesized and utilized for removal of Methylene blue (MB) from textile effluents. The photostability of the new photocatalyst was confirmed.

It had been suggested that the alginate fibrils more efficiently stabilize the metal oxides than alginate beads do. The explanation is that fibers possess reactive and flexible cavity to restrictly stabilize the metal oxide nanoparticles. For example, CaAlg polymer fibers have showed higher productivity in decolonization of methyl orange from wastewater (Zhu et al., 2012). In another study by Reveendran and Ong studied, NaALg-TiO_2 composite thin film utilized for removal of congo red (CR) as an anionic diazo dye (Reveendran and Ong, 2018). Based on their previous study, the removal of malachite green greatly improved through integration of TiO_2 in the polymer network of thin films. TiO_2 nanoparticles can be well dispersed and stabilized in sodium alginate (NaALg) as well as calcium alginate (CaAlg) aqueous solution. Zhao in 2014 and Reveendran in 2018, prepared TiO_2/CaAlg composite to degrade methyl orange. The composite thin films were fabricated by immobilizing TiO_2 in sodium alginate matrix through crosslinking with divalent metal ions. The important point is that recycled TiO_2 retained the same photocatalytic efficiency. In fact TiO_2/calcium alginate composite film represented better photocatalytic activity compared to TiO_2/calcium alginate composite beads (Zhao et al. 2014; Reveendran and Ong, 2018).

According to reported results (Harikumar et al. 2013) the chemical and structural features of dyes directly affect on the degradation rate by TiO_2 that is attributed to their surface area and adsorption of light by dyes. Calcium alginate hydrogel suffers from weak mechanical stiffness that made it at risk of breaking up in aqueous solutions. To solve this weakness problem, Wei in 2016, prepared really robust cross-linked hydrogels using alginate and acrylamide. The prepared polyacrylamide/calcium alginate/TiO_2 (PAM/CA/T) film revealed excellent flexibility and toughness that didn't break up after swelling in NaCl solutions and showed better absorption capability as well as reusability (Wei et al. 2016).

Numerous studies were performed on hybrid photocatalysts and immobilization on on Ca-Alg gel for elimination of heavy metal cations (Harikumar et al., 2013, Esmat et al.

2017). In one study (Kanakaraju et al., 2017) TiO_2/ZnO–CaAlg composite was prepared through integration of calcium alginate beads and hybrid photocatalyst TiO_2/ZnO. Those TiO_2/ZnO–CaAlg beads proved higher photocatalytic property in comparison to bare calcium alginate beads in removal of Cu ion contaminants.

A hybrid photocatalytic/ultrafiltration membrane was developed through dispersing TiO_2 photocatalysts and stabilizing into the porous matrix of CaAlg fibers (Papageorgiou et al. 2012). The fibers increased the absorption capacity of the photocatalytic membrane film that caused up to three folds enhancement of the Methyl Orange removal efficiency from polluted water by the porous Ca Alg/TiO_2 fibers.

Surface functionalization of nanoparticles with noble metals (eg. Au, Ag) has increased the minimum wavelength of photoactivity to visible region. Buaki-Sogo (2013) prepared TiO_2 supported Au NPs immobilized on alginate beads as photosensitizer that presented high photoactivity at visible light for generation of H_2 gas from mixture of methanol and water. Photocatalytic efficiency of Au/TiO_2 samples increased successfully by eight times using this novel biopolymer templating method (Buaki-Sogo et al. 2013).

Alginate–based composites appeared with high reactivity, perfect mechanical stability and compatibility, which introduced alginate as a desirable support for photo-degradation processes.

2.3. Cellulose

Cellulose $(C_6H_{10}O_5)_n$ as the principal structural component of plants and most abundant carbohydrate is mainly obtained from plants. This simple polysaccharide is a long unbranched chain made up of glucose units that are hold together by B- β-(1-4)-glycosidic linkages to form cellobiose as the building block of cellulose chain (Klemm et al. 1998; Klemm et al. 2005). Each glucose unit contains three hydroxyl groups on C2, C3 and C6 (Ummartyotin and Manuspiya, 2015). Figure 4 exhibits the structural chemical formula of cellulose.

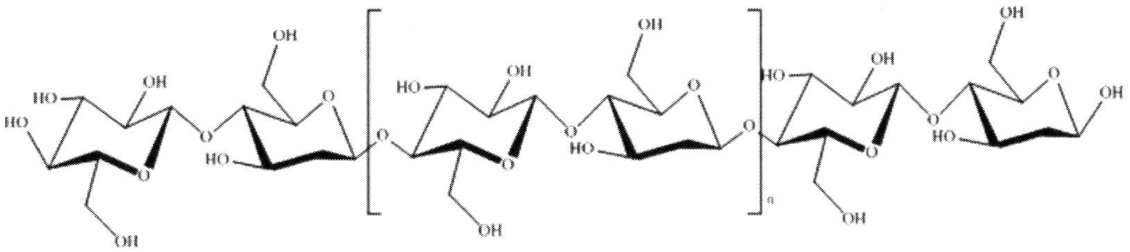

Figure 4. Structural formula of cellulos. Reprinted from *Renewable and Sustainable Energy Reviews*, Vol 41, (Ummartyotin and Manuspiya 2015), "A critical review on cellulose: From fundamental to an approach on sensor technology", Pages 402-412, Copyright (2018), with permission from Elsevier.

Cellulose fibrils consist of crystalline and amorphous regions. Upon disconnection of crystalline regions from amorphous sections, nanocellulose is produced (Tang et al. 2017). The nanocelluloses have distinctive characteristics such as hydrophobicity, large surface area with ease of surface modification. Since nanocelluloses are produced from different origins through different processing conditions, two main classes including cellulose nanocrystals (CNCs) and cellulose nanofibrils (CNFs) have been obtained both are nanosized cellulose fibers with superior physiochemical properties and mechanical stability than the origin cellulose (Moon et al., 2011). However, CNCs and CNFs have different size in nanometer regime and different composition. Cellulose nanofibril (CNFs) particles are finer cellulose fibrils and contain both amorphous and crystalline region.

Cellulose nanocrystals (CNCs) obtained through acid hydrolysis of wood fibers is rigid particles having up to hundreds of nanometers length. The active hydroxyls on the linear cellulose nanocrystls has made them suitable candidate for development of functional nanomaterials (Lima and Borsa, 2018). There are evidences from studies that CNCs as active adsorbents are utilized in water purification processes and as support for nanoparticles as photocatalyst.

Abundant hydroxyl groups on the glucose monomers of the cellulose chain are reactive sites for chemical modification of cellulose surface. The hydroxyl groups provide a cationic nature to the cellulose nanocrystals, therefore, cellulosic materials are applied as alternative adsorbent to remove acid dyes of textile effluent (Kumar et al. 2015), cationic dyes (Zhou et al. 2013), heavy metals like PbII (Lu et al., 2016) and Cr (VI) (Li et al. 2012). The first report on capability of CNCs in adsorption of cationic dyes mainly methylene blue was released by He (He et al. 2013). Thereafter, Batmaz in 2014, improved the adsorption capability of CNCs through surface modification (Batmaz et al. 2014)that raised the charge density of polymer. Enhancement of the surface adsorption capacity is possible through modification of raw cellulose employing different methods such as physical adsorption (Kumar et al. 2017), chemical modifications (Trejo-O'reilly et al. 1997), crosslinking or surface functionalization with noble metals (Snyder et al. 2013).

Modification is associated with incorporation of chemical moieties to the cellulose surface for enhancement of binding affinity to cellulose nanomaterial to prepare desirable materials (Hokkanen et al. 2016). Modification of cellulosic materials has been achieved by both organic and inorganic functional groups. In this way, the potential applications of cellulosic materials in specific fields maybe expanded (Lam et al. 2012, Wang et al. 2014).

Inorganic materials have similarly been used for surface modification of nanocrystalline cellulose due to their specific catalytic properties. Accordingly, metal oxide nanoparticles have attracted great attention among scientists. The deposition of these nanoparticles such as TiO_2 on the surface of nanocrystalline cellulose prepared hybrid materials with photocatalytical characteristic (Yu et al. 2000) that have been extensively used in oxidation and reduction processes and photo degradations of dyes including methylene blue or methyl orange. The photocatalyst was prepared through immobilization of TiO_2 in cellulosic matrix. For example (Nagaokaa et al. 2002) examined photodegradation of solid organic materials using TiO_2-carbon composite prepared from cellulose/TiO_2 composites. Furthermore, photodegradation of phenol was examined when irradiated with UV light (Zeng et al. 2010). In both studies, the photocatalyst was prepared thorugh immobilization of TiO_2 in cellulosic matrices.

However, the close contact of photcatalysts and organic pollutants simply affect on degradation of cellulose and limits the long-term applications of these composite materials. It is well known that, TiO_2 may generate active radical oxygen species when exposed to UV irradiation. The generated radicals perform photodegradation along with cleavage of cellulose chain (Marques et al. 2006). Having these limitations in mind, surface modification of cellulose and nanoparticles improves properties of the TiO_{2}-cellulose composites.

CNCs are able to form stable colloidal dispersions to produce 2D films or 3D composites. The 3D composite of cellulose hydrogel with increased surface area can increase the adsorption capacity of polymer. Nanocellulose – based aerogels are stable and have large surface area with high porosity. They are renewable, biocompatible, biodegradable, thus, might be applied for many purposes in chemical and biological applications (Long et al. 2018). Yang in 2014, fabricated cellulose hydrogel through crosslinking of the cellulose with epichlorohydrin, NH_4OH and the TiO_2 nanoparticles. Increasing the surface area has increased the adsorption capacity of cellulose (Yang et al. 2014). In the first place, the methyl orange in the aqueous solution was easily adsorbed on the cellulose gel because of its porous structure. Degradation of the adsorbed methyl orange has stimulated adsorption of MO from the solution resulting in photodegradation by TiO_2 nanoparticles. In another study (Kettunen et al. 2011), surface modification caused decline in inherent hydrophilicity of CNs and improved tendency to hydrophobic compounds and anionic contaminants. The consequence is elimination of organic materials from wastewater. In this experiment, entangled native cellulose nanofibrils formed highly porous aerogel network, which is a superabsorbent. Since TiO_2-coated nanocellulose aerogels showed photocatalytic activity, it can be used for water purification. In similar experiment, the cellulose nanofibrils aerogel functionalized with a hydrophobic but oleophilic coating, such as TiO_2. The resulting material is able to absorb oil when is floated on water (Korhonen et al. 2011). Modification of nanocrystaline materials with positively charged functional groups allows collection and removal of anionic impurities (Wang et al. 2016). Addition of amine groups to crystalline nanocellulose resulted in complete elimination of chromium (VI) from solution.

Water resistant microfibrillated cellulose (MFC)—polyamide-amine-epichlorohydrin (PAE)—titanium dioxide (TiO_2) composites were prepared. PAE was first added to a

MFC suspension, followed by TiO_2 addition. Here, PAE is used both as wet strength agent to consolidate the MFC structure and as a retention aid for the TiO_2 NPs. The prepared composite was tested for decontamination of methyl orange (MO) aqueous solutions under UV light. MO is an azo dye of relatively high toxicity and poor biodegradability which provides a good reference for waste water residues from the textile and dying industries. Results show that only TiO_2 nanoparticles (NPs) were contributed to photocatalytic activity. The composites had the same photocatalytic efficiency even after 3 times reproducible experiments that it confirms the reproducibility and stability of the composite. Further, the key feature of this composite is that it does not produce any residual contamination because no loss of TiO_2 Nps was monitored during the photocatalysis process (Garusinghe et al. 2018).

A group of scientists synthesized graphen oxide -TiO_2 nanocrystals hybrid materials whenTiO_2 nanoparticles were tightly immobilized on graphene oxide sheets followed by calcination at appropriate temperature. The graphene oxide-TiO_2 hybrid material performed well when exposed to near UV light. Then bacterial cellulose polymer employed as biomatrix for GO-TiO_2 nanoparticles to prepare GO-TiO_2/BC composite that holds antibacterial properties (Liu et al., 2017).

So far, a number of noble-metal-doped semiconductor composites were fabricated to expand their photocatalytic activity into the visible region (Chan and Barteau, 2005). For example, Synder in 2013, fabricated TiO_2-CNF composite films including Au–TiO_2 and Ag–TiO_2 nanoparticles in ultraviolet and/or visible light for degradation of methylene blue (MB) as recalcitrant organic compounds in water (Snyder et al., 2013). The results proved the recyclability of fabricated films and showed that composites were able to degrade methylene blue by 75% in UV-Vis spectrum. Morawski (2013), also provided TiO_2/nitrogen-modified cellulose composites with high UV-Vis light absorption (Morawski et al., 2013). The composite was prepared thorough hydrogen bonding of $NTiO_2$ and microcrystalline cellulose in ionic liquid (Janus et al. 2008).

3. APPLICATIONS

Photocatalysis has been considered as the promising method in overcoming global barriers to water recycling and water reuse. The photocatalysis process benefits from many advantages including oxidizing strength, reasonable operational temperature, and sustainability associated with the abundant sunlight resource. Photocatalytic reactions induced by ultraviolet and/or visible light irradiation were utilized for decontamination of water resources. Trace amount of any kind of pollutants in water resources is a serious treat for human health; hence, semiconductor metal-oxide photocatalysts would be promising and effective method in removal of various complicated pollutants.

3.1. Photocatalytic Degradation of Organic Pollutants

Heterogeneous photocatalysts have been utilized to accelerate photodegradation of organic contaminants of industrial chemicals and pharmaceuticals in water and wastewater. Removal of dyes from textile effluents has always been significant issue of interest in last decades (Vandevivere et al. 1998). Dye stuffs are an important part of varieties of wastewaters because many industries mainly textiles, papers, leathers and printing, discharge huge quantity of hazardous synthetic dyes in water effluents. Thus treatment of wastewaters before discharging into water resources is vital (Holkar et al. 2016).

Dyestuffs are very persistent in environment and their decomposition in nature is generally slow because they are manufactured with thermal, chemical, and photolytical stability (Forgacs et al. 2004). Azo dyes are synthetic chemicals utilized in variety of industries (Shaul et al. 1991). Majority of azo dyes have been described with low biodegradability and solubility and high toxicity. The planar N = N- double bond is commonly linked to aromatic rings such as benzene or naphthalene and when the double bond is cleaved decoloration of dyes happens.

Several physical and chemical approaches have been examined to achieve satisfactory decolorization and decontaminating such effluents. Among them adsorption is the most practical technique (Cestari et al. 2004). But it ends up in secondary pollution which needs to be processed further. Natural biopolymers (Twu et al. 2003) have been reported as very successful and efficient adsorbents for organic dyes (Ding et al. 2009), metal ions (Fu and wang, 2011) and inorganic pollutants (Xie et al. 2013). For example, organic molecules are adsorbed by chitosan through the electrostatic attraction of the –NH_2 groups and solutes, while stable coordination of metal ions and chitosan is explained through chelating groups of the chitosan (Chiou et al. 2004). The problem is that adsorption cannot be considered as a complete treatment method because this process ends up in secondary pollution which needs to be processed further. Furthermore, adsorption technique just transfer one form of waste to another form e.g., liquid to solid (Twu et al. 2003;Vijaya et al. 2008). Nevertheless, oxidative degradation has been extensively applied for treatment of complicated pollutants to more simple and non-toxic biodegradable molecules. Oxidative degradation has already demonstrated a good and efficient rate of degradation for organic compounds (Sahel, 2007;AbuTariq et al. 2008).

Photocatalysis has also been introduced as the more promising technique for elimination of dyes from wastewater (Khataee et al. 2009). This method utilize strong oxidizing species such as •OH and •O radicals to decompose the macromolecules (Rauf and Ashraf, 2009). The macromolecule is often entirely decomposed into water and carbon dioxide. Photoactive semiconductors (Chen and Ray, 2001) specifically TiO_2, ZnO, Fe_2O_3 and/or CdS have been reported very useful in decolorization of water resources by decolorization of the dye solutions. Titanium dioxide (TiO_2) (Gayya and

Abdullah, 2008) has attracted extensive attention since it has potential in removal of organic contaminants (Mahmoodi et al. 2006).

TiO_2 biocompatible composites have widely been used for removal of dye. Sodium Alginate/titania nanoparticle for removal of textile dyes (Mahmoodi et al. 2011), Ca-alginate/titanium dioxide beads (Alg/TiO_2) for cationic dyes (Lam et al. 2017) and methylene blue (Chen and Zhang, 2008), alginate/carboxymethyl cellulose/TiO_2 nanocomposite hydrogel for Congo Red dye (Molly et al. 2016), TiO_2/chitosan for removal of monoazo dye (Zainal et al. 2009), Benzopurpurin (BP) (Zubieta et al. 2008) and Pt/TiO_2/alginate hydrogels for removal of phenolic compounds (Andreozzi et al. 2000; Chen et al. 2008; Chen et al. 2008; Jawad and Nawi, 2012; Nawi and Sheilatina, 2012) have been reported. The TiO_2-biopolymer composite photocatalysts have shown faster mineralization rate of contaminant in environmental treatments. The photocatalyst systems, therefore, have been considered as an environmentally friendly method through oxidization of pollutant macromolecules. Hence, TiO_2 and natural biopolymers can complement each other with their own advantages when the polymer as support for metal oxides photocatalysts (Zhu et al. 2013).

Moreover, contaminant pharmaceutical wastes are contaminants occurring in trace amounts and consists different group of organic chemicals harmful for human health and other living organisms. Many studies are focused on identifying the pharmaceuticals and residues of pharmaceuticals in industrial wastewaters (Kanakaraju et al. 2017). Majority of pharmaceuticals are soluble in water, while some others do not breakdown in the environment for long time depending on the environment media. Pharmaceutical effluents have a complex physiochemical nature since they are mixtures of various types of pharmaceuticals and other contaminants, thus, advanced treatment namely oxidative photocatalysis technique have been applied to their degradation (Tijani et al. 2016). However, the reports revealed that behavior of the drug compounds toward TiO_2 photocatalysis is complex. Chitosan and chitosan-based polymers have been introduced as effective and environmentally friendly adsorbent for removal of pharmaceuticals and the residues from polluted water (Kyzas et al. 2013). Effective processes for eliminating pharmaceuticals using TiO_2-aliginate beads have been reported (Sarkar et al. 2015).

3.2. Photocatalytic Disinfection of Biological Pollutants

Biological contaminants in water resources include microorganisms such as viruses and bacteria as well as pathogens (Montgomery and Elimelech, 2007;Shannon et al. 2008). The presence of these types of pollutants in water resources is a permanent threat for environment and humans health and the safety of drinking waters and water resources is an important issue (Bryant et al. 1992, Byrne et al. 2015). To remove these contaminants, water is disinfected commonly through chemical methods by addition of chlorine. However, hazardous and carcinogenic disinfection byproducts (DBP) (Sadiq and Rodriguez, 2004) are potentially produced during water chlorination. Moreover, development of resistance to chlorine has been reported (How et al. 2017). Even though advances in disinfection materials and methods have been effective, many of them need costly chemicals and equipment. Hence, development of alternative water treatment methods for effective removal of waterborn infections for providing safe and sustainable water sources is of great significance. Photocatalysis (Glover et al. 2006) is a recent approach, a feasible alternative, and a promising application for the removal of pollutant from the environment using sunlight. So far, photocatalysis (Markowska-Szczupak et al. 2011, Dimitroula et al. 2012) has been considered as a recent, alternative and promising method and TiO_2 (Matsunaga et al. (1985); (1988) has been studied as preferred photocatalyst (Sunada et al. 1998; Li et al. 2008; Tiwari et al. 2008).

The intention of photocatalysis process is to generate and utilize the reactive oxygen species (ROS) mainly hydroxyl free radical (-°OH) (Maness et al. 1999) (Huang et al. 2005) to destroy organic and biological molecules that cannot be oxidized by the conventional oxidants. They are converted to carbon dioxide and water vapor through complete mineralization of compounds (Jacobe et al. 1998). Hydroxyl radicals are nonselective species that are able to react with varieties of contaminants since they are attached to surface of hydrated metal ions to suppress the activity of pathogens.

By far, the microorganism *E. coli*, has been largely studied in drinking water systems (Sökmen et al. 2001; Rincón and Pulgarin, 2004). Cho and his coworkers (2005), published the first report on linear correlation between presence of reactive radicals and inhibition of E. coli in decontamination of water supplies (Cho et al. 2005). As provided by the study (Jacobe et al. 1998), organic molecules constitute around 96% of dry weight of microbe's cells. They are complex molecules that are degraded by reactive oxygen species. Maximum 1% of dry weight is inorganic ions (Knabner-Kogel, 2002) and the remaining of the dry weight includes monomer precursors as well as sugars. Findings of most of the researches have revealed that oxidation of aforementioned organic constitutes and inorganic ones for destruction of cell membrane happens when microorganisms exposed to photocatalytic situations (Wei et al. 1994; Largeau and De Leeuw, 1995).

Despite of all advantages, photocatalytic treatment has its challenges. The properties that make metal oxides effective catalyst could be improved through doping with metal ions (Reddy et al. 2007; Pradeep and Anshup, 2009; Liga et al. 2011) non-metal ions (Liu et al. 2007) and inorganic oxides (Markowska-Szczupak et al., 2011). However, the preferred doping agents are non-metal ions i.e., C, S, N, P that are not toxic and their applications has not been limited. As a result of doping, light absorption capacity of TiO_2 has shown a shift from UV region to visible region of spectrum. The antimicrobial activity of the photocatalysts partly depends to uniformly dispersing of nanoparticles and avoiding their aggregation during application. Natural Biopolymers with antimicrobial activity have been utilized as suitable support for semiconductors in in treatment of wastewaters (Rabae et al., 2003). Decolonization ability of TiO_2/biopolymer composites support disinfection of the water resources (Albarelli et al., 2009; Gilpavas et al. 2014). For example, chitosan has shown inhibitory affect against different microorganism (Rhim et al. 2006; Raafat et al. 2008; Jarmila and Vavríkova, 2011). The chitosan with positive nature may attract bacteria with opposite charge (Holappa et al., 2006) while chitosan has the ability to chelate metal ions to enhance its antimicrobial activity (Nehra et al., 2018) and controls production of toxic materials and growth of bacteria. For example, assay of Ag nanoparticles-chitosan composite revealed that composite has higher antimicrobial activity than that of each constituents (Pinto et al. 2012). According to study (Zhang et al. 2017), TiO_2 nanoparticles was uniformly dispersed into chitosan to prepare chitosan-TiO_2 film that showed significant antimicrobial activity against four strains, i.e., *Escherichia coli*, *Staphylococcus aureus*, *Candida albicans*, and *Aspergillus Niger* irradiated with visible light for food pathogenic microbes. Additionally, TiO_2- calcium alginate beads were successfully utilized for the photocatalytic inhibition of pathogenic microorganisms (Gilpavas et al., 2014). Furthermore, chitosan with different molecular weight (MW) (No et al., 2002) has shown different reaction to diverse pathogen. The microbial action is promoted with lower degree substituted chitosan at lower pH <6 because at acidic conditions the density of protonated amino groups and correspondingly positive charge of chitosan has increased (Rabae et al., 2003). In spite of all findings, disinfection mechanism of biopolymers for the targeted purposes should be more realized (Kong et al., 2010).

Recently, cellulosic fabric loaded with TiO_2 nanoparticles have been utilized for the development of antimicrobial fabrics (Khan et al., 2015; Khafagha et al., 2016). In addition to this, a thin film has also been prepared from the regenerated bacterial cellulose (RBC) -TiO_2 nanocomposites for the inhibition of *E. coli* bacterial growth. Moreover, titanium dioxide (TiO_2) with antibacterial properties was applied to hydrophobic and hydrophilic dressing materials for minimizing the development of serious wound infections (Haugen and Lyngstadaas 2016). In addition, the nanocomposite dressing prepared from combinations of poly(N-vinylpyrrolidone) (PVP)/chitosan/TiO_2 has showed excellent antimicrobial ability, excellent hydrophilic nature, high swelling properties and good biocompatibility make it suitable for wound healing dressing (Archana et al. 2013; Bui et al., 2017).

Recently, research on antimicrobial food packaging has increased. Variety of antimicrobial compounds has been evaluated in the form of films and edible forms as well as synthetic polymers. Moreover, food packaging films incorporating TiO_2 NPs have been commonly used. These TiO_2-based nanocomposites (Othman et al., 2014) and chitosan-TiO_2 composite films (Zhang et al., 2017) have presented antibacterial property. photocatalytic paper sheets have also been introduced for mineralization of the wastewater (Rehim et al., 2016). Paper sheets composed of TiO_2/NaAlg nanocomposites revealed high antibacterial effect against *salmonella typhimurium* that made it suitable for food packaging applications.

3.3. Photocatalytic Degradation of Inorganic Pollutants

Removal of inorganic contaminants including heavy metals and residual ions from contaminated water has been a complicated environmental issue. The toxic and non-biodegradable inorganic ions such as copper (Cu^{2+}), nickel (Ni^{2+}) cadmium (Cd^{2+}), lead (Pb^{2+}), and zinc (Zn^{2+}) and their derivatives have dangerous effects on environment and human health (Karvelas et al., 2003, Hua et al., 2012). Therefore, many researches focused on development of suitable adsorbents for heavy metal ions. Adsorption has known as a cost effective and eco-friendly technique (Crini, 2005). Biopolymer-based materials are emerging, easy available and environmentally-friendly adsorbents (Yu et al., 2000; Özer et al., 2005; Sud et al., 2008).

Chitosan and its derivatives that contain reactive functional moieties possess distinctive chelating and adsorption abilities for heavy metals in solutions. There is significant increase in reports on utilization of chitosan as adsorbent for heavy metals in polluted waters (Babel and Kurniawan, 2003). Dragon (2014) studied chitosan hydrogels as adsorbent for heavy metals. Jiang (Jiang et al., 2014) reported high capacity of crosslinked chitosan for Cu^{+2}. Karthik and Meenakshi (2015) reviewed the ability of chitosan-grafted-polyaniline composites in removal of chromium (VI) ions (Karthik and Meenakshi, 2015). The adsorption capacity of chitosan definitely enhanced after grafting (Dragan, 2014), crosslinking (Wang et al., 2013) and impregnation (Kuang et al. 2013).

The carboxylic groups of alginate polymers enhanced its affinity to heavy metals (Fourest and Volesky 1995). For example, calcium – alginate hydrogel beads utilized for encapsulation of iron-based nanoparticles (CaAlg-Ni/Fe beads) and the composite showed significant capacity for removal of Cu^{+2} (Kuang et al., 2013). Moreover, Showky in 2011, prepared Alg-montmorilloni (clay) composite indicating a high affinity to Pb^{+2} (Showky, 2011). There has been many efforts devoted to efficiently elimination of metals through photocatalysis process (Pekakis et al., 2006; Choi et al., 2007). However, limitations of photocatalytic methods have been reported. As an instance, release of doped metal ions in water has cause serious treats to environment. The supported photocatalysts have been more practical in heavy metal removal while biopolymers reported as highly efficient support for metal oxide nanoparticles. Biopolymers are porous with larger surface area. For example, there are many studies on elimination of Cr(VI), Cd(II), Cr(III), Cu(II), and Co(II) utilizing TiO_2-immobilized on CaALg beads (Wu, Wei et al. 2012). Hybrid catalyst TiO_2/ZnO–CaAlg beads have been applied for elimination of Cu^{+2} (Kanakaraju et al. 2017)

According to the following reactions, mechanism of photocatalytic process is illustrated based on inorganic species and metal-ion reduction eq (1)- (4)(Kabra et al. 2004):

$$TiO_2 + hv(UV) \rightarrow TiO_2\,(e + h^+) \quad (1)$$

$$h^+ + H_2O \rightarrow HO\bullet + H^+ \quad (2)$$

$$h^+ + OH^- \rightarrow HO\bullet \quad (3)$$

$$Cu^{2+} + 2e^- \rightarrow Cu^0 \quad (4)$$

The h^+ represents strong oxidation agent that upon reaction with water produces hydroxyl radicals as powerful oxidant. Then metal cation Cu^{+2} adsorbed onto adsorbent e.g., CaAlg is reduced to elemental Cu^0. Sun and coworkers studied visible-light induced photocatalysis of Cr(VI) using P25 TiO_2 (Sun et al., 2005). Photoreduction of aqueous hexavalent chromium reported as highly efficient and without generation of secondary pollution (Qiu et al., 2012).

In spite of all findings, elimination of heavy metals from water resources is still remained demanding and requires more investigations.

CONCLUSION

Heterogeneous semiconductor photocatalysis has been introduced as effective and sustainable method for treatment of varieties of organic, inorganic and biological contaminants due to their presence in the water resources and environment. The importance of this technology as partly has been evidenced by numerous published reports in this area. Photocatalysis is demonstrated as inexpensive and environmentally friendly technology for water treatment due to utilizing sunlight irradiations, while it has been used successfully in many developing countries to remove pathogens in water supplies.

Semiconductor metal oxide nanoparticles specifically TiO_2 nanoparticles are widely used for efficient management of wastewaters as a consequence of high activity and stability during irradiation. The emergence of TiO_2 immobilized onto the natural biopolymers such as cellulose, chitosan, alginate etc. enhanced their properties and applications. The new composites with higher activity and mechanical properties, excellent biodegradability and compatibility are desirable candidates for photodegradation processes. Now it is possible to prepare photocatalysts with enhanced and beneficial activities for specific applications to solve many serious environmental and pollution challenges. Despite the improved performance and tremendous potential of biopolymer-based photocatalysts in the past decade, this technology suffers from some essential problems and limitations that hinder its commercialization. The main limitations are post-treatment for recovery of nanoparticles after oxidative process, enhancement of photocatalysis in visible region of spectrum and uniform distribution in aqueous suspension and biopolymer matrix. In addition, surface modification has been the reason of extension of the absorption edge of catalysts toward visible region of spectrum.

This chapter provides a comparative evaluation of different natural biopolymers which have been introduced as alternative support for metal oxide photocatalysts mainly TiO_2 NPs. The effectiveness of these biopolymers has been illustrated by their practical applications in various decontamination processes. However, more extensive benefits of biopolymer-based metal oxides photocatalysts could be attained after eliminating their inherent limitations. Many studies have been carried out with the scientific communities to overcome the limitations, however, further research is required to expand the application of the oxidative nanoparticle composites in photodegradation of pollutants.

ACKNOWLEDGMENT

The authors would like to acknowledge the financial support by DIP-2015-028 research funds.

REFERENCES

Abdelwahab N.A. and Helaly F.M. 2017 "Simulated Visible Light Photocatalytic Degradation of Congo Red by Tio_2 Coated Magnetic Polyacrylamide Grafted Carboxymethylated Chitosan." *Journal of Industrial and Engineering Chemistry* 50: 162-171.

Abu Tariq A., Faisal M.M., Saquib M. Andmuneer M. 2008 "Heterogeneous Photocatalytic Degradation of an Anthraquinone and A Triphenylmethane Dye Derivative in Aqueous Suspensions of Semiconductor." *Dyes and Pigments* 76: 358-365.

Acharyulu S., Gomathi T. and Sudha P.N. 2013 "Physico-Chemical Characterization of Cross Linked Chitosan-Polyacrylonitrile Polymer Blends." *Der Pharmacia Lettre* 5:354-363.

Ajmal A., Majeed I., Malik R.N., Idriss H. and Nadeem M.A. 2014 "Principles and Mechanisms of Photocatalytic Dye Degradation on Tio_2 - Based Photocatalysts: A Comparative Overview." *Rsc Advances* 70: 37003-37026.

Albarelli J., Santos D.T., Murphy S. and Oelgemo-Ller M. 2009 "Use of Ca–Alginate as A Novel Support for Tio_2 Immobilization in Methylene Blue Decolorisation." *Water Science and Technology* 60: 1081-1087.

Alinsafi A., Evenou E.M., Abdulkarim N.M., Zahraa O., Benhammou A., Yacoubi A. and Nejemddin A. 2007 "Treatment of Textile Industry Wastewater by Supported Photocatalysis." *Dyes Pigments* 74: 439-445.

Andreozzi R., Caprio V., Insola A., Longo G. and Tufano V. 2000 "Photocatalytic Oxidation of 4-Nitrophenol in Aqueous Tio_2 Slurries: An Experimental Validation of Literature Kinetic Models." *Journal of Chemical Technology and Biotechnology* 75:131-7.

Aranaz I., Mengibar M. and Harris R. 2009 "Functional Characterization of Chitin and Chitosan." *Current Chemical Biology* 3: 203-320.

Archana D., Singh B.K., Dutta J. and Dutta P.K. 2013 "In Vivo Evaluationof Chitosan-Pvp-Titanium Dioxide Nanocomposite as Wound Dressing Material." *Carbohydrate Polymmers* 95: 530-539.

Asahi R., Morikawa T., Ohwaki T., Aoki K. and Taga Y. 2001 "Visible Light Photocatalysis in Nitrogen -Doped Titanium Oxides." *Science*293:269-271.

Aslani P. and Kennedy R.A. 1996 "Studies on Diffusion in Alginate Gels, I. Effects of Cross-Linking with Calcium or Zinc Ions on Diffusion of Acetaminophen." *Journal of Controlled Release* 42: 75-82.

Azlan K., Saimeb W.N.W. and Ken L. 2009 "Chitosan and Chemically Modified Chitosan Beads for Acid Dyes Sorption." *Journal of Environmental Sciences* 21: 296-302.

Babel S. and Kurniawan T.A. 2003 "Low-Cost Adsorbents for Heavy Metals Uptake from Contaminated Water: A Review." *Journal of Hazardous Materials* 97: 219-243.

Batmaz R., Mohammed N., Zaman M., Minhas G. and Berry R.M. 2014. "Cellulose Nanocrystals as Promising Adsorbents from the Removal of Cationic Dyes." *Cellulose* 21:1655-1665.

Bezbaruah A.N., Shanbhogue S.S., Simsek S. and Khan E. 2011. "Encapsulation of Iron Nanoparticles in Alginate Biopolymer for Trichloroethylene Remediation." *Journal of Nanoparticle Research* 13: 6673-6681.

Bryant E.A., Fulton G.P. and Budd G.C. 1992 *"Disinfection Alternatives for Safe Drinking Water."* Environmental Engineering Series. New York, Van Nostrand Reinhold.

Buaki-Sogo M., Serra M., Primo A., Alvaro M. and Garcia H. 2013 "Alginate as Template in the Preparation of Active Titania Photocatalysts." *Chemcatchem* 5: 513-518.

Bui. V.K.H., Park D. and Lee Y. 2017 "Chitosan Combined with Zno, Tio_2, and Ag Nanoparticles for Antimicrobial Wound Healing Applications: A Mini Review of the Research Trends." *Polymers* 9:1-21.

Byrne J.A., Dunlop P.S.M. and Hamilton J.W.J. 2015 "A Review of Heterougenous Photocatalysis for Water and Surface Disinfection." *Molecules* 20: 5574-5615.

Cestari A.R., Vieira E.F.S., Santos A.G.P.D., Mota J.A. and Almeida V.P.D. 2004 "Adsorption of Anionic Dyes on Chitosan Beads. The Influence of the Chemical Structures of Dyes and Temperature on the Adsorption Kinetics." *Journal of Colloid and Interface Science* 280: 380-386.

Chan S.C. and Barteau M.A. 2005 "Preparation of Highly Uniform Ag/Tio_2 and Au/Tio_2 Supported Nanoparticle Catalysts by Photodeposition." *Langmuir* 21: 5588-5595.

Chen A., Liu S., Chen C., and Chen C. 2008 "Comparative adsorption of Cu(II), Zn(II), and Pb(II) ions in aqueous solution on the crosslinked chitosan with epichlorohydrin." *Journal of Hazardous Materials* 154: 184–191.

Chen A.H., Liu S.C. and Chen C.Y. 2008. *Comparative Adsorption of Cu(II), Zn(II).*

Chen D. and Ray A.K. 2001 "Removal of Toxic Metal Ions from Wastewater by Semiconductor Photocatalysis." *Chemical Engineering Science* 56: 1561-1570.

Chen J.Y., Zhou P.J., Li J.L. and Wang Y. 2008 "Studies on the Photocatalytic Performance of Cuprous Oxide/Chitosan Nanocomposites Activated by Visible Light." *Carbohydrate Polymers* 72: 128–132.

Chen P. and Zhang X. 2008 "Fabrication of Pt/Tio_2 Nanocomposites in Alginate and Their Applications to the Degradation of Phenol and Methylene Blue in Aqueous Solutions." *Clean Soil Air Water* 36: 507-511.

Chiou M., P. Ho and Li H. 2004 "Adsorption of Anionic Dyes in Acid Solutions Using Chemically Cross-Linked Chitosan Beads." *Dyes and Pigments* 60: 69-84.

Choi H., Stathatos E. and Dionysiou D.D. 2007 "Photocatalytic Tio_2 Films and Membranes for the Development of Efficient Wastewater Treatment and Reuse Systems." *Desalination* 202: 199-206.

Cho M., Chung H., Choi W. and Yoon J. 2005 "Different Inactivation Behavior of Ms-2 Phase and E. Coli in Tio_2 Photocatalytic Disinfection." *Applied Environmental Microbiology* 71: 270-275.

Crini G. 2005 "Recent Developments in Polysaccharide-Based Materials Used as Adsorbents in Wastewater Treatment." *Progress in Polymer Science* 30: 38-70.

Cunff. J. Le, Tomasic V. and Gomzi Z. 2015 "Preparation and Photoactivity of the Immobilized Tio_2/Chitosan Layer." *Chemical Engineering Transactions* 43: 856-870.

Devi L.G. and Kavitha R. 2013 "A Review on Non Metal Ion Doped Titania for the Photocataly*tic* Degradation of Organic Pollutants Under Uv/Solar Light: Role of Photodegenerated Charge Carrier Dynamics in Enhancing the Activity." *Applied Catalysis B: Environmental* 140-141:559-587.

Dimitroula H., Daskalaki V., Frontistis Z., Kondarides D., Panagiotopoulou P. and Xekoukoulotakis N. 2012 "Solar Photocatalysis for the Abatement of Emerging Micro Contaminants in Wastewater: Synthesis, Characterization and Testing of Various Tio_2 Samples." *Applied Catalysis B: Environmental* 117-118: 283-291.

Ding Y., Zhao Y. and Tao X. 2009 "Assembled Alginate/Chitosan Microshells for Removal of Organic Pollutants." *Polymer* 50: 2841-2846.

Donati I., Holtan S., Morch Y.A., Borgogna M., Dentini M. and Skjak-Bræk G. 2005 "New Hypothesis on the Role of Alternating Sequences in Calcium-Alginate Gels." *Biomacromolecules* 6:1031-1040.

Dong F., Guo S., Wang H., Li X. and Wu Z. 2011 "Enhancement of the Visible Light Photocatalytic Activity of C-Doped Tio_2nanomaterials Prepared by A Green Synthetic Approach." *Journal of Physical Chemistry* 115: 13285-13292.

Dragan E.S. 2014 "Design and Applications of Interpenetrating Polymer Network Hydrogels: A Review." *Chemical Engineering Journal* 243: 572-590.

Esmat M., Farghali A.A., Khedr M.H. and El-Sherbiny I.M. 2017 "Alginate-Based Nanocomposites for Efficient Removal of Heavy Metal Ions." *International Journal of Biological Macromolecules* 55: 214-223.

Essawy A.A.S., El-Nggar S.M., El-Nggar, A.M. 2017 "Wastewater Remediation by Tio_2-Impregnated Chitosan Nano-Grafts Exhibited Dual Functionality: High Adsorptivity and Solar-Assisted Self-Cleaning." *Journal of Photochemistry and Photobiology B: Biology* 173: 170-178.

Fabiyi M.E. and Skelton R.L. 2000 "Photocatalytic Mineralization of Methylenen Blue Using Buoyant Tio_2-Coated Polystyrene Beads." *Journal of Photochemistry and Photobiology* A 132: 121-128.

Forgacs E., Cserhati T. and Ores G. 2004 "Removal of Synthetic Dyes Wastewaters: A Review." *Environmental International* 30: 953-971.

Fourest E. and Volesky B. 1995 "Contribution of Sulfonate Groups and Alginate to Heavy Metal Biosorption by the Dry Biomass of Sargassum Fluitans." *Environmental Science and Technology* 30: 277-282.

Fu F. and Wang Q. "Removal of Heavy Metal Ions from Wastewaters: A Review." *Journal of Environmental Managment* 92: 407-418.

Garusinghe U.M., Raghuwanshi V.S., Batchelor W. and Garnier G. 2018 "Water Resistant Cellulose – Titanium Dioxide Composites for Photocatalysis." *Scientific Reports* 8:2306:1-13.

Gayya U.I. and Abdullah A.H. 2008 "Heterogeneous Photocatalytic Degradation of Organic Contaminants Over Titanium Dioxide: A Review of Fundamentals, Progress and Problems." *Journal of Photochemistry and Photobiologyc: Photochemistry Reviews* 9: 1-12.

Gilpavas E., Acevedo J., Lopez L.F. and Gomez-Garciab M.A. 2014 "Solar and Artificial Uv Inactivation of Bacterial Microbes Byca-Alginate Immobilized Tio_2 Assisted by H2o2 Using Fludized Bed Photoreactor." *Journal of Advanced Oxidation Technologies* 17: 343-351.

Gjipalaj J. and Alessandri I. 2017 "Easy Recovery, Mechanical Stability, Enhanced Adsorption Capacity and Recyclability of Alginate-Based Tio_2 Macrobead

Photocatalysts for Water Treatment." *Journal of Environmental and Chemical Engineering* 5: 1763-1770.

Glover S., Gomez L.A., Reyes K. and Leal M.T. 2006 "A Practical Demonstration of Water Disinfection Using Tio_2 Films and Sunlight." *Water Research* 40: 3274-3280.

Gopalakannan V. and Viswanathan N. 2015 "Synthesis of Magnetic Alginate Hybrid Beads for Efficient Chromium (Vi) Removal." *International Journal of Biological Macromolecule* 72: 862-867.

Grant G.T., Morris E.R., Rees D.A., Smith P.J.C. and Thom D. 1973 "Biological Interactions between Polysaccharides and Divalent Cations: The Egg-Box Model." *Febs Letters* 32: 195-198.

Guibal E. 2004 "Interactions of Metal Ions with Chitosan-Based Sorbents: A Review." *Separation and Purification Technology 38*: 43-74.

Haarstrick A., Kut O.M. and Heinzle E. 1996 "Tio_2-Assisted Degradation of Environmentally Relevant Organic Compounds in Wastewater Using A Novel Fluidized Bed Photoreactor." *Environmental Science and Technology* 30:817-824.

Hamalainen C., H.S. Mard and Cooper A.S. 1972 "Comparison of Application Techniques for Deposition of Resins in Cotton Fibres." *American Dyestuff Reporter* 71: 30-38.

Hamdine M., Heuzey M. and Begin A. 2005 "Effect of Organic and Inorganic Acids on Concentrated Chitosan Solutions and Gels." *International Journal of Biological Macromolecules* 37: 134-142.

Hamedi H., Moradi S., Hudson S.M., Tonelli A. E. 2018. Chitosan based hydrogels and their applications for drug delivery in wound dressings: A review. *Carbohydrate Polymers* 199(1), 445-460

Harikumar P., Litty J. and Dhanya A. 2013 "Photocatalytic Degradation of Textile Dyes by Hydrogel Supported Titanium Dioxide Nanoparticles." *Journal of Environmental Engineering and Ecological Science* 2:1-9.

Hashimoto K.I. and Fujishima H. 2005 "Tio_2 Photocatalysis: A Historical Overview and Future Prospects." *Japanese Journal of Applied Physics* 44:8269-8285.

Haugen H.J. and Lyngstadaas S.P. 2016 "Antibacterial Effects of Titanium Dioxide in Wounds." 439-450.

Hay I.D., Rehman Z.U., Ghafoor A. and Rehm B.H.A. 2010 "Bacterial Biosynthesis of Alginate." *Journal of Chemical Technology and Biotechnology* 85: 752-759.

He D., Li Y., Wang I., Wu J. and Yang Y. 2017 "Carbon Wrapped and Doped Tio_2 Mesoporous Nanostructure with Efficient Visible Light Photocatalysis for No Removal." *Applied Surface Science* 391: 318-325.

He X., Male K.B.N., Esterenko P.N., Brabazon D. and Luong J. 2013 "Adsorption and Desorption of Methylene Blue on Porous Carbon Monoliths and Nanocrystalline Cellulose." *Acs Applied Material Interfaces* 5: 8796-8804.

Herrmann J.M. 1999 "Heterogenous Photocatalysis: Foundamentals and Applications to the Removal of Various Types of Aqueous Pollutants." *Catalysis Today* 53: 115-129.

Hokkanen S., Bhatnagar A. and Sillanpaa M. 2016 "A Review on Modification Methods to Cellulose-Based Adsorbents to Improve Adsorption Capacity." *Water Research* 91: 156-173.

Holappa J., Hjalmarsdottir M., Másson M., Rúnarsson O., Asplund T. and Soininen P. 2006 "Antimicrobial Activity of Chitosan N-Betainates." *Carbohydrate Polymer* 65: 114-118.

Holkar C.R., Jadhav A.J. and Pinjari D.V. 2016 "A Critical Review on Textile Wastewater Treatments: Possible Approaches." *Journal of Environmental Managment* 182: 351-366.

How Z.T., Kristiana I., F. Busetti, Linge K.L. and Joll C.A. 2017 "Organic Chloramine in Chlorine –Based Disinfected Water Systems: A Review." *Journal of Environmenta*l Science 58: 2-18.

Hua M., Zhang S., Pan B., Zhang W., L.V. and Zhang Q. 2012 "Heavy Metal Removal from Water/Wastewater by Nanosized Metal Oxides: A Review." *Journal of Hazardous Materials* 211-212: 317-331.

Huang L., Li D., Lin Y., Wei M., Evans D. and Duan X. 2005 "Controllable Preparation of Nano-Mgo and Investigation of Its Bactericidal Properties." *Journal of Inorganic Biochemistry* 99: 968-993.

Jacobe W., Maness P., Wolfrum E., Blake D. and Fennell J. 1998 "Mineralization of Bacteria Cell Mass on A Photocatalytic Surface in Air." *Environmental Science and Technology* 32: 2650-2653.

Janus M., Choin J. and Morawski A.W. 2008 "Azo Dyes Decomposition on New Nitrogen- Odified Anatase Tio_2 with High Adsorptivity." *Journal of Hazardous Materials* 166: 1-5.

Jarmila V. and Vavríková E. 2011 "Chitosan Derivatives with Antimicrobial, Antitumour and Antioxidant Activities--A Review." *Current Pharmaceutical Design* 17: 3596-3607.

Jawad A.H. and Nawi M.A. 2012 "Oxidation of Crosslinked Chitosan-Epichlorohydrine Film and Its Application with Tio_2 for Phenol Removal." *Carbohydrate Polymers* 90: 87-94.

Jiang W., Chen X., Pan B., Zhang Q., Teng L. and Liu L. 2014 "Spherical Polystyrene-Supported Chitosan Thin Film of Fsat Kinetics and High Capacity for Copper Removal." *Journal of Hazardous Material* 276: 295-301.

Kabra, K., R. Chaudhary and R.L. Sawhney (2004). "Treatment of Hazardous Organic and Inorganic Compounds through Aqueous-Phase Photocatalysis: A Review." *Industrial Engineering and Chemical Research* 43: 7683-7696.

Kanakaraju D., Shantini R. and Lim Y.C. 2017 "Combined Effects of Adsorption and Photocatalysis by Hybrid Tio_2/Zno-Calcium Alginate Beads for the Removal of Copper." *Journal of Environmental Science* 55: 214-223.

Karthik R. and Meenakshi S. 2015 "Removal of Pbii and Cd II Ions from Aqueous Solution Using Polyaniline Grafted Chitosan." *Chemical Engineering Journal* 263: 168-177.

Karvelas M., Katsoyiannis A. and Samara C. 2003 "Occurance and Fate of Heavy Metals in the Wastewater Treatment Process." *Chemosphere* 53: 1201-1210.

Kettunen M., Houbenov.N, Nykanen A. and Ruokolainen J. 2011 "Photoswitchable Superabsorbency Based on Nanocellulose Aerogels." *Advanced Functional Materials* 21: 510-517.

Khafagha M.R., Ali H.E. and El-Naggar A. 2016 "Antimicrobial Finishing of Cotton Fabrics Based on Gamma Irradiated Carboxymethyl Cellulose/Poly (Vinyl Alcohol)/Tio_2 Nanocomposites." *Journal of the Textile Institute* 107: 766-773.

Khan S., Ui-Islam M., Khattak W.A. andul Lah M. 2015 "Bacterial Cellulose-Titanium Nanocomposites: Nanostructural Characteristics, Antibacterial Mechanism, and Biocompatibility." *Cellulose* 22: 565-579.

Khataee A, Pons M. and Zahraa O. 2009 "Photocatalytc Degradation of Three Azo Dyes Using Immobilized Tio_2 Nanopartcles on Glass Plates Activated by Uv Light Irradiation: Influence of Dye Molecular Structure." *Journal of Hazardous Materials* 168: 451-457.

Kim Y., Neudeck C. and Walsh D. 2010 "Biopolymer Templating as Synthetic Route to Functional Metal Oxide Nanoparticles and Porous Sponges." *Polymer Chemistry* 1: 272-275.

Klemm D., Philpp B., Heinze T., Heinze U. and Wagenknecht W. 1998 *"Comprehensive Cellulose Chemistry." Volume 1: Fundamentals and Analytical Methods*, Wiley-Vch Verlag Gmbh.

Klemm D., Heublein B., Fink H. and Bohn A. 2005 "Cellulose: Fascinating Biopolymer and Sustainable Raw Material." *Angewandte Chemi* 44: 3358-3393.

Knabner-Kogel I. 2002 "The Macromolecular Organic Composition of Plant and Microbial Residues as Inputs to Soil Organic Matter." *Soil Biology and Biotechnology* 34: 139-162.

Kong M., Chen X., Xing K. and Park H. 2010 "Antimicrobial Properties of Chitosan and Mode of Action: A State of the Art Review." *International Journal of Food Microbiology* 144: 51-63.

Korhonen J.T., Kettunen M., Ras R.H.A. and Ikkala O. 2011 "Hydrophobic Nanocellulose Aerogels as Floating, Sustainable, Reusable, and Recyclable Oil Absorbents." *Acs Applied Materials and Interfaces* 3: 1813-1816.

Kuang Y., Du J., Zhou R., Chen Z., Megharaj M. and Naidu R. 2013 "Preparation of Triethylene –Tetramine Grafted Magnetic Chitosan for Adsorption of Pb II Ion from Aqueous Solutions." *Journal of Hazardous Materials* 260: 210-219.

Kumar G.N., Kumar V., Misra N. and Varshney L. 2015 "Cellulose Based Cationic Adsorbent Fabricated Via Radiation Grafting Process for Treatment of Dyes Waste Water." *Carbohydrate Polymers* 132: 444-451.

Kumar R., Sharma R.K. and Singh A.P. 2017 "Cellulose Based Grafted Biosorbents - Journey from Lignocellulose Biomass to Toxic Metal Ions Sorption Applications - A Review." *Journal of Molecular Liquids* 232: 62-93.

Kyzas G., Kostoglou M. and Lazaridas N. 2013 "Environmental Friendly Technology for the Removal of Pharmaceutical Contaminants from Wastewater Using Modified Chitosan Adsorbents." *Chemical and Engineering Journal* 222: 248-258.

Lam E., Male K.B., Chong J.H., Leung A.C.W. and Leung J.H.T. 2012 "Applications of Functionalized and Nanoparticle-Modified Nanocrystalline Cellulose." *Trends in Biotechnology* 30: 283-290.

Lam W., Chong M., Horri B.A. and Tey B. 2017 "Physicochemical Stability of Calcium Alginate Beads Immobilizing Tio_2 Nanoparticles for Removal of Cationic Dye Under Uv Irradiation." *Applied Polymers* 134:1-8.

Largeau C. and De Leeuw J.W. 1995 "Insoluble, Nonhydrolyzable, Aliphatic Macromolecular Constituents of Microbial Cell Walls." *Advances in Microbial Ecology*. J.J.G. Boston, Ma, Springer. 14.

Leopore G.P., Persaud L. and Langford C.H. 1996 "Supporting Titanium Dioxide Photocatalysts on Silica Gel and Hydrophobically Modified Silica Gel." *Journal of Photochemistry & Photobiology A: Chemistry* 98: 103-111.

Li C., Zhang Y., Peng J., Wu H., Li J. and Zhai M. 2012 "Adsorption of Cr(Vi) Using Cellulose Microsphere-Based Adsorbent Prepared by Radiation-Induced Grafting." *Radiation Physics and Chemistry* 81: 976-970.

Li Q., Mahendra S., Lyon D., Brunet L., Liga M. and Li D. 2008 "Antimicrobial Nanomaterials for Water Disinfection and Microbial Control: Potential Applications and Implications." *Water Research* 42: 4591-4602.

Li Q., Su H.J. and Tan T.W. 2008 "Synthesis of Ion-Imprinted Chitosan–Tio_2 Adsorbent and Its Multi-Functional Performances." *Biochemical Engineering Journal* 38: 212-218.

Li Y., Wang W., Zhan Z. and Woo M. 2010 "Photocatalytic Reduction of Co_2 with H2o on Mesophorous Silica Supported Cu/Tio_2 Catalysts." *Applied Catalysis B: Environmental* 100: 386-392.

Liga M., Bryant E., Colvin V. and Li Q. 2011 "Virus Inactivation by Silver Doped Titanium Dioxide Nanoparticles for Drinking Water Treatment." *Journal of Photochemistry and Photobiology A: Chemistry* 190: 94-100.

Lima M.M.D.S. and Borsa R. 2018 "Rodlike Cellulose Microcrystals: Structure, Properties, and Applications." *Macromolecular Rapid Communications 39*: 771-787.

Liu L., Yang X., Ye L., Xuea D., Liu M., Jia S. and Hou Y. 2017 "Preparation and Characterization of A Photocatalytic Antibacterial Material: Graphene Oxide/Tio_2/Bacterial Cellulose Nanocomposite." *Carbohydrate Polymers* 174: 1078-1086.

Liu Y., Li J., Qiu X. and Burda C. 2007 "Bactericidal Activity of Nitrogen-Doped Metal Oxide Nanocrystals and the Influence of Bacterial Extracellular Polymeric Substances (Eps)." *Journal of Photochemistry and Photobiology A: Chemistry* 190: 94-100.

Long L., Weng Y.X. and Wang Y.Z. 2018 "Cellulose Aerogels: Synthesis, Applications, and Prospects." *Polymers* 10:623-29.

Lu J.J., Liu R.N., Wang C., Ouyang Y.F. 2016 "Magnetic Carboxylated Cellulose Nanocrystals as Adsorbent for the Removal of Pb(II) from Aqueous Solution." *International Journal of Biological Macromolecules* 93: 547-556.

Magalhaes P.A., Nunes L. and Mendes O.C. 2017 "Titanium Dioxide Photocatalysis: Foundamnetals and Application on Photoinactivation." *Review on Advance Sciences* 51: 91-129.

Mahmoodi N.M., Arami M., Limaee N., Gharanjig K. and Nourmohammadian F. 2007 "Nanophotocatalysis Using Immobilized Titanium Dioxide Nanoparticle: Degradation and Mineralization of Water Containing Organic Pollutant: Case Study of Butachlor." *Materials Research Bulliten* 42: 797-806.

Mahmoodi N.M., Arami M., Limaee N.Y. and Tabrizi N.S. 2006 "Kinetics of Heterogeneous Hotocatalytic Degradation of Reactive Dyes in an Immobilized Tio_2 Photocatalytic Reactor." *Journal of Colloid and Interface Science* 295: 159-164.

Mahmoodi N.M., Hayati B., Arami M. and Bahrami H. 2011 "Preparation, Characterization and Dye Adsorption Properties of Biocompatible Composite(Alginate/Titania Nanoparticle." *Desalination* 275: 159-193-110.

Mahmoud M.E., Gehan M., Abdel-Aal N.H., Fekry N.A. and Maher M.O. 2018 "Imprinting "Nano-Sio2-Crosslinked Chitosan-Nano-Tio_2" Polymeric Nanocomposite for Selective and Instantaneous Microwave-Assisted Sorption of Hg(II) and Cu(II)." *Acs Sustainable Chemical Engineering* 6: 4564-4573.

Majeti N.V. and Kumar R. 2000 "A Review of Chitin and Chitosan Applications." *Reactive and Functional Polymers* 46: 1-27.

Maness P., Smolinska S., Blake D., Huang Z., Wolfrum E. and Jacobe W.A. 1999 "Bactericidal Activity of Photocatalytic Tio_2 Reaction: Toward an Understanding of Its Killing Mechanism." *Applied Environmental Microbiology* 65: 4094-4098.

Markowska-Szczupak A., Ulfig K. and Morawski A. 2011 "The Application of Titanium Dioxide for Deactivation of Bioparticulates: An Overview." *Catalysts Today* 169: 257-269.

Marques P., Trindade T. and Neto P.C. 2006 "Titanium Dioxide/Cellulose Nanocomposites Prepared by A Controlled Hydrolysis Method." *Composites Science and Technology* 66: 1038-1044.

Marschall R. and Wang L. 2014 "Non –Metal Doping of Transition Metal Oxides for Visible –Light Photocatalysis." *Catalysis Today* 225: 111-135.

Matsunaga T., Tomoda R., Nakajima T. and Wake H. 1985 "Photoelectrochemical Sterilization of Microbial Cells by Semiconductor Powders." *Fems Microbiology Letters* 29: 211-214.

Molly T., Naikoo G.A., Din Sheikh M.U. and Bano M. 2016 "Effective Photocatalytic Degradation of Congo Red Dye Using Alginate/Carboxymethyl Cellulose/Tio_2 Nanocomposite Hydrogel Under Direct Sunlight Irradiation." *Journal of Photochemistry and Photobiology A: Chemistry* 327: 33-43.

Montgomery M. and Elimelech M.V2007 "Water and Sanitation in Developing Countries: Including Health in the Equation." *Environmental Science and Technology* 17:9741-9748.

Moon R.J., Martini A., Nairn J., Simonsen J., Youngblood J. 2011. "Cellulose nanomaterials review: structure, properties and nanocomposites". *Chem. Soc. Rev.* 40 (2011) 3941-3994

Morawski A.W., Ewelina K., Jacek P., Kordala R. and Pernak J. 2013 "Cellulose-Tio_2 Nano-Composite with Enhanced Uv–Vis Light Absorption." *Cellulose* 20: 1293-1300.

Morch Y., Donati I., Strand B. and Skjåk-Bræk G. 2006 "Effect of Ca2+, Ba2+, and Sr2+ on Alginate Microbeads." *Biomocromolecules* 7: 471-480.

Nagaokaa S., Hamasakic Y., Ishiharab S., Negata M.I., K. and Ihara H. 2002 "Preparation of Carbon/Tio_2 Microsphere Composites from Cellulose/Tio_2 Microsphere Composites and Their Evaluation." *Journal of Molecular Catalysis A: Chemical* 177: 255-263.

Naseeruteen F., Hamid N., Suah F., Ngah W. and Mehamod F. 2018 "Adsorption of Malachite Green from Aqueous Solution by Using Novel Chitosan Ionic Liquid Beads." *International Journal of Biological Micromolecules* 107:1270-1277.

Nasikhudin, M., Diantoro A., Kusumaatmaja A. and Triyana K. 2018 "Stabilization of Pva/Chitosan/Tio_2 Nanofiber Membrane with Heat Treatment and Glutaraldehyde Crosslink." *Materials Science and Engineering* 367: 213-220.

Nawi M.A. and Sheilatina S.S. 2012 "Photocatalytic Decolourisation of Reactive Red 4 Dye by an Immobilised Tio_2/Chitosan Layer by Layer System." *Journal of Colloid and Interface Science* 372: 80-87.

Nawi M.A., Jawad A.H., Sabar S. and Ngah W.S.W. 2011 "Immobilized Bilayer Tio_2/Chitosan System for the Removal of Phenol Under Irradiation by A 45 Watt Compact Fluorescent Lamp." *Desalination* 280: 288–296.

Nehra P., Chauhan R., Garg N. and Verma K. 2018 "Antibacterial and Antifungal Activity of Chitosan Coated Iron Oxide Nanoparticles." *Journal of Biomedical Science* 75: 13-18.

Ngah W.S.W., Ariff N., Hashim A. and Megat Hanafiah M.A.K. 2010 "Malachite Green Adsorption Onto Chitosan Coated Bentonite Beads: Isotherms, Kinetics and Mechanism." *Clean Soil Air Water* 38: 394-400.

Ngah W.S.W., Teong L.C. and Megat Hanafiah M.A.K. 2011 "Adsorption of Dyes and Heavy Metal Ions by Chitosan Composites: A Review." *Carbohydrate Polymer* 83: 1446–1456.

Nita I., Iorgulescu M. and Spiroiu M.F. 2007 "The Adsorption Og Heavy Metal Ions on Porous Calcium Alginate Micrcoparticles." *Analele Universitatii Din Bucuresti – Chimie, Anul Xvi (Seri Noua)* 1: 59-67.

No H., Park N., Lee S. and Meyers S. 2002 "Antibacterial Activity of Chitosan and Chitosan Oligomers with Different Molecular Weights." *International Journal of Food Microbiology* 74: 65-72.

Othman S.H., Abd Salam N.R. and Zainal N. 2014 "Antimicrobial Activity of Tio_2 Nanoparticles-Coated Film for Potential Food Packaging Applications." *International Journal of Photoenergy* 2014:1-6.

Özer A., Özer D. and Ekiz H.I. 2005 "The Equilibrium and Kinetic Modelling of the Biosorption of Copper(II) Ions on Cladophora Crispata." *Adsorption* 10:317-326.

Papageorgiou S.K., Katsaros F.K., Favvas E.P., Em Romanos G., Athanasekou P.C., Beltsios K.G., Tziall O.I. and Falaras P. 2012 "Alginate Fibers as Photocatalyst Immobilizing Agents Applied in Hybrid Photocatalytic/Ultrafiltration Water Treatment Processes." *Water Resources* 46: 1858-1872.

Pathakoti K., Morrow S., Han C., Pelaez M., He X., Dionysiou D.D. and Hwang H.M. 2013 "Photoinactivation of Escherichia Coli by Sulfur-Doped and Nitrogen–Fluorine-Codoped Tio_2 Nanoparticles Under Solar Simulated Light and Visible Light Irradiation." *Environmental Science and Technology* 47: 9988-9996.

Pawar S.N. and Edgar K.J. 2012 "Alginate Derivatization: A Review of Chemistry, Properties and Applications." *Biomaterials* 33: 3279-3305.

Pekakis P.A., Xekoukoulotakis N.P. and Mantzavinos D. 2006 "Treatment of Textile Dyehouse Wastewater by Tio_2 Photocatalysis." *Water Research* 40: 1276-1286.

Pinto R.J.B., Fernandes S.C.M. and Freire C.S.R. 2012 "Antibacterial Activity of Optically Transparent Nanocomposite Films Based on Chitosan or Its Derivatives and Silver Nanoparticles." *Carbohydrate Research* 348: 77-83.

Pozzo, R.L., Baltanas M.A., Cassano A.E. 1997, Supported Titanium Oxide as Photocatalyst in Water Decontamination. *Catal. Today* 39: 219–231

Pradeep T. and Anshup 2009 "Nobel Metal Nanoparticles for Water Purification: A Critical Review." *Thin Solid Films* 517: 6441-6479.

Qiu R., Zhang D., Diao Z., Huang X. and He C. 2012 "Visible Light Induced Photocatalytic Reduction of Cr(Vi) Over Polymer –Sensitized Tio_2 and Its Synergism with Phenol Oxidation." *Water Research* 46: 2299-2306.

Raafat D., Bargen K., Hass A. and Sahl H. 2008 "Insights into the Mode of Action of Chitosan as an Antimicrobial Compound." *Applied Environmental Microbiology* 74: 3764-3773.

Rabae E., Badawy M., Stevens C., Smagghe G. and Steurbaut W. 2003 "Chitosan as Antimicrobial Agent: Applications and Mode of Action." *Biomicromolecules* 4: 1457-1465.

Rauf M.A. and Ashraf S.S. 2009 "Fundamental Principles and Application of Heterogeneous Photocatalytic Degradation of Dyes in Solution." *Chemical Engineering Journal* 151: 10-18.

Reddy M., Venugopal A. and Subrahmanyam M. 2007 "Hydroxypetite-Supported Ag–Tio_2 as Escherichia Coli Disinfection Photocatalyst." *Water Research* 41: 379-386.

Rehim M.H.A., El-Samahy M.A., Badawy A.A., MOhram M.E. 2016. "Photocatalytic activity abd abtimicrobial properties of paper sheets modified with TIO2/sodim alginate nanocomposites". *Carbohydrate Polymers* 148, 194-199

Reveendran G. and Ong S.T. 2018 "Application of Experimental Design for Dyes Removal in Aqueous Environment by Using Sodium Alginate-Tio_2 Thin Film." *Chemical Data Collections* 15-16: 32-40.

Rhim J., Hong S., Park H. and Ng P. 2006 "Preparation and Characterization of Chitosan Based Nanocomposite Films with Antimicrobial Activity." *Journal of Agriculture and Food Chemistry* 54: 5814-5822.

Rincón A. and Pulgarin C. 2004 "Effect of Ph, Inorganic Ions, Organic Matter and H2o2 on E. Coli K12 Photocatalytic Inactivation by Tio_2 Implications in Solar Water Disinfection." *Applied Catalysis B: Environmental* 51: 283-302.

Sadiq R. and Rodriguez M. 2004 "Disinfection By-Products in Drinking Water and Predictive Models for Their Occurance: A Review." *Science of the Total Environment* 321: 21-46.

Sahel K.P., Chermette N., Bordes H., Derriche C. and Guillard Z. 2007 "Photocatalytic Decoloration of Remazol Black 5 (Rb5) and Procion Red Mx-5b—Isotherm of Adsorption, Kinetic of Decoloration and Mineralization." *Applied Catalysis B: Environmental* 77: 100-109.

Salehi R., Arami M., Mahmoodi N.M., Bahrami H. and Khorramfar S. 2010 "Novel Biocompatible Composite (Chitosan–Zinc Oxide Nanoparticle): Preparation, Characterization and Dye Adsorption Properties." *Colloids and Surfaces B* 80: 86-93.

Sarkar S., Chakraborty S. and Bhattacharjee C. 2015 "Photocatalytic Degradation of Phamceutical Wastes by Alginate Supported Tio_2 Nanoparticles in Packed Bed Photoreactor(Pbpr)." *Ecotoxicology and Environmental Safety* 121: 263-270.

Sarrouh B.F., Santos D.T. and Silva S.S. 2007 "Biotechnological Production of Xylitol in A 35 Three-Phase Fluidized Bed Bioreactor with Immobilized Yeast Cells in Ca-Alginate Beads." *Biotechnology Journal* 2: 759–763.

Shannon M., Bohn P., Elimelech M., Georgiadis J., Mariñas B. and Mayes A. 2008 "Science and Technology for Water Purification in the Coming Decades." *Nature* 452: 301.

Shao Y., Cao C., Chen S. and He M. 2015 "Investigation of Nitrogen Doped and Carbon Species Decorated Tio_2 with Enhanced Visible Light Photocatalytic Activity by Using Chitosan." *Applied Catalysis B: Environmental* 179: 344-351.

Shaul, G., T. Holdsworth, C. Demspsey and K. Dostal (1991). *Chemosphere* 22(1-2): 107-119.

Showky H.A. 2011 "Improvement of Water Quality Using Alginate/Montmorillonite Composite Beads." *Journal of Applied Polymer Science* 119: 2371-2378.

Sikorski P., Mo F., Skjåk-Bræk G. and Stokke B. 2007 "Evidence for Egg-Box-Compatible Interactions in Calcium-Alginate Gels from Fiber X-Ray Diffraction." *Biomacromolecules* 8: 2098-2103.

Simo G., Fernandez E. and Crespo J. 2017 "Research Progress in Coating Techniques of Alginate Gel Polymer for Cell Encapsulation." *Carbohydrate Polymers* 170: 1-14.

Snyder A., Bo Z., Moon R., Rochet J. and Stanciu L. 2013 "Reusable Photocatalytic Titanium Dioxidecellulose Nanofber Flms." *Journal of Colloid and Interface Science* 399: 92-98.

Sökmen M., Candan F. and Smer Z. 2001 "Disinfection of E. Coli by the Ag–Tio_2/Uv System: Lipidperoxidation." *Journal of Photochemistry and Photobiology A: Chemistry* 143: 241-244.

Song, J., Wang X., Y. Bu Y. and Zhang J. 2017 "Photocatalytic Enhancement of Floating Photocatalyst: Layer-By-Layer Hybrid Carbonized Chitosan and Fe-N-Codoped Tio_2 on Fly Ash Cenosphere." *Applied Surface Science* 391: 236-250.

Stasinakis A.S. 2008 "Use of Selected Advanced Oxidation Processes (Aops) for Wastewater Treatment – A Mini Review." *Global Nest Journal* 10: 376-385.

Su W., Zhang Y., Li Z., Wu L., Wang X., Li J. and Fu X. 2008 "Multivalency Iodine Doped Tio_2: Preparation, Characterization, Theoretical Studies, and Visible-Light Photocatalysis." *Langmuir* 24: 3422-3428.

Sud D., Mahajan G. and Kaur M.P. 2008 "Agricultural Waste Material as Potential Adsorbent for Sequestering Heavy Metal Ions from Aqueous Solutions – A Review." *Bioresource Technology* 99: 6017-6027.

Sun B., Reddy E.P. and Smirniotis P.G. 2005 "Visible Light Cr(Vi) Reduction and Organic Chemical Oxidation by Tio_2 Photocatalysis." *Environmental Science and Technology* 39: 6251-6259.

Sunada K., Kikuchi Y., Hashimoto K. and Fujishima A. 1998 "Bactericidal and Detoxification Effects of Tio_2 Thin Film Photocatalyst." *Environmental Science Technology* 32: 726-728.

Tang J., Sisler J., Grishkewich N. and Tam K. 2017 "Functionaliztion of Cellulose Nanocrystals for Advanced Applications." *Journal of Colloid and Interface Science* 494: 397-409.

Tennakone K., Tilakaratne C.T.K. and Kottegoda I.R.M. 1995 "Photocatalytic Degradation of Organic Contaminants in Water with Tio_2 Supported on Polythene Films." *Journal of Photochemistry and Photobiology.A* 87: 177-179.

Tijani J.O., Fatoba O.O., Babajide O.O. and Petril L.F. 2016 "Pharmaceuticals, Endocrine Disruptors, Personal Care Products, Nanomaterials and Perfluorinated Pollutants: A Review." *Environmental and Chemical Letters* 14: 27-49.

Tiwari D., Behari J. and Sen P. 2008 "Application of Nanoparticles in Wastewater Treatment." *World Applied Sciences Journal* 3: 417-433.

Torimoto T., Okawa Y. and Takeda N. 1997 "Effect of Activated Carbon Content in Tio_2-Loaded Activated Carbon on Photodegradation Behaviors of Dichloromethane." *Journal of Photochemistry & Photobiology A: Chemistry* 103: 153-157.

Trejo-O'reilly J., Cavaille J. and Gandini A. 1997 "The Surface Chemical Modification of Cellulosic Fibres in View of Their Use in Composite Materials." *Cellulose* 4: 305-320.

Twu Y., Huang H., Chang S. and Wang S. 2003 "Preparation Sorption Activity of Chitosan/Cellulose Blend Beads." *Carbohydrate Polymers* 54: 425-430.

Ummartyotin S. and Manuspiya H. 2015 "A Critical Review on Cellulose: From Fundamental to an Approach on Sensor Technology." *Renewable and Sustainable Enegy Reviews* 41: 402-412.

Vakili M., Rafatullah M.D., Salamatinia B.A., Ibrahim M.H., Tank. B. and Gholami Z. 2014 "Application of Chitosan and Its Derivatives as Adsorbents for Dye Removal from Water and Wastewater: A Review." *Carbohydrate Polymers* 113: 115–130.

Vandevivere P., Bianchi R. and Verstraete W. 1998 "Review: Treatment and Reuse of Wastewater from the Textile Wet-Processing Industry: Review of Emerging Technologies." *Journal of Chemical Technology and Biotechnology* 72:289-302.

Vijaya Y., Popuri S., Boddu V. and Krishnaiah A. 2008 "Modified Chitosan and Calcium Alginate Biopolymer Sorbents for Removal of Nickel (II) Through Adsorption." *Carbohydrate Polymers* 72: 261-271.

Vipin A.K., Hu B. and Fugetsu B. 2013 "Prussian Blue Caged in Alginate/Calcium Beads as Adsorbent for Removal of Cesium Ions from Contaminated Water." *Journal of Hazardous Materials* 33: 3279-3305.

Walalawela N.G.A. 2014 "Photoactive Chitosan: A Step Toward A Green Strategy for Pollutant Degradation." *Photochemistry and Photobiology* 90: 1216–1218.

Wang Y., Yadav S., Heinlein T., Konjik V., Breitzke H., Buntkowsky G., Schneider J.J. and K. Zhang K. 2014 "Ultra-Light Nanocomposite Aerogels of Bacterial Cellulose and Reduced Graphene Oxide for Specific Absorption and Separation of Organic Liquids." *Rsc Advances* 4: 21553-21558.

Wang X., Deng W., Xie Y. and Wang C. 2013 "Selective Removal of Mercury Ions Using A Chitosan –Poly(Vinyl Alcohol) Hydrogel Adsorbent with Three-Dimentional Network Structure." *Chemical Engineering Journal* 228: 232-242.

Wang Y., Zhao L., Peng H., Wu J., Liu Z. and Guo X. 2016 "Removal of Anionic Dyes from Aqueous Solutions by Cellulose-Based Adsorbents: Equilibrium, Kinetics, and Thermodynamics." *Journal of Chemical & Engineering Data* 61: 3266-3276.

Wei C., Lin W., Zainal Z., Williams N., Zhu K. and Kruzic A. 1994 "Bactericidal Activity of Tio_2 Photocatalyst in Aqueous Media: Toward A Solar Assisted Water Disinfection System." *Environmental Science and Technology* 28: 934-938.

Wei S., Zhang X., Zhao K. and Fu Y. 2016 "Preparation, Characterization, and Photocatalytic Degradation Properties of Polyacrylamide/Calcium Alginate/Tio_2 Composite Film." *Polymer Composites* 37(4): 1292-1301.

Woan K., Pyrgios G. and Sigmund W. 2009 "Photocatalytic Carbon-Nanotube-Tio_2 Composites." *Advanced Material* 21: 2233-2239.

Wu N., Wei H. and Zhang N. 2012 "Efficienct Removal of Heavy Metal Ions with Biopolymer Template Synthesized Mesoporous Titania Beads of Hundreds of Micrometers Size." *Environmental Science and Technology* 46: 419-425.

Xie J., Li C. and Wu D. 2013 "Chitosan Modified Zeolite as A Versatile Adsorbent for the Removal of Different Pollutants from Water." *Fuel* 103: 480-485.

Yang F., Song X., Yan L. and Wei W. 2014 "Photocatalysis Degradation of Azo Dye Using Nanotio$_2$-Coated Porous Cellulose Gel: Enhancement by Adsorption and Its Self-Clean Characteristic." *Micro and Nano Letters* 9: 193-197.

Yang J., Xie Y. and He W. 2011 "Research Progress on Chemical Modification of Alginate: A Review." *Carbohydrate Polymers* 84: 33-39.

Yu B., Zhang Y., Shukla A., Shukla S.S. and Dorris K.L. 2000 "The Removal of Heavy Metal from Aqueous Solutions by Sawdust Adsorption — Removal of Copper." *Journal of Hazardous Materials* 80: 33-42.

Yu J., Zhao X. and Zhao Q. 2000 "Effect of Surface Structure on Photocatalytic Activity of Tio_2 Thin Films Prepared by Sol-Gel Method." *Thin Solid Films* 1-2: 7-14.

Zainal Z., Hui L.K., Hussein M.Z., Abdullah A.H. and Hamadneh I.R. 2009 "Characterization of Tio_2-Chitosan/Glass Photocatalyst for the Removal of A Monoazo Dye Via Photodegradation-Adsorption Process." *Journal of Hazardous Materials* 164: 138–145.

Zainal Z., Lee K.H., Hussein M.Z., Taufiq-Yap Y.H., Abdullah A.H. and Ramli I. 2005 "Removal of Dyes Using Immobilized Titanium Dioxide Illuminated by Fluorescent Lamps." *Journal of Hazardous Material B* 125: 113-120.

Zaleska-Medynska A. 2018 *"Metal Oxide-Based Photocatalysis: Fundamentals and Prospects for Application, Metal Oxides."* Elsevier: ISBN 978-0-12-811634-0.

Zargar V., Asghari M. and Dashti A. 2015 "A Review on Chitin and Chitosan Polymers: Structure, Chemistry, Solubility, Derivatives and Applications." *Chembioeng Reviews* 2: 204-226.

Zeng J., Liu S., Cai J. and Zhang L. 2010 "Tio$_2$ Immobilized in Cellulose Matrix for Photocatalytic Degradation of Phenol Under Weak Uv Light Irradiation." *Journal of Physical Chemistry C* 114: 7806-7811.

Zhang L., Zeng Y. and Cheng Z. 2016 "Removal of Heavy Metal Ions Using Chitosan and Modified Chitosan: A Review." *Journal of Molecular Liquids* 214: 175-191.

Zhang Y., Chen Y., Westerhoff P., Hristovski K. and Crittenden J.C. 2008 "Stability of Commercial Metal Oxide Nanoparticles in Water." *Water Research* 42: 2204-2212.

Zhang Y., Ma H., Yi M., Shen Z. and Yu X. 2017 "Magnetron –Sputtering Fabrication of Noble Metal Nanodots Coated Tio$_2$ Nanoparticles with Enhanced Photocatalytic Performance." *Material and Design* 125: 94-99.

Zhang Y., Ma H., Yi M., Shen Z. and Yu X. 2017 "Magnetron –Sputtering Fabrication of Noble Metal Nanodots Coated Tio$_2$ Nanoparticles with Enhanced Photocatalytic Performance." *Material and Design* 125: 94-99.

Zhao K., Feng L. and Lin H.F. 2014 "Adsorption and Photocatalytic Degradation of Methyl Orange Imprinted Composite Membrane Using Tio$_2$/Calcium Alginate Hydrogel as Matrix." *Catalysis Today* 236: 127-134.

Zhou Y., Zhang M., Hu X., Wang X., Niu J. and Ma T. 2013 "Adsorption of Cationic Dyes on A Cellulose-Based Multicarboxyl Adsorbent." *Journal of Chemical Engineering Data* 58: 413-421.

Zhu H., Jiang R. and Xiao L. 2013 "Cds Nanocrystals/Tio$_2$/Crosslinked Chitosan Composite: Facile Preparation, Characterization and Adsorption-Photocatalytic Properties." *Applied Surface Science* 273: 661-669.

Zhu H., Jiang R., Fu Y., Guan Y., Yao J., Xiao L. and Zeng G. 2012 "Effective Photocatalytic Decolorization of Methyl Orange Utilizing Tio$_2$/Zno/Chitosan Nanocomposite Films Under Simulated Solar Irradiation." *Desalination* 286: 41-48.

Zubieta C.E., Messina P.V., Luengo C., Dennehy M., Pieroni O. and Schulz P.P. 2008 "Reactive Dyes Remotion by Porous Tio$_2$-Chitosan Materials." *Journal of Hazardous Materials* 152: 765-777.

In: Photocatalysis
Editors: P. Singh, M. M. Abdullah, M. Ahmad et al.
ISBN: 978-1-53616-044-4

Chapter 5

PHOTOCATALYTIC PERSPECTIVES OF NANOMATERIALS FOR ENVRIONMENTAL PROTECTION

***Mohit Yadav*[1,2], *Seema Garg*[2] *and Amrish Chandra*[3]**
[1]Amity Institute of Nanotechnology, Amity University, Noida, India
[2]Department of Chemistry, Amity Institute of Applied Sciences, Amity University, Noida, India
[3]Amity Institute of Pharmacy, Amity University, Noida, India

ABSTRACT

In the past few decades, nanomaterials-based photocatalysis, with the virtues of easy operation and high efficiency towards the degradation of toxic contaminants, has shown a promising solution for environmental remediation. Apart from environmental protection, nanotechnology has also played a significant role in numerous advancements and industry divisions i.e., data innovation, vitality, ecological science, medicine, surveillance and security, transportation, and food and agriculture, etc. In the present scenario, nanotechnology is involved in various branches of science i.e., material science, chemistry, physics, and biotechnology/biology to develop novel nanomaterials that possess unique properties compared to their bulk form. This chapter abridges the wide range of applications of nanomaterials in recent times and upcoming future advances.

INTRODUCTION

Nanotechnology is an expansive and interdisciplinary territory of innovative work movement that has been emerging progressively worldwide over the years. It deals with the manipulation of matter at the nanoscale and the elementary components which establish the framework of this technology are known as nanomaterials. The range of the nanomaterials varies from 1-100 nanometers (nm) and they possess unique chemical, optical, electrical, physical and biological properties in comparison to their bulk size. Nanomaterials have been studied and applied extensively in a wide range of application because of their vast surface to volume proportion and higher chemical stability in the chemical reactions (Balbus et al., 2007; Hasan, 2015).

Nanomaterials and their composites can be synthesized by tuning and modulating their band structure to display unique chemical, physical, optical, magnetic, electronic, and biological properties. Researchers across the globe with different research backgrounds i.e., information technology, chemistry, physics, biology, metrology, engineering, or any other fields are working extensively and cooperatively, which is setting a trademark in the nanotechnology's research breakthroughs. The basic virtues of nanotechnology involves the manipulation of matter atthe nanoscale through synthesis, structural characterization, analysis of the properties measurements, and modeling. Such traits of nanotechnology prove to be beneficialin identifying and analyzing the usual characteristics of nanomaterial resulted from their intermediate relationship between the bulk macroscopic level and the nanoscale level. To make optimum utilization ofone of kind properties of nanomaterials, the research and development has been focused to gather many insights in the understanding and modernization of the nanomaterial-based systems and devices (Cao, 2004; Lu et al., 2007; Akbarzadeh et al., 2012; Tjong, 2013).

Nanotechnology encompasses a wide variety of the materials and systems with varying characteristics including shape, constituents and structural design, which endows its remarkable ability. The nanotechnology also serves as a base for researchers, engineers and other professionals to make optimum use of such unique properties of nanomaterials obtained by gaining a manipulative control over the morphological and other features at the nanoscale. Further advancement in the field of nanotechnology can also provide a learning curve for young scientists to effectively invest and analyze these nanoscale devices for fulfilling the requirements ofsustainable development. However, it is necessary to keep-up the durability of boundaries between nanomaterials and incorporate these "nanostructures" at the micro/macroscopic level. Moreover, the characteristics of nanomaterials are generally observed to inevitable from characteristics that are usually identified at the bulk size. The most probable variations in characteristics and mechanism of nanomaterials specifically rely on the results obtained from their basic nature, rather than the individual decrease in the order of their magnitude dimension. These interpretations comprise of size confinement, prevalence of quantum mechanics, and interfacial occurrences, which are coherent to the electronic conformations and orbital illustrations (Hulteen and Martin, 1997; Xia et al., 2003; Gupta and Gupta, 2005; Chen and Mao, 2007; Huang et al., 2009).

After five decades of extensive research on exploitation and exploration of nanotechnology, it became much more practical for researchers to control, modify, and tune the properties of nanomaterials by various means of techniques, such as functionalization, doping with other metal or no-metal ions, grafting via later adjustment or In-situ decorations. It is well established that the magnitude of the nanostructures can be either be decreased or increased to meet the nanoscale optimization, which allows the advancement in the novel products with unique properties, such as carbon-based nanomaterials (graphene, carbon nanotubes), quantum confinement of nanostructures (nanowires and quantum dots), thin-films, laser emitters and DNA-based nanostructures. This novel advancement of nanomaterials and nanoscale devices represents an innovator age for revolutionizing future science and technology (Patzke et al., 2002; Koch, 2003; Rolison, 2003). Moreover, we can discover and make optimum use of the underlying applications which are shown in the Figure 1.

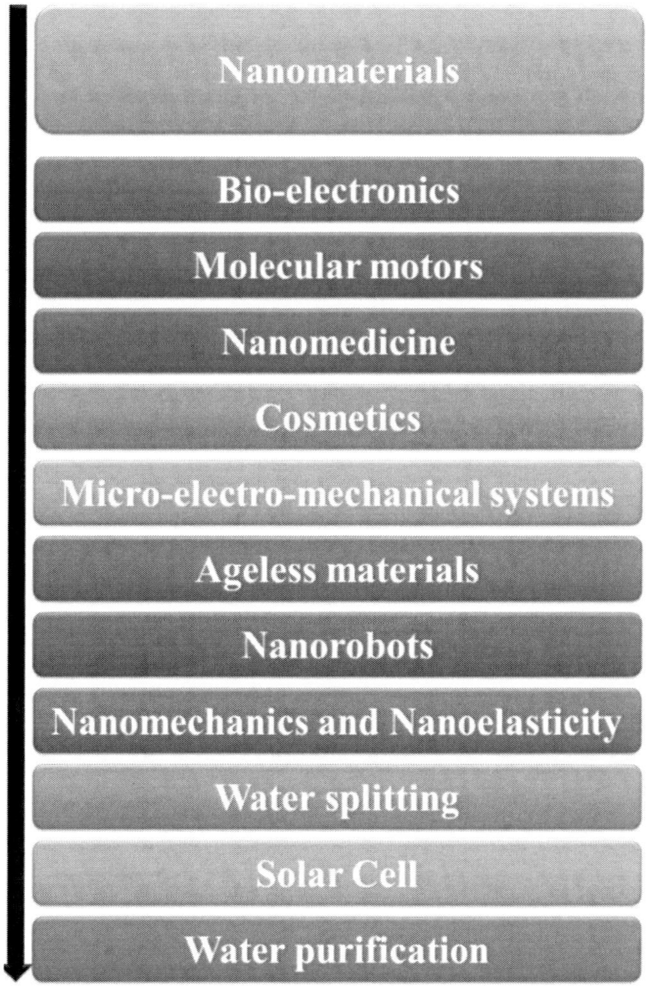

Figure 1. Applications of the nanomaterials.

Clean Energy Production and Storage Applications

With the growing population the energy crisis is also increasing and the environmental safety is also becoming a major issue. Scientists across the globe are trying to find ways to develop a clean, economically viable and renewable source of energy, which can act as an alternate for current energy sources to diminish the energy consumption, and reduce the detrimental effects to the environment. As discussed earlier, nanotechnology has already established its root in numerous applications and a lot of work is going on to maximize the usage of nanomaterials in the future applications. New kind ofsolar cells offer a much efficient and cheaper alternative to many other sources that are designed to convert sunlight into electricity. Unlike discrete panels, the solar cells in the nano regime are easy to install and can be stored as flexible rolls. Benefitting from these advantages, a lot of research is going on in developing a system comprising of thin-film solar electric panels attached on a computerized operating case, and agile piezoelectric nanowire interlaced into clothing to power electronic devices, such as mobile phones, portable fan and computers, etc. (Li et al., 2009; Low et al., 2015; Najim et al., 2015; Jalaja et al., 2016; Mobasser and Firoozi, 2016; Pratsinis, 2016; Sabet et al., 2016).

Newly developed batteries possess innumerable advantages over the customary batteries, for instance, light weight, low flammable rate, higher charging efficiency, large power density and higher charge holding capacity for example, lithium-sulfur with a graphene wrapper battery is known to be one of the future batteries (Figure 2). These batteries are economically viable and profitable due to their high specific energy density in comparison to the previously existing batteries such as lithium-ion (Chen et al., 2014; Rabbani et al., 2016).

Researchers have also been working on development of safe and lightweight fuel tanks for the hydrogen storage and is expected to be applied in future devices on a large scale. Nanomaterials also, hold the application as a catalyst to comprehend fuel cells as potential alternatives for conventional and current transportation technologies at cheaper rates. Numerous nanotechnology-based projects have been currently operating to transform waste heat and other alternatives into a defined power source, which could be utilized on a large scale for various applications, such as energy efficient lighting systems to cut off the large consumption of energy for illumination, transportation sector, advanced electronics, nano-engineered low-friction lubricants for fans, pumps and other machinery items, nano-coated glass for highly efficient light absorption to complement heating and cooling devices and fast rechargeable emergency lanterns with high illumination efficiency (Mishra et al., 2012; Tarafdar et al., 2013; Fan et al., 2016; Sadeghi et al., 2016).

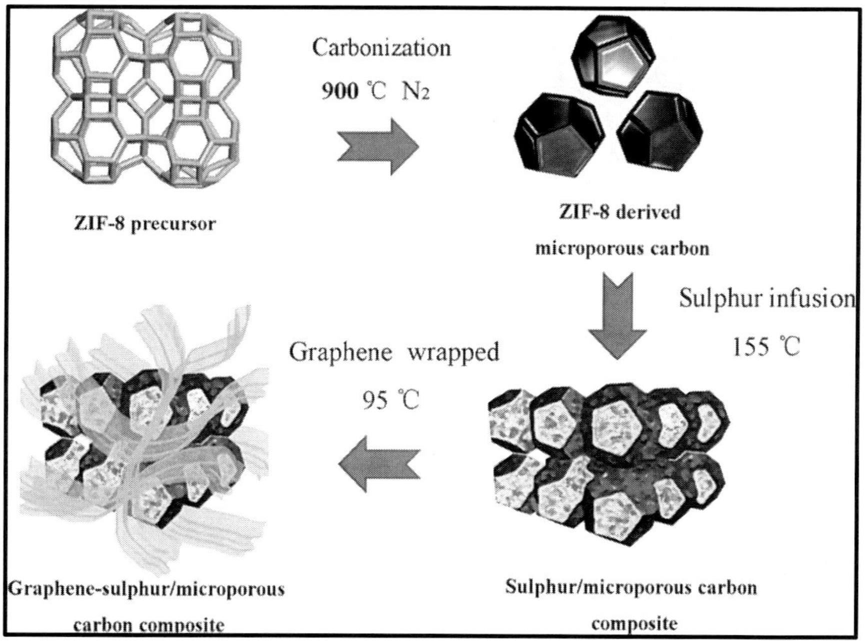

Figure 2. Fabrication scheme for 3-D hierarchical structured graphene-sulfur/carbon ZIF8-D composite (Chen et al.). Reprinted with permission from APL Materials, AIP.

BIOSENSORS AND MEDICAL APPLICATIONS

Nanomaterials potentially hold the tendency to reform a wide range of therapeutic and biotechnology devices in a way that they are more customized, compact, less expensive, safe and easy to regulate. The biosensors that comprise of nanocomponents, such as nanowires, nano-cantilevers, etc. can be used extensively for early stage recognition of hereditary and molecular orientations, thus offering the possibility to distinguish unique molecular signals related with malignancy. The nanoparticles in these biosensors assist as multifunctional medication with high stereospecificity towards cancer cells; thereby, reducing the threat to normal tissues (Yashveer et al., 2014; Ng et al., 2015; Weiss, 2015). Researchers are developing an imaging tool to detect the quantity of a nanomaterial-antibody complex that amasses specifically in the plague, which can be monitored as the treatment would progress. Similarly, gold (Au) nanomaterials can be employed for early recognition of Alzheimer's disease. Research empowering influences, for example, microfluidic chip-based nano-labs are designed to monitor and manipulate the particular cells, and nanoscale tests to monitor the activity of cells and individual molecules depending on their mobility in the surroundings (Boisseau and Loubaton, 2011; Taha et al., 2013; Schulte et al., 2014; Adam et al., 2015). Research is in progress to avail nanotechnology to goad the development of nerve cells, for example, in ruptured brain cells or spinal cord. In one strategy, a nanostructured gel-type material is used to fill the gap between the existing cells, which inflicts the formulation of new cells. Lab-on-chip is a typical example of nano-lab monitoring and manipulation of cells, and also analyzes the cell movements. For example, polymerase chain reaction (PCR) is realized on a chip for the DNA analysis as shown in Figure 3.

Quantum dots (QDs) are another typical example of exotic nanomaterials that hold the great application in enhancing the biological imaging up to 1000 times larger than many conventional dyes used in medicinal diagnostics such as MRIs. These quantum dots under UV-light irradiation have the tendency to transmit wide-ranging bright colors, which can be applied to detect and distinguish particular section of cells and their biological actions. Streptavidin-covered QDs are profoundly used as the most appropriate ones, in light of the fact that they can be effortlessly conjugated to various monetarily accessible biotinylated antibodies. The QD-antibody acting agent conjugate can be utilized for immunofluorescent bioimaging by two general methodologies. In the first method, a streptavidin-coated QD is conjugated to a biotinylated prime antibody, which perceives specifically focused on the antigen. The second method for a QD-assisted identification of antigens includes primary and secondary antibodies. The essential immune response focuses on the antigen and is then perceived by a biotinylated secondary immune response, which ties a streptavidin-coated QD (Figure 4) (Morganti et al., 2011; Yordanov and Dushkin, 2011; Milliron, 2014; Schnitzenbaumer and Dukovic, 2014; Tam et al., 2014; Raspa et al., 2016).

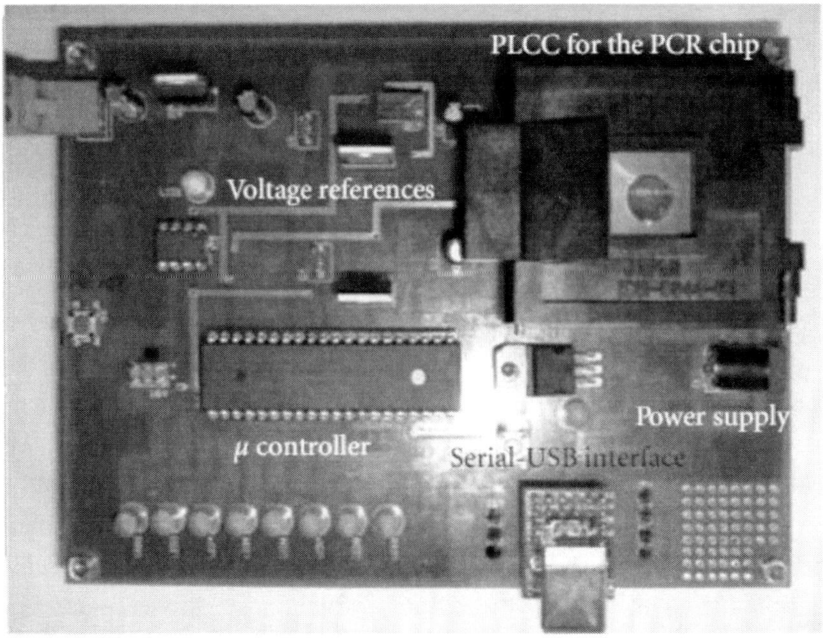

Figure 3. Image of a realized PCR electronic board (Morganti et al.). Reprinted with permission from *Journal of Sensors*, Hindawi.

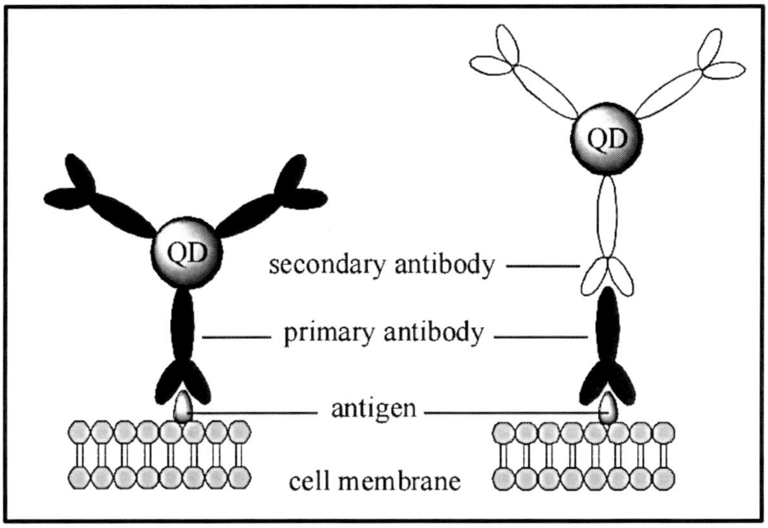

Figure 4. Schematic of QDs application for immunofluorescent bioimaging using biotinylated primary antibody (left) or a combination of a primary antibody and a biotinylated secondary antibody (right) (Yordanov et al.). Reprinted with permission from University Press, Sofia University, St. kliment Ohridski.

FOOD AND AGRICULTURE

Asia is known to be the biggest continent on Earth, where nearly 50% of the global population lives and the majority of the population living in developing countries such as Pakistan, Bhutan, India, Nepal, etc. suffer hygienic of food and fresh water. The possible reason for such conditions can be speculated as a result of the ecological imbalance; on the other hand, the developed countries have been managing their food and freshwater resources. The developing countries are mostly driven on finding solutions to make drought and pest resistant crops, which can advance the yield. Concurrently, the developed countries are mostly driven on developing and strengthening the food industry as per the consumer demand, which is considered to hike the business. One of the prime examples of such investments are of the food industries in the UK, which are exponentially developing almost at an annual growth rate of 5.3%, the major reason behind this hike is the high demand for fresh food. In addition, some other sectors such as energy, information, and communication technology industries have been well-advertised and recognized (Allen and Cullis, 2004; Dreher, 2004; Wallace et al., 2004; Kim et al., 2014a).

In this regard, nanotechnology has been utilized on a large scale and numerous products have been manufactured that are already available in the market, such as self-cleaning windows, antiseptic/bacterial dressings, sunscreens, scratch free paints for cars, and stain-resistant fabrics. United States Department of Agriculture in 2003 was the first to address the implementation of nanotechnology to the agricultural and food industries. Their inspiration was driven on the fact that nanotechnology will revolutionize the entire concept of the food industry in various aspectssuch as production, processing, packaging, transportation, and consumption. With time, food and agriculture industries have faced certain challenges, such as increasing stipulations for healthy, food safety; rising concerns with respect to the diseases, and unexpected dangers to agricultural and fishery products with varying weather designs. Moreover, developing an agricultural budget is a tough challenge and a perplex course, which involves the intersection of distinctive virtues of science (Ahmed et al., 2016; Wong et al., 2016; Yetisen et al., 2016).

The implications of new aspects of nanomaterials can modernize the agricultural and food industry. With the technological advancements in nanotechnology would initiate the development of new tools, which could be utilized on a large scale for various treatments, such as rapid disease detection, molecular treatment of diseases, boosting the potential of plants for the absorption of nutrients, etc. In addition, the concept of smart sensors and smart delivery systems would certainly give big support for the agricultural industries to eradicate the crop pathogen and other viruses. In the near future, exotic nanomaterials in the form of nanocomposites and catalysts would be developed, which promisingly would enhance the ability of pesticides and herbicides. In addition, future research on the exotic nanomaterials would also help in protecting the environment via alternative sources of energy supplies, filters, and catalysts for pollution control. Controlled Environment Agriculture (CEA) is a well-known rural technique generally utilized as a part of the USA, Europe, and Japan, which effectively uses present-day innovation for crop administration and management. This technique is a progressed and severe type of agriculture, where development of plant inside a well-controlled condition takes place to realize the agricultural practices (Lead et al., 2018; Nurulhuda et al., 2018; Pourmohammadi et al., 2018).

A lot of work has been going on to develop nanoscale devices with unique characteristics, which can transform the current agricultural system into a "smart" futuristic system. The concept of such nanoscale devices is based on the identification and prediction of the plant health issues before being noticed by the farmers. The implication of such nanoscale devices will allow a real-time response of the varying situations and accordingly will take extreme measures for remedial action. However, even if these devices fail to analyze the environmental conditions, they will surely act as an alarm, which will alert the farmers for the upcoming problems. In such a way, these nanoscale devices will perform a dual nature by acting as a protective and timely threatening system. In addition, these devices will also be able to deliver the chemicals in a well-organized and targeted way, similar to how nano-medicine are applied for drug delivery (Sun et al., 2014; Zhang et al., 2014).

The development of nanomedicines allows high precision treatments, such as targeted drug delivery (to a specific tissue and organ) and cancer in animals. Nanoscale technologies in the form of nanoencapsulation and controlled drug release have transformed the utility of pesticides and herbicides on a large scale, and also preventing the wastage. Many industries design certain formulations with encapsulation of small sized (50-100 nm) nanoparticles, which can easily dissolve in water as compared to the conventional and existing ones. Some other companies are aiming to develop oil and water based nanoparticle suspensions and nano-emulsions containing pesticides or herbicides in the nano-range of 200-400 nm. Such suspension type composites can be easily fused with various matrixes, for example; creams, liquid, gels, etc. and most importantly such suspensions will have multiple synergistic implications for treatment, prevention or conservation of the gathered product. The future research likewise means to develop such devices which will help in consumption of water, pesticides, and fertilizers to the plants more effectively that will simultaneously help in pollution reduction and enhance the agriculture to be more sustainable. Agriculture lays the foundation of most under-developed and developing nations, with over 60% general population dependent on it. In addition, creating enhanced frameworks for checking natural conditions and conveying supplements or pesticides as suitable, nanotechnology can boost our understanding of the science of various products, and therefore, conceivably advance the nutritional qualities and yields (Aureli et al., 2018; Gil et al., 2018; van Riper et al., 2018).

TRANSPORTATION APPLICATIONS

Nanoengineering and reformation of cementation materials, such as concrete, steel asphalt, etc. offers promising traits as far as enhancing the execution, durability, and lifespan of transportation and highway infrastructure segments along with economic viability. New frameworks may fuse creative aptitudes into conventional foundation materials, for example, the capacity to produce or emit energy. Nanoscale sensors and gadgets may offer economic viability for incessant structural performance and monitoring of rails, tunnels, bridges, pavements, and parking structures over time. In addition, new advanced technology may likewise bolster an improved transportation foundation that can co-relate with vehicle-based frameworks to enable drivers to keep up path position, avoid accidents, modify set out courses to circumnavigate clog, and other such exercises (Naqvi et al., 2009; Wang et al., 2010; Green and Klein, 2011; Reddy and Sivakumar, 2013; Bai and Li, 2014; Lin et al., 2014; Wada, 2014; Agzenai et al., 2015; Hernandez-Bautista et al., 2017; Li et al., 2017; De la Varga et al., 2018; Gheibi et al., 2018; Moon et al., 2018).

Environmental Remediation

Over the decades, the rapid industrialization has led to a proportional increase in the usage of toxic chemicals which are discharged in the environment without any proper treatment. Most of the toxic chemicals are organic compounds which are hard to degrade because of their complex chemical structure and tendency to accumulate in the water streams for longer period of time. The organization of damaged soil and groundwater is a striking environmental concern. The continuous rise in the variety of contaminants in soils, residue, surface and ground waters influences the human health around the world. Nanotechnology in this field has already shown promising applications and a lot more can be explored (Wang et al., 2014; Xue et al., 2014; Simon et al., 2018).

Nanotechnology can be classified into four main measures toward environmental remediation: (1) remediation, (2) fortification, (3) preservation, and (4) enrichment. However, out of these four, remediation is considered to be the most quickly developing classification, while both fortification and preservation represent the principle virtue of nanotechnology application, and enrichment of ecological conditions make a short portion but also an important aspect of nanotechnology application. Nanomaterials can be used in air purification, wastewater treatment processes, as mesoporous components for green science, and in semiconductor photocatalysis, etc. (Yang and Westerhoff, 2014; Arts et al., 2015; Ullah et al., 2015).

Under favorable circumstances, the properties of the nanomaterials can be tuned and modified for degradation organic pollutants either via photocatalysis or adsorption methods. For a sustainable environment, proper implementation of nanotechnology can be effectively realized at the lab-scale. However, to execute the large scale implementation of nanotechnology certain parameters have to be addressed, such as affirmation of their viability, welfare in the field, cost-effectiveness and long-term stability, etc. The present research aims to overcome the limitations of traditional remediation techniques and develop an efficient method for degradation of the contaminants concentration noticeable in the air, soil, and water (Ushiroda et al., 2005; Karimi and Navidbakhsh, 2014; Manzano et al., 2014; Karimi-Maleh et al., 2018).

Until now, various nanomaterials have been applied for environmental remediation, such as silver (Ag), TiO_2, ZnO, SiO_2, iron (Fe), CNT (carbon nanotubes), chitosan, etc. However, out of these nanomaterials, TiO_2 has turned out to be the most efficient nanomaterial due to its high efficiency for the disintegration of pollutants. TiO_2 has displayed remarkable ability in different roles such as in sensing applications i.e., biological, chemical, and numerous gas sensing (H_2, CO, NO_x, SO_x, etc.), and other remedial purposes.Furthermore, TiO_2 has also shown great application as a reducing agent for the metals into their respective zero oxidation states, ultimately helpful for wastewater treatment. Moreover, the disintegration of the pollutants by TiO_2 results in non-toxic byproducts, thereby, reducing the secondary pollution. Although, extensive research has been envisaged on TiO_2 but its ability to perform under natural sunlight is still a matter of concern, as the UV radiations are nearly 4% of the total solar radiations, hence, it tends to underperform when utilized under visible light source that accounts for more than 42% of the total solar spectrum. Likewise, the other UV light driven nanoparticles and photocatalysts such as ZnO, SiC, ZnS, Fe_2O_3, SnO_2, etc. face the same photocatalytic confinement (He et al., 2005; Dai et al., 2011; Liu et al., 2012).

Therefore, advanced and revised strategies need to be initiated and explored to counter these future challenges. One of the major approaches that are expected to be executed in the future is the applicability of clean and green technology for the synthesis of nanomaterials on a large scale. If TiO_2 can be synthesized by a green route such that its band gap potential decreases simultaneously that would be a great achievement, and its applicability to perform under natural sunlight would be enhanced. To overcome the limitations of TiO_2, various visible light driven nanophotocatalysts have been studied extensively, and one of them is the bismuth oxyhalides (BiOX) (Garg et al., 2018a; Garg et al., 2018b; Garg et al., 2018c; Garg et al., 2019). They are non-hazardous and possesses high chemical stabilty and unique optical properties. However, the study on the bismuth oxyhalides and various other nanophotocatalysts is still in the early stage, and a lot more is still expected in the upcoming decades (Li et al., 2010; Yang et al., 2010; Su et al., 2012; Kim et al., 2014b; Mahlambi et al., 2015).

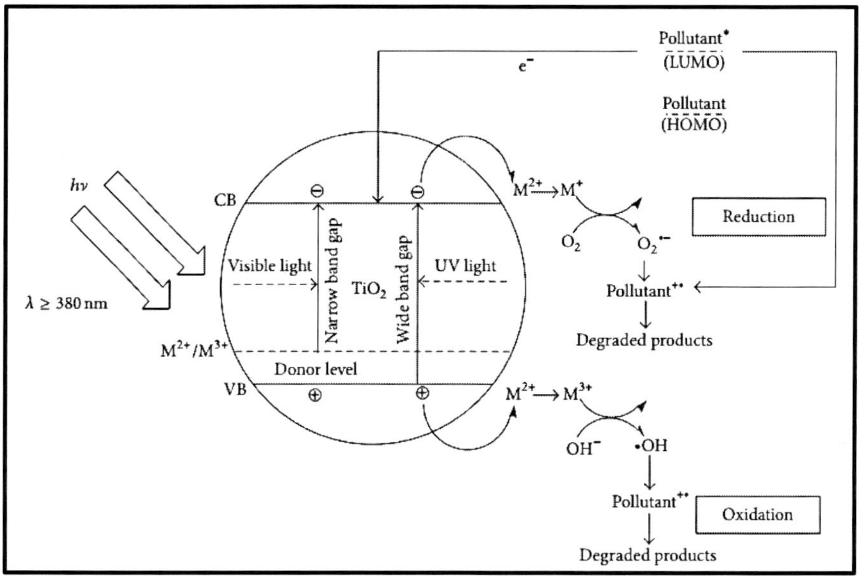

Figure 5. Photocatalysis mechanism of TiO_2 (Mahlambi et al.). Reprinted with permission from *Journal of Nanomaterials*, Hindawi.

Conclusion

In summary, nanotechnology is still in its early times and vast possible outcomes can be speculated in the form of futuristic nanotechnology-based products that will overcome all the limitations of the conventional and modern day techniques. The use of nanomaterials will ease the workload and will provide a well-organized and less time-consuming framework for numerous applications, for example, food and horticulture, mechanical, building materials, transportation, electrical building and medicine, etc. In spite of the fact that replication of the natural framework is a standout amongst the most encouraging zones of this innovation, researchers are as yet endeavoring to get a handle on their bewildering complexities. Furthermore, nanotechnology is a quickly developing field of research in which the novel photocatalytic properties of nanomaterials can be explored for the advantage of industrialization, environmental remediation and numerous other sectors, which can possibly improve the life-cycle rate of building infrastructure to make the future more comfortable and luxurious.

References

Adam, M., Wang, Z., Dubavik, A., Stachowski, G. M., Meerbach, C., Soran-Erdem, Z., Rengers, C., Demir, H. V., Gaponik, N. & Eychmüller, A. (2015). Liquid–Liquid Diffusion-Assisted Crystallization: A Fast and Versatile Approach Toward High Quality Mixed Quantum Dot-Salt Crystals. *Advanced Functional Materials*, *25*, 2638-2645.

Agzenai, Y., Pozuelo, J., Sanz, J., Perez, I. & Baselga, J. (2015). Advanced self-healing asphalt composites in the pavement performance field: mechanisms at the nano level and new repairing methodologies. *Recent Patents on Nanotechnology*, *9*, 43-50.

Ahmed, S., Ahmad, M., Swami, B. L. & Ikram, S. (2016). A review on plants extract mediated synthesis of silver nanoparticles for antimicrobial applications: A green expertise. *Journal of Advanced Research*, *7*, 17-28.

Akbarzadeh, A., Samiei, M. & Davaran, S. (2012). Magnetic nanoparticles: preparation, physical properties, and applications in biomedicine. *Nanoscale Research letters*, *7*, 144.

Allen, T. M. & Cullis, P. R. (2004). Drug delivery systems: entering the mainstream. *Science*, *303*, 1818-1822.

Arts, J. H., Hadi, M., Irfan, M. A., Keene, A. M., Kreiling, R., Lyon, D., Maier, M., Michel, K., Petry, T., Sauer, U. G., Warheit, D., Wiench, K., Wohlleben, W. & Landsiedel, R. (2015). A decision-making framework for the grouping and testing of nanomaterials (DF4nanoGrouping). *Regulatory Toxicology and Pharmacology, 71*, 26.

Aureli, R., La-Marta, J., Grossi, A.B., Della Pia, E.A., Esteve-Garcia, E., Wulf-Andersen, L. & Thorsen, M. (2018). A novel glucuronoxylan hydrolase produced by fermentation is safe as feed additive: Toxicology and tolerance in broiler chickens. *Regulatory Toxicology and Pharmacology, 25*, 30246-30240.

Bai, C. & Li, Y.(2014). Time series analysis of contaminant transport in the subsurface: applications to conservative tracer and engineered nanomaterials. *Journal of Contaminant Hydrology, 164*, 153-162.

Balbus, J.M., Maynard, A.D., Colvin, V.L., Castranova, V., Daston, G.P., Denison, R.A., Dreher, K.L., Goering, P.L., Goldberg, A. M. & Kulinowski, K. M.(2007). Meeting report: hazard assessment for nanoparticles—report from an interdisciplinary workshop. *Environmental Health Perspecives,115*, 1654.

Boisseau, P. & Loubaton, B. (2011). Nanomedicine, nanotechnology in medicine. *Comptes Rendus Physique,12*, 620-636.

Cao, G. (2004). Nanostructures & nanomaterials: synthesis, properties & applications. Imperial college press. (doi.org/10.1142/7885).

Chen, R., Zhao, T., Tian, T., Cao, S., Coxon, P.R., Xi, K., Fairen-Jimenez, D., Vasant Kumar, R.& Cheetham, A.K.(2014). Graphene-wrapped sulfur/metal organic framework-derived microporous carbon composite for lithium sulfur batteries. *APL Materials, 2*, 124109.

Chen, X.& Mao, S. S.(2007). Titanium dioxide nanomaterials: synthesis, properties, modifications, and applications. *Chemical Reviews, 107*, 2891-2959.

Dai, G., Yu, J. & Liu, G.(2011). Synthesis and enhanced visible-light photoelectrocatalytic activity of p−n junction BiOI/TiO2 nanotube arrays. *The Journal of Physical Chemistry C,115*, 7339-7346.

De la Varga, I., Munoz, J.F., Bentz, D.P., Spragg, R.P., Stutzman, P. E. & Graybeal, B. A.(2018). Grout-Concrete Interface Bond Performance: Effect of Interface Moisture on the Tensile Bond Strength and Grout Microstructure. *Construction and Building Materials, 170*, 747-756.

Dreher, K. L.(2004). Health and environmental impact of nanotechnology: toxicological assessment of manufactured nanoparticles. *Toxicology Sciences, 77*, 3-5.

Fan, W., Shi, J. & Bu, W. (2016). Engineering Upconversion Nanoparticles for Multimodal Biomedical Imaging-Guided Therapeutic Applications. *Advances in Nanotheranostics I*. Springer, pp. 165-195.

Garg, S., Yadav, M., Chandra, A., Gahlawat, S., Ingole, P. P., Pap, Z. & Hernadi, K. (2018a). Plant leaf extracts as photocatalytic activity tailoring agents for BiOCl towards environmental remediation. *Ecotoxicology and Environmental Safety, 165*, 357-366.

Garg, S., Yadav, M., Chandra, A. & Hernadi, K.(2019). A Review on BiOX (X= Cl, Br and I) Nano-/Microstructures for Their Photocatalytic Applications. *Journal of Nanoscience and Nanotechnology,19*, 280-294.

Garg, S., Yadav, M., Chandra, A., Sapra, S., Gahlawat, S., Ingole, P., Todea, M., Bardos, E., Pap, Z.& Hernadi, K. (2018b). Facile Green Synthesis of BiOBr Nanostructures with Superior Visible-Light-Driven Photocatalytic Activity. *Materials,11*, 1273.

Garg, S., Yadav, M., Chandra, A., Sapra, S., Gahlawat, S., Ingole, P. P., Pap, Z. & Hernadi, K. (2018c). Biofabricated BiOI with enhanced photocatalytic activity under visible light irradiation. *RSC Advances,8*, 29022-29030.

Gheibi, M., Karrabi, M., Shakerian, M. & Mirahmadi, M.(2018). Life cycle assessment of concrete production with a focus on air pollutants and the desired risk parameters using genetic algorithm. *Journal of Environmental Health Science and Engineering, 16*, 89-98.

Gil, J.D.B., Reidsma, P., Giller, K., Todman, L., Whitmore, A. & van Ittersum, M. (2018). Sustainable development goal 2: Improved targets and indicators for agriculture and food security. *Ambio, 28*, 018-1101.

Green, C. G. & Klein, E.G.(2011). Promoting active transportation as a partnership between urban planning and public health: the columbus healthy places program. *Public Health Reports, 1*, 41-49.

Gupta, A. K. & Gupta, M.(2005). Synthesis and surface engineering of iron oxide nanoparticles for biomedical applications. *Biomaterials, 26*, 3995-4021.

Hasan, S.(2015). A review on nanoparticles: their synthesis and types. *Research Journal of Recent Sciences* ISSN 2277: 2502.

He, C., Li, X., Xiong, Y., Zhu, X. & Liu, S. (2005). The enhanced PC and PEC oxidation of formic acid in aqueous solution using a Cu–TiO2/ITO film. *Chemosphere, 58*, 381-389.

Hernandez-Bautista, E., Sandoval-Torres, S., de, J. C. B.P. F. & Bentz, D. P.(2017). Modeling Heat and Moisture Transport in Steam-Cured Mortar: Application to Aashto Type Vi Beams. *Construction and Building Materials, 151*, 186-195.

Huang, X., Neretina, S.& El-Sayed, M.A. (2009). Gold nanorods: from synthesis and properties to biological and biomedical applications. *Advanced Materials, 21*, 4880-4910.

Hulteen, J. C. & Martin, C. R.(1997). A general template-based method for the preparation of nanomaterials. *Journal of Material Chemistry, 7*, 1075-1087.

Jalaja, K., Naskar, D., Kundu, S. C. & James, N. R.(2016). Potential of electrospun core–shell structured gelatin–chitosan nanofibers for biomedical applications. *Carbohydrates and Polymers, 136*, 1098-1107.

Karimi-Maleh, H., Bananezhad, A., Ganjali, M.R., Norouzi, P.& Sadrnia, A. (2018). Surface amplification of pencil graphite electrode with polypyrrole and reduced graphene oxide for fabrication of a guanine/adenine DNA based electrochemical biosensors for determination of didanosine anticancer drug. *Applied Surface Sciences*, *441*, 55-60.

Karimi, A. & Navidbakhsh, M.(2014). Material properties in unconfined compression of gelatin hydrogel for skin tissue engineering applications. *Biomedical Technology*, *59*, 479-486.

Kim, D., Wu, X., Young, A. T. & Haynes, C. L.(2014a). Microfluidics-based *in vivo* mimetic systems for the study of cellular biology. *Accounts of Chemical Research*, *47*, 1165-1173.

Kim, W. J., Pradhan, D., Min, B. K. & Sohn, Y.(2014b). Adsorption/photocatalytic activity and fundamental natures of BiOCl and BiOClxI1− x prepared in water and ethylene glycol environments, and Ag and Au-doping effects. *Applied Catalysis B: Environmental*, *147*, 711-725.

Koch, C.(2003). Top-down synthesis of nanostructured materials: Mechanical and thermal processing methods. *Reviews on Advanced Materials Science*, *5*, 91-99.

Lead, J.R., Batley, G.E., Alvarez, P.J.J., Croteau, M.N., Handy, R.D., McLaughlin, M. J., Judy, J. D. & Schirmer, K.(2018). Nanomaterials in the environment: Behavior, fate, bioavailability, and effects-An updated review. *Environmental Toxicology and Chemistry*, *37*, 2029-2063.

Li, G., Zhang, D., Yu, J. C. & Leung, M. K.(2010). An efficient bismuth tungstate visible-light-driven photocatalyst for breaking down nitric oxide. *Environmental Science & Technology*, *44*, 4276-4281.

Li, X., Cai, W., Colombo, L. & Ruoff, R. S.(2009). Evolution of graphene growth on Ni and Cu by carbon isotope labeling. *Nano Letters*, *9*, 4268-4272.

Li, X., Grasley, Z.C., Garboczi, E. J. & Bullard, J. W.(2017). Simulation of the Influence of Intrinsic C-S-H Aging on Time-Dependent Relaxation of Hydrating Cement Paste. *Construction and Building Materials*, *157*, 1024-1031.

Lin, J., Yuan, Y., Zhou, J. & Gao, J. (2014). Methodology to improve design of accelerated life tests in civil engineering projects. *PLoS One*, *9*. (DOI:10.1371/journal.pone.0103937).

Liu, L., Liu, Z., Bai, H. & Sun, D. D.(2012). Concurrent filtration and solar photocatalytic disinfection/degradation using high-performance Ag/TiO 2 nanofiber membrane. *Water Research*, *46*, 1101-1112.

Low, J., Yu, J. & Ho, W. (2015). Graphene-based photocatalysts for CO2 reduction to solar fuel. *The journal of physical chemistry letters*, *6*, 4244-4251.

Lu, A.H., Salabas, E.e. L. & Schüth, F. (2007). Magnetic nanoparticles: synthesis, protection, functionalization, and application. *Angewandte Chemie International Edition*, *46*, 1222-1244.

Mahlambi, M.M., Ngila, C. J. & Mamba, B. B.(2015). Recent Developments in Environmental Photocatalytic Degradation of Organic Pollutants: The Case of Titanium Dioxide Nanoparticles-A Review. *Journal of Nanomaterials*, *29*.

Manzano, S., Poveda-Reyes, S., Ferrer, G. G., Ochoa, I. & Hamdy Doweidar, M.(2014). Computational analysis of cartilage implants based on an interpenetrated polymer network for tissue repairing. *Computational Methods and Programs Biomedicine*, *116*, 249-259.

Milliron, D. J.(2014). Quantum dot solar cells: The surface plays a core role. *Nature materials*, *13*, 772.

Mishra, Y., Chakravadhanula, V., Hrkac, V., Jebril, S., Agarwal, D., Mohapatra, S., Avasthi, D., Kienle, L.& Adelung, R. (2012). Crystal growth behaviour in Au-ZnO nanocomposite under different annealing environments and photoswitchability. *Journal of Applied Physics*, *112*, 064308.

Mobasser, S. & Firoozi, A. A.(2016). Review of nanotechnology applications in science and engineering. *Journal of Civil Engineering and Urbanism*, *6*, 84-93.

Moon, H., Ramanathan, S., Suraneni, P., Shon, C.S., Lee, C. J. & Chung, C. W.(2018). *Revisiting the Effect of Slag in Reducing Heat of Hydration in Concrete in Comparison to Other Supplementary Cementitious Materials. Materials*, *11*. (doi: 10.3390/ma11110184).

Morganti, E., Collini, C., Potrich, C., Ress, C., Adami, A., Lorenzelli, L.& Pederzolli, C. (2011). A micro polymerase chain reaction module for integrated and portable DNA analysis systems. *Journal of Sensors*, *7*.

Najim, M., Modi, G., Mishra, Y.K., Adelung, R., Singh, D.& Agarwala, V. (2015). Ultra-wide bandwidth with enhanced microwave absorption of electroless Ni–P coated tetrapod-shaped ZnO nano-and microstructures. *PCCP*, *17*, 22923-22933.

Naqvi, A.A., Garwan, M.A., Maslehuddin, M., Nagadi, M.M., Al-Amoudi, O.S., Khateeb ur, R.& Raashid, M. (2009). Prompt gamma analysis of fly ash, silica fume and Superpozz blended cement concrete specimen. *Applied Radiation and Isotopes*, *67*, 1707-1710.

Ng, C.K., Mohanty, A.& Cao, B. (2015). Biofilms in Bio-Nanotechnology: Opportunities and Challenges. *Bio-Nanoparticles: Biosynthesis and Sustainable Biotechnological Implications*, 83-100.

Nurulhuda, K., Gaydon, D.S., Jing, Q., Zakaria, M.P., Struik, P. C. & Keesman, K.J. (2018). Nitrogen dynamics in flooded soil systems: an overview on concepts and performance of models. *Journal of the Science of Food and Agriculture*, *98*, 865-871.

Patzke, G.R., Krumeich, F.& Nesper, R. (2002). Oxidic nanotubes and nanorods—anisotropic modules for a future nanotechnology. *Angewandte Chemie International Edition*, *41*, 2446-2461.

Pourmohammadi, K., Abedi, E., Hashemi, S.M.B. & Torri, L. (2018). Effects of sucrose, isomalt and maltodextrin on microstructural, thermal, pasting and textural properties of wheat and cassava starch gel. *International Journal of Biological Macromolecules*, *26*, 172.

Pratsinis, S.E. (2016). Overview-Nanoparticulate Dry (Flame) Synthesis & Applications. *UNE*, 13-15.

Rabbani, M. M., Ahmed, I. & Park, S. J. (2016). Application of nanotechnology to remediate contaminated soils. *Environmental Remediation Technologies for Metal-Contaminated Soils*. Springer, pp. 219-229.

Raspa, A., Pugliese, R., Maleki, M.& Gelain, F. (2016). Recent therapeutic approaches for spinal cord injury. *Biotechnology and Bioengineering*, *113*, 253-259.

Reddy, M.L.& Sivakumar, S. (2013). Lanthanide benzoates: a versatile building block for the construction of efficient light emitting materials. *Dalton Transactions*, *42*, 2663-2678.

Rolison, D.R. (2003). Catalytic nanoarchitectures--the importance of nothing and the unimportance of periodicity. *Science*, *299*, 1698-1701.

Sabet, M., Hosseini, S., Zamani, A., Hosseini, Z.& Soleimani, H. (2016). Application of Nanotechnology for Enhanced Oil Recovery: A Review. *Defect & Diffusion Forum*, *367*, 149-156.

Sadeghi, R., Ansari, S., Uzun, S., Bozkurt, F., Gezer, P., Karimi, M.& Kokini, J. (2016). Nanobiotechlogy: Applications in Food Science and Engineering. *UNE*, 13-15.

Schnitzenbaumer, K.J.& Dukovic, G. (2014). Chalcogenide-Ligand Passivated CdTe Quantum Dots Can Be Treated as Core/Shell Semiconductor Nanostructures. *The Journal of Physical Chemistry C*,*118*, 28170-28178.

Schulte, P., Geraci, C., Murashov, V., Kuempel, E., Zumwalde, R., Castranova, V., Hoover, M., Hodson, L.& Martinez, K. (2014). Occupational safety and health criteria for responsible development of nanotechnology. *Journal of Nanoparticles Research*, *16*, 2153.

Simon, G., Gyulavári, T., Hernadi, K., Molnár, M., Pap, Z., Veréb, G., Schrantz, K., Náfrádi, M.& Alapi, T. (2018). Photocatalytic ozonation of monuron over suspended and immobilized TiO2–study of transformation, mineralization and economic feasibility. *Journal of Photochemistry and Photobiology A: Chemistry*, *356*, 512-520.

Su, M., He, C., Sharma, V.K., Asi, M.A., Xia, D., Li, X. z., Deng, H.& Xiong, Y. (2012). Mesoporous zinc ferrite: synthesis, characterization, and photocatalytic activity with H 2 O 2/visible light. *Journal of Hazardous Material*, *211*, 95-103.

Sun, Q., Dai, L., Nan, C. & Xiong, L. (2014). Effect of heat moisture treatment on physicochemical and morphological properties of wheat starch and xylitol mixture. *Food Chemistry*, *143*, 54-59.

Taha, M.R., Khan, T.A., Jawad, I.T., Firoozi, A.A.& Firoozi, A.A. (2013). Recent experimental studies in soil stabilization with bio-enzymes-a review. *Electronic Journal of Geotechnical Engineering*,*18*, 3881-3894.

Tam, R.Y., Fuehrmann, T., Mitrousis, N.& Shoichet, M.S. (2014). Regenerative therapies for central nervous system diseases: a biomaterials approach. *Neuropsychopharmacology*, *39*, 169.

Tarafdar, J., Sharma, S.& Raliya, R. (2013). Nanotechnology: Interdisciplinary science of applications. *African Journal of Biotechnology*, *12*, 125-129.

Tjong, S. C. (2013). *Nanocrystalline materials: their synthesis-structure-property relationships and applications*. Newnes.

Ullah, F., Othman, M.B., Javed, F., Ahmad, Z.& Md Akil, H. (2015). Classification, processing and application of hydrogels: A review. *Material Science and Engineering C-Material for Biological Applications*, *57*, 414-433.

Ushiroda, S., Ruzycki, N., Lu, Y., Spitler, M.& Parkinson, B. (2005). Dye sensitization of the anatase (101) crystal surface by a series of dicarboxylated thiacyanine dyes. *Journal of American Chemical Society*, *127*, 5158-5168.

van Riper, C.J., Browning, M., Becker, D., Stewart, W., Suski, C.D., Browning, L.& Golebie, E. (2018). Human-Nature Relationships and Normative Beliefs Influence Behaviors that Reduce the Spread of Aquatic Invasive Species. *Environmental Management*, *28*, 018-1111.

Wada, M. (2014). [Civil engineering education at the Imperial College of Engineering in Tokyo: an analysis based on Ayahiko Ishibashi's memoirs]. *Kagakushi Kenkyu*, *53*, 49-66.

Wallace, L.A., Emmerich, S.J.& Howard-Reed, C. (2004). Source strengths of ultrafine and fine particles due to cooking with a gas stove. *Environmental Science and Technology*, *38*, 2304-2311.

Wang, T., Yang, G., Liu, J., Yang, B., Ding, S., Yan, Z.& Xiao, T. (2014). Orthogonal synthesis, structural characteristics, and enhanced visible-light photocatalysis of mesoporous Fe 2 O 3/TiO 2 heterostructured microspheres. *Appled Surface Science*, *311*, 314-323.

Wang, Y., Li, Y., Kim, H., Walker, S.L., Abriola, L. M. & Pennell, K. D.(2010). Transport and retention of fullerene nanoparticles in natural soils. *Journal of Environmental Quality*, *39*, 1925-1933.

Weiss, P.S. (2015). *Where are the Products of Nanotechnology?* ACS Publications. (DOI: 10.1021/acsnano.5b02224).

Wong, M.H., Misra, R.P., Giraldo, J.P., Kwak, S.Y., Son, Y., Landry, M.P., Swan, J.W., Blankschtein, D.& Strano, M.S. (2016). Lipid Exchange Envelope Penetration (LEEP) of Nanoparticles for Plant Engineering: A Universal Localization Mechanism. *Nano Letters*, *16*, 1161-1172.

Xia, Y., Yang, P., Sun, Y., Wu, Y., Mayers, B., Gates, B., Yin, Y., Kim, F.& Yan, H. (2003). One-dimensional nanostructures: synthesis, characterization, and applications. *Advanced Materials*, *15*, 353-389.

Xue, C., Xia, J., Wang, T., Zhao, S., Yang, G., Yang, B., Dai, Y.& Yang, G. (2014). A facile and efficient solvothermal fabrication of three-dimensionally hierarchical BiOBr microspheres with exceptional photocatalytic activity. *Materials Letter*, *133*, 274-277.

Yang, L., Luo, S., Li, Y., Xiao, Y., Kang, Q.& Cai, Q. (2010). High efficient photocatalytic degradation of p-nitrophenol on a unique Cu2O/TiO2 pn heterojunction network catalyst. *Environmental science & technology*, *44*, 7641-7646.

Yang, Y.& Westerhoff, P. (2014). Presence in, and release of, nanomaterials from consumer products. *Advances in Experimental Medicine and Biology*, *811*, 1-17.

Yashveer, S., Singh, V., Kaswan, V., Kaushik, A.& Tokas, J. (2014). Green biotechnology, nanotechnology and bio-fortification: perspectives on novel environment-friendly crop improvement strategies. *Biotechnology and Genetic Engineering Reviews*, *30*, 113-126.

Yetisen, A. K., Qu, H., Manbachi, A., Butt, H., Dokmeci, M. R., Hinestroza, J. P., Skorobogatiy, M., Khademhosseini, A. & Yun, S. H. (2016). Nanotechnology in Textiles. *ACS Nano*, *10*, 3042-3068.

Yordanov, G. & Dushkin, C. (2011). *Quantum Dots for Bioimaging Applications: Present Status and Prospects.*

Zhang, Y., Rempel, C. & Liu, Q. (2014). Thermoplastic starch processing and characteristics-a review. *Critical Reviews in Food Science and Nutrition*, *54*, 1353-1370.

In: Photocatalysis
Editors: P. Singh, M. M. Abdullah, M. Ahmad et al.
ISBN: 978-1-53616-044-4

Chapter 6

CHALLENGES AND INFLUENCING FACTORS OF NANOPARTICLES FOR PHOTOCATALYSIS: A CLASSICAL APPROACH IN THEIR SYNTHESIS

Prashant Hitaishi[1], Rohit Verma[1,*], Parul Khurana[2] and Sheenam Thatai[1,†]

[1]Amity Institute of Applied Sciences, Amity University, Noida, India
[2]Department of Chemistry, G. N. Khalsa College of Arts, Science and Commerce, University of Mumbai, Mumbai, India

ABSTRACT

Nanotechnology has revolutionized the world with its multidisciplinary applications. In the nanoscale domain, materials show enhanced electrical, optical and magnetic properties because nanostructures exhibit unique physical, chemical, and biological properties. Nanostructures show unique characteristics as compared with the parent bulk material such as the improved surface area to volume ratio, high reactive, catalytic and antimicrobial activity, large thermal conductivity, magnetic behavior etc. Nanostructures are used across the world for different applications such as photo-catalyst, solar cells, medicines, pharmaceuticals, targeted drug delivery, cosmetics, sensors, fuel for energy storage, optical and electronic devices, etc. Magnetic nanostructures are used for water treatment to remove toxic and heavy metals from it. This chapter focuses on the classical approaches used for the synthesis of nanomaterials and it further covers the challenges and factors influencing their characteristics. Synthesis process of nanostructures play a very crucial role in nanotechnology, so our main aim is to describe the synthesis methods

* Corresponding Author's Email: rverma85@amity.edu.
† Corresponding Author's Email: sheenam@gmail.com

of materials with superior properties at comparatively lower cost. Particle size, surface morphology, quantity, reactivity and all other physical, chemical properties depend on the synthesis mechanism used. The main objective to study the synthesis processes is to minimize the unwanted by-products left in the environment and maximizes the product yield.

Keywords: nanoparticle, nanostructures, synthesis, top-down, bottom-up, cvd, mechanical milling, laser ablation, nanolithography, sputtering, thermal decomposition, sol-gel, pyrolysis, biosynthesis, spinning.

1. Introduction

Nanoparticles (NPs) are defined as the particles with a size in the nanometer range (1 nm – 100 nm) (Ealia et al. 2017). Nanostructures show different chemical and physical characteristics as compared to the identical bulk material. Metallic nanostructure exhibit unique optical, chemical, mechanical and magnetic properties like higher mechanical strength, specific surface areas optical properties, specific magnetizations, and low melting points (Horikosi et al. 2013). Optical properties include the reflection, absorption and light penetration capabilities of the nanostructure (Ealia et al. 2017). Chemical properties of nanostructures made them a good candidate for sensors and detector applications due to their high reactivity and sensitivity towards the atmosphere, light, heat, and moisture (Ealia et al. 2017).

Nanotechnology has received great attention in past decades. The properties of the bulk material are size independent whereas NPs shows size-dependent properties (Horikosi et al. 2013). Metal nanostructures such as Silver (Ag), Aluminium (Al), Gold (Au), and Iron (Fe) has already been reported useful for different applications. Nanostructures are widely used for medical applications such as antimicrobial (Guzman et al. 2012), catalysis (Jiang et al. 2005), fluorescence (Xu et al. 2010) are for diagnosis (Choi et al. 2007). Aluminum nanostructures are used in various technologies for coating purpose including solar cells, sensors, mirrors, and an optical device. It is also used to raise the reaction energy, blast rates, and flame temperature. Hence, useful in explosive industries and is expected to be used in aerospace applications as a propellant in the near future (Ghorbani and Hamid 2014). ZnO nanostructures are non-toxic, low-cost and absorb the large fraction of solar spectrum hence used as a photocatalyst in the photodegradation of carbon-based organic pollutants (Ong et al. 2018). Au nanoparticles (>10 nm) show photocatalytic property and are used for water splitting into H_2 and O_2 (Tian et al. 2018). TiO_2 and ZnO nanostructures reflect the UV rays and are transparent to visible light, hence used in sunscreens, cosmetics, transparent window panels and glasses (Ealia et al. 2017). Au nanostructures are proved to be useful in medical science for the detection and diagnosis of cancer cells. These are also useful to identify and differentiate the different classes of bacteria. Au nanostructures are capable to detect the DNA in a sample, hence used in DNA fingerprinting (Tomar. 2013).

Major challenges in the development of NPs include:

- Synthesis of high-quality NPs with high purity.
- Cost effective synthesis process to ensure more efficient production of NPs at a commercial level.
- Better control over the whole synthesis process to enhance the properties of NPs.

Generally, two main approaches are used for the synthesis of nanomaterials, named Top-down & Bottom-up. For commercial and research applications, NPs have been synthesized mainly by three methods namely Physical, Chemical and Biological processes (Ealia et al. 2017).

These methods of preparing NPs include Crushing, Grinding, Physical vapor deposition, Chemical vapor deposition (CVD), Sol-gel method, a Plasma method, Pulsed Laser ablation, Electrodeposition, Hydrothermal, Combustion method etc., each process have some advantage as well as a disadvantage over other processes.

Nano-size particles have a larger surface area and so they are highly reactive and toxic (Vithiya et al. 2011). Such small size particles have the ability to penetrate the human body at the cellular level (Vishwakarma et al. 2010). Biosynthesis of NPs is an alternative green method to reduce the adverse effects of conventionally used chemical methods. In this process, fungi, plants, yeast, bacteria etc. are used and no chemical byproducts left in the atmosphere hence, it is an eco-friendly approach (Vithiya et al. 2011).

This chapter describes the classical synthesis methods of nanostructures, challenging factors and their influence on the properties of nanostructures.

2. Synthesis Methods of Nanostructures

The synthesis of nanomaterials is the initial development of nanoscience. It has been a challenging task to synthesize the nanomaterials efficiently, that means it should be cost-effective, simple and with a high product yield. It has been observed that the synthesis of nanostructures is a challenging task because the shape and size of NPs depend on many physical and chemical conditions such as:

1. Viscosity and time
2. Concentration of medium
3. Temperature and humidity
4. Quality and purity of solvents used in the synthesis

5. Physical environment present or created externally including gases, vacuum, pressure etc.

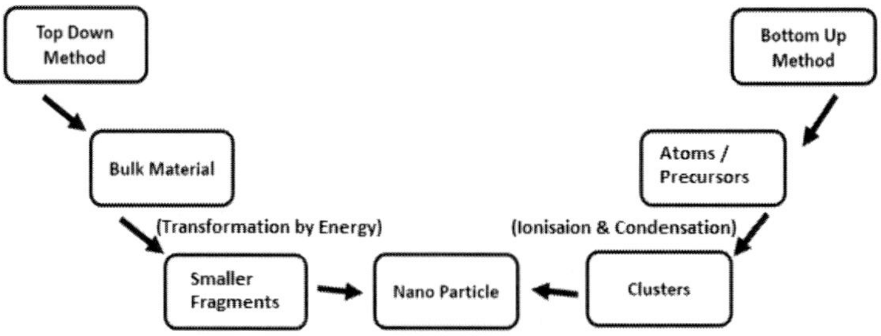

Figure 1. Schematic representation of the synthesis of NPs by top-down and bottom-up approach (Ealia et al. 2017).

Nanostructures can be synthesized from zero dimension to multi-dimension depending on the approach and method used for the synthesis. As shown in Figure 1, two approaches are followed for the preparation of NPs. The first is Top-Down (breakdown) technique and second is Bottom-Up (build-up) technique.

3. Top-Down Technique: Synthesis of Nanostructures from Bulk Material

Top-Down is a synthesis approach in which process starts initially with the bulk material and continues with the objective reduce to particle size in the nanometer range. Bulk material is broken into smaller fragments under externally applied energy. This externally applied energy can be a chemical, thermal, mechanical or another form of energy. The main disadvantage of this approach is that it introduced in the surface structure (Thakkar et al. 2010) and the nanostructured formed are non-uniform in size. Particles are formed in a wide size range and to maintain the particle size expensive equipment are used in methods such as focused ion beam sputtering and lithography synthesis. The top-down approach is considered for particles having size more than 100 nm. Most commonly used top-down methods are tabulated in Table 1.

3.1. Mechanical Milling

This process is used for the synthesis of different types of materials such as amorphous, nanocrystalline alloys and metal/non-metal nanocomposite materials. In this process, milling of different powder is mixed to form one (single) powder, followed by the annealing of compound powder in an inert atmosphere. The main objective of ball milling is the reduction of particle size and blending of different particles in a new phase. Dense steel or tungsten carbide metal balls have used that rolls down the surface of the chamber and produce an impact on the powder beneath them.

Table 1. Different methods of synthesis of nanostructures (Ealia et al. 2017)

Category	Method	NPs
Top-down method	Mechanical milling	Metal, oxide, and polymer-based
	Nanolithography	Metal-based
	Laser ablation	Carbon and metal oxide based
	Sputtering	Metal-based
	Thermal decomposition	Carbon and metal oxide based
Bottom-up method	Sol-gel	Carbon, metal and metal oxide based
	Spinning	Organic polymers
	Chemical Vapor Deposition (CVD)	Carbon and metal based
	Pyrolysis	Carbon and metal oxide based
	Biosynthesis	Organic polymers and metal based

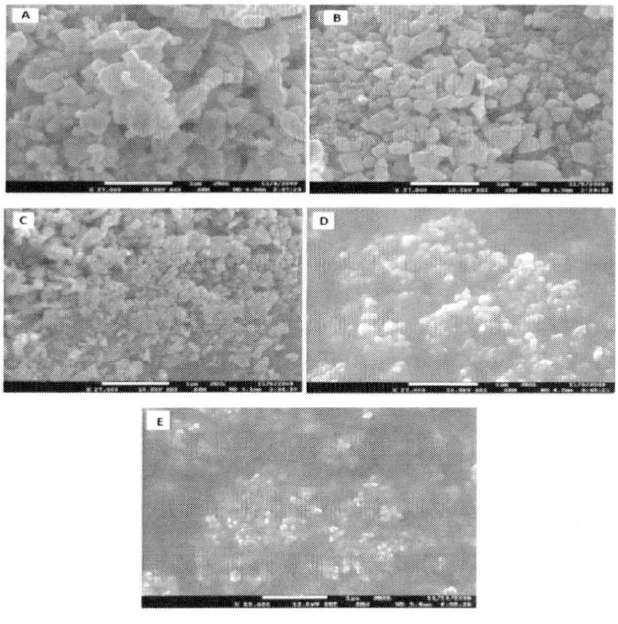

Figure 2. Scanning electron microscope images of ZnO samples before and after milling: (A) as purchased. (B, C, D, and E) for the samples ball milled for 2, 10, 20 and 50 hours respectively (Salah et al. 2011).

The yield and quality of nanostructure depend on the dry and wet type of milk used, milling speed, size and distribution of the balls, temperature and time. Figure 2 shows the effect of milling time on ZnO nanostructures, synthesized using mechanical ball milling. Figure 2 (A) shows the SEM image of the ZnO sample (un-synthesized as purchased) having particles of different sizes and shapes. The size of particles is large and covers the range 0.1-0.6 μm but after ball milling for 2, 10, 20, 50 hours the particles size becomes smaller and uniform as shown in Figure 2 (B, C, D, E) respectively (Salah et al. 2011). The basic mechanics in the process is that particles during the milling get trapped between the milling balls and undergo deformation and fracture process. Deformation causes the change in particle shape and fracture decreases the particle size. High energy ball milling is more effective than traditional ball milling. In traditional ball milling particles are fractured and milled to reduce the size, while high-energy ball milling (HEBM) continues for a longer time and helps to make them chemically active. Different factors such as rotation speed, the mass of balls, impact angle, and impact velocity can be controlled in HEBM to form better products in comparison to the ordinary milling process.

Synthesis of ZnO NPs using HEBM has already been reported earlier (Salah et al. 2011). The impact of variation of milling time and speed has been studied and reported (Salavati-Niasari et al. 2013). Figure 2 shows the effect of milling speed and time on the size of particles. With the increase in speed and time, particle size decreases. Milling time has a direct impact on the particle size as it reduces from 600 nm to 30 nm and the lattice constant c get increased from 5.204 to 5.217 Å with the increase in milling time from 2 to 50 hrs (Salah et al. 2011). Using ball milling the synthesis of Silica nanostructures from Rice husk ash has been reported. After 6 hours, ball milling of the silica nanostructures having size 70 nm was found in powdered form. During synthesis decrease in particle size has been observed with the increase in ball milling speed or time.

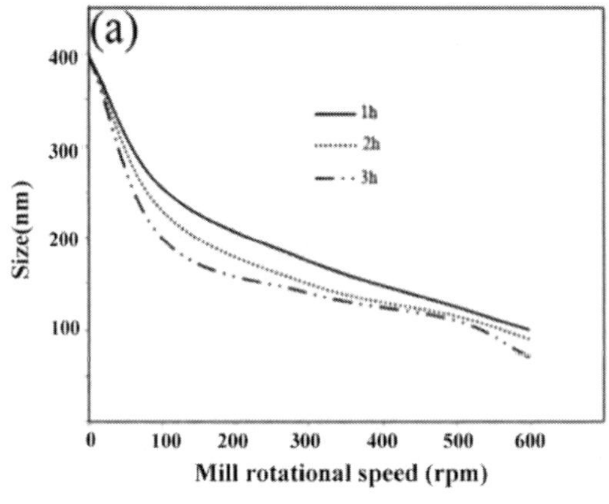

Figure 3. The effect of milling time and rotation speed on silica NPs size synthesized from rice husk ash (Salavati-Niasari et al. 2013).

3.2. Nanolithography

It is a process of transferring a predefined and required pattern from one material to another with the objective to modify the surface of the second material to nanometer resolution. Several lithographic techniques are Photolithography, Electron beam lithography, X-ray lithography etc. In the photolithography, process photoresist is applied to the bulk material and a mask is used to expose the light selectively on the desired surface of the material. Depending on the type of photoresist used, a partial area of the material is etched away and the desired material remains with a modified nanostructured surface. In photolithography, the highly energetic optical source and x-rays are used. This method is used to synthesis the metal-based nanostructure of desired shape and size.

3.3. Laser Ablation

Laser ablation is a process used to modify the surface of a material by irradiating it with a laser beam. Depending on the intensity of the beam required a pulsed or continuous laser beam can be used to ablate the material. The depth of laser beam interaction depends on the optical properties of materials and laser pulse. The target material is placed in an intermediate medium or phase that can be a vacuum, liquid or gaseous medium. The energy of incident laser light absorbed by the target material causes either thermal or chemical reaction. This method does not require expensive setup. This is a simple and more versatile method. The amount of total mass ablated from the target in each laser pulse is known as ablation rate. Synthesis of Al nanostructures has been reported (Baladi et al. 2010), in which Al targets were kept in ethylene glycol, acetone, and ethanol. Synthesis of Al NPs in the acetone medium was fine in comparison to other mediums used. In the acetone medium, the size of particles varies from 10 to 100 nm and the maximum particles have a diameter of 30 nm. Zn-Al layered double hydroxides and Mg-Al nanostructures synthesized by laser ablation in the liquid medium has been reported (Hur et al. 2009). The layer thickness of the Zn-Al structure was found to be 6.0 nm with an average diameter of 500 nm. The ablation rate decreases in the colloidal solution because the high concentration of particles in the colloidal solution blocks the path of incident laser and the energy is partially absorbed instead of interacting with the target surface (Baladi et al. 2010). The schematic representation of the synthesis of nanostructures by laser ablation is shown in Figure 4.

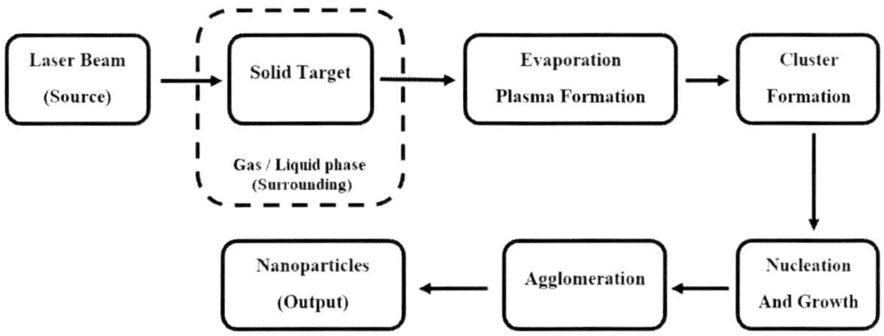

Figure 4. Schematic representation of nanostructures synthesized by the laser ablation method.

Synthesis of nanostructures by Pulse Laser Ablation (PLA) is further divided into two phases based on their state i.e., gas or liquid:

3.3.1. Gas Phase PLA

In this process a chamber is filled either with pure inert gas or a mixture of inert gases and the solid target is placed in it. The main role of inert gas is to control the plasma expansion during the ablation process. The plasma produced in the gaseous atmosphere by using highly intense laser beam is used to fulfill the requirement of high temperature for the synthesis of nanostructures (Scott et al. 2002). PLA is used for the synthesis of carbon and metal oxide based nanostructures.

3.3.2. Liquid Phase PLA

In this process, a chamber or a reactor is filled with a liquid and the solid target is kept in it. The laser beam incident on the solid target and vaporizes it to form an intermediate reactive product that reacts with the liquid filled in the reactor to form nanostructured products. The intensity, wavelength, and power of incident laser beam are the important parameters along with the solid target and precursor liquid used in the synthesis process (Scott et al. 2002). The quality and nanoparticle properties vary with all these parameters. It is a cost-effective, simple synthesis process and does not require an expensive system.

Finally, the nanostructures formed are collected by precipitation, filtration, and evaporation process.

3.4. Sputtering

Sputtering is a physical vapor deposition (PVD) technique, used for the deposition of thin films with a thickness of nm to µm. It is a kind of non-thermal vaporization technique in which deposition takes place in a vacuum chamber or filled with background gas. In sputtering, a target material is bombarded with highly energetic ions and ejected particles get deposited on the substrate surface (Ealia et al. 2017). Generally, the target is attached to the cathode and opposite charged, highly energetic ions collide with the target. Ion-target interaction is a complex phenomenon depends on the ion energy and ion to target mass ratio.

The advantage of sputtering is that it is possible to sputter one material on multiple targets of different materials at the same time. Sputtering is used to synthesize metal nanostructures using metal thin films or plates. Synthesis process depends on the distance between the target and the substrate (Asanithi et al. 2012). Depending on this distance and mean free path, there are two possibilities either formation of thin film or nanostructure. If the mean free path is greater than the target-substrate distance, there is no collision between sputtered atoms and a continuous thin film is obtained. Oppositely, if the mean free path is less than the target- substrate distance, multiple collision take place and nanostructures are formed. Due to collisions in the mean free path before reaching the substrate, sputtered atoms condense rapidly and grow into nanostructure in the vapor phase. Synthesis of Ag nanostructure, using DC magnetron sputtering has been reported (Asanithi et al. 2012) and there is an inverse relation in the size of NPs and target-substrate distance. The size of NPs decreases with the increase in target-substrate distance. Schematic representation of DC sputtering technique used for the formation of thin-film is shown in Figure 5.

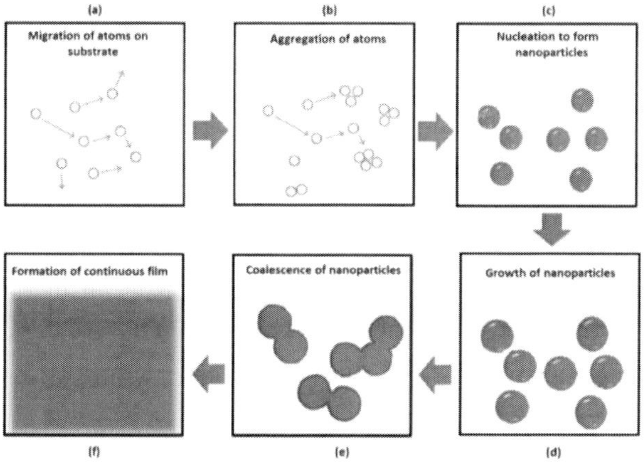

Figure 5. Schematic representation of the formation of thin-film using a DC sputtering technique (Asanithi et al. 2012).

3.5. Thermal Decomposition

Decomposition is a process in which large substances are broken down into small ones. In thermal decomposition, external heat is supplied to break the bulk target into the nanostructure. Temperature is an important factor in this process that affects the characteristics of the product formed. The specific temperature at which decomposition takes place is known as decomposition temperature (Ealia et al. 2017). External heat supplied to the system is used to break the chemical bonds. The schematic representation of thermal decomposition is shown in Figure 6.

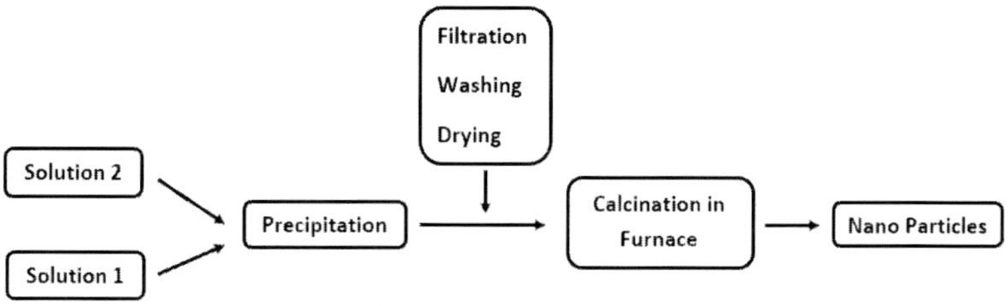

Figure 6. Schematic representation of the synthesis of NPs by the thermal decomposition method.

Synthesis of magnetic $Co_{0.4}Zn_{0.6}Fe_2O_4$ nanostructures using thermal decomposition method has been reported. In this process metal acetylacetonate in solvents and fatty acids having high boiling point temperature are used to synthesize NPs (Sharifi et al. 2016). Magnetic particles have a unique feature to respond in the presence of magnetic force. Magnetic properties of materials have attracted them to find applications in medical science.

Now a day's magnetic NPs have been used in magnetic resonance imaging (MRI), drug targeting, bioseparation etc. Synthesis of core-shell Si-Fe oxide nanostructures has been reported using thermal decomposition approach. SiO_2 spheres having 244 nm diameter are used to deposit iron oxide nanostructure. Iron (III) acetylacetonate, diphenyl ether is used to synthesize the core-shell NPs in the presence of SiO_2 by thermal decomposition technique (Kishore et al. 2012).

Various applications of magnetic core-shell nanostructures are wastewater treatment, immunological assays, immobilization of biomolecules etc. Synthesis of spherical shaped ZnO NPs by thermal decomposition method has been reported in which $Zn_4(SO_4)(OH)_6(0.5H_2O)$ used as a precursor. This method has been reported as a convenient, cheap and simple method to produce ZnO NPs at industrial scale (Darezereshki et al. 2011).

4. Bottom-Up: Synthesis of Nanostructures from Atomic Level

The Bottom-up approach is also known as a constructive method. It is a synthesis method of NPs from the atomic level to nano-sized particles. In this process initially, the building blocks of nanostructures are formed and later assembled to form the nanostructures (Thakkar et al. 2010).

4.1. Sol Gel

Sol-gel is a bottom-up method, widely used for the synthesis of nanostructures. It is a very simple technique, doesn't require expensive equipment, and has a low processing temperature (room temperature) and almost all type of NPs can be synthesized from this method. The sol-gel method can be used for the deposition of high quality, homogenous films on different types of substrates. In this method, solid particles (~ 0.1-1 nm) are suspended in liquid and known as a sol. These particles are small in size and have negligible weight hence, unaffected by the gravitational forces and the only interaction between the particles is Coulomb force and short-range van der Wall's attraction. There two phenomena in the processing i.e., flocculation and coagulation. In the flocculation process, bonding between the sol particles is reversible and in the coagulation process, it is irreversible. Sol particles grow up to macroscopic dimension and transform into a gel (Shahjahan et al. 2017, Azlina et al. 2016, Livage et al. 1989, Sui et al. 2012).

Steps of Sol-Gel Process:

In the thin film deposition and nanostructure synthesis process multiple steps are involved.

1. *Hydrolysis:* The general chemical equation for the hydrolysis process is

$$M(OR) + H_2O \rightarrow M(OH) + ROH \quad (1)$$

This reaction shows that during hydrolysis, hydroxyl ions get attached to the metal atom. Here, R is a proton or another ligand, M is metal and ROH is alcohol. The rate and time duration of the hydrolysis process depends on the availability of catalyst and water.

2. *Condensation and Polymerization:* Condensation reactions partially take place in two steps.

$$M(OH) + M(OH) \rightarrow M_2O + H_2O \rightarrow \text{Water condensation} \quad (2)$$

$$M(OR) + M(OH) \rightarrow M_2O + ROH \rightarrow \text{Alcohol condensation} \quad (3)$$

These reactions will continue to grow and form large particles containing metal and the process is known as the method of polymerization. Hydrolysis and condensation reaction undergoes at the same time.

3. *Gelation:* The gel is an intermediate form of solid and liquid particles in the form of colloids having a particle size of about 1μm particle. In the process of gelation, charged species present in the solution get adsorbed on the surface and develops a surface charge. This charge separation leads to the formation of a new opposite charged layer just below the charged surface layer. Sol particles interact by the attractive van der Waals force and repulsive short range force (Born repulsion). The total energy of the system becomes negative when particles cross the energy barrier and lead to the formation of solid fractal aggregates. These solid aggregates extend and get coagulated throughout the sol. And this transformation of sol into solid is known as sol-gel transition or gelation.
4. *Aging:* Aging is a process of formation of giant molecules where small polymers formed in the gelation process attracted to form the main network. It is further

followed with the phase transformation and condensation, which leads to the shrinking of some gels through transportation of liquid from the pores present in the gel.

5. *Drying:* Drying and evaporation lead to the shrinking of the gel structure due to some chemical or physical process such as capillary action. Drying process leads to the formation of a porous gel. Capillary action is observed more effective for small size pores that cause more stress and leads to the cracking of film.
6. *Heat treatment:* The gels or films should be annealed to remove the residual water to obtain the desired material. During annealing, the volatile components that have the capability to escape more rapidly remove out of the final product and finally resulting in the shrinkage of films. The bonding between film and substrate become stronger due to the process of shrinkage of films.

Factors affecting the product yield and characteristics are

1. sol's concentration
2. chemical equilibrium
3. pH value
4. temperature
5. time

Sol-gel process has been used and reported for the synthesis of Ag nanostructures, having size 6-22 nm (Shahjahan et al. 2017). The chemical reaction was carried out in the presence of CH_3COONa and a reducing agent named hydrazine at room temperature. Synthesis of SiO_2 NPs using the sol-gel process has been reported (Azlina et al. 2016), in which acetic acid was used as a catalyst and distilled water was used as the hydrolyzing agent. The effect of calcination temperature (600-700°C) and aging time (2-6 hrs) has been studied.

4.2. Spinning

Spinning process is very simple and convenient, used for the synthesis of organic polymer nanostructures. The reactor chamber consists of a spinning rotating disc. This process is also known as the spinning method. The disc used in the chamber is coupled with a temperature-controlled heater to maintain the temperature required during the process. The reactor is generally filled with an inert gas to avoid unwanted chemical reactions in the chamber.

The liquid precursors and water are placed on the rotating / spinning disc as a result molecule get coagulated to form solid-like structure. This intermediate product further precipitated, collected and dried to form NPs (Ealia et al. 2017). The characteristics of NPs depends on various parameters such as speed of the rotating disc, liquid/precursor ratio, the flow rate of liquid at which fluid is injected into the chamber, the surface of disk used, temperature maintained during the process (Mohammadi et al. 2014).

4.3. Chemical Vapour Deposition (CVD)

It is a thermal process in which substrate or target is heated at a high temperature and reactants in vapor or gas phase undergo a chemical reaction and get deposited on the heated substrate (Ealia et al. 2017). It is known as CVD because a chemical reaction takes place when the gases meet the heated substrate. CVD system comprises of a deposition chamber with heater, gas supply system, vacuum system, and gas exhaust system. A thermal CVD system requires a high activation temperature above 900 °C, while in a plasma CVD activation energy is provided by the plasma and deposition takes place at lower temperature 300 and 700 °C. NPs formed by the CVD system are in highly pure form, maintains size uniformity and distribution. But the main disadvantages of the CVD system are the by-product gases exhausted by the system are highly toxic in nature and CVD system is very expensive.

4.4. Pyrolysis

In this process, liquid or vapor precursor are injected into the furnace at high temperature and pressure. Precursors are burned using flame in the furnace to form intermediate by-products in a gaseous form which further processed to form final NPs. In the furnace, precursors can be heated using direct flame, LASERs, and plasma to evaporate the reactants (D'Amato et al. 2013). This process is commercially used for the synthesis of carbon and metal oxide based nanostructures (Ealia et al. 2017).

4.5. Biosynthesis

Biological ways of synthesis are the alternative ways over chemical and physical methods (Vithiya et al. 2011). Green or Biosynthesis of the nanoparticle is an eco-friendly, non-hazardous approach of synthesis. It is carried out using different types of microorganisms and plants (Shankar et al. 2004). The synthesis of Cu nanostructures using cumin seed extracts has been reported (Rajesh et al. 2018). It is a green approach to produce metal nanostructures using bio-molecules.

Different research groups have reported the synthesis technique and applications of biosynthesis. Synthesis of Au-Ag nanostructures has been reported by the group Nestor et al. in which green tea (Camellia sinensis) extracts are used as a reducing and stabilizing agent (Vilchis-Nestor et al. 2008). Synthesis of Ag nanostructure using organism, Bacillus lichenformis has been reported by the group Kalishwaralal et al., 2008.

Three different sources used in the biosynthesis:

a. Use of microorganisms (Klaus et al. 1999) like fungi, yeast, bacteria, etc.
b. Use of plant extracts (Shankar et al. 2004) or enzymes (Willner et al. 2006).
c. Use of templates like DNA, membranes, virus etc.

Advantages:

a. Eco-friendly
b. Cost-effective
c. Biocompatible
d. NPs produced are free from impurities.

5. Size and Shape-Dependent Properties of Nanostructures

5.1. Quantum Confinement

Electronic and optical properties of a material change as we go in the smaller size, typically in the nanometer range (< 10 nm). In nanometer range, bandgap increases because electrons and holes are confined in the critical quantum range known as exciton Bohr radius (Yoffe 1993). In a quantum confined structure, the motion of both the charge carriers is confined in two to zero dimension by the potential barriers (Miller 1984). Due to dimension confinement of particles in 1, 2 and 3 dimensions, particles are categorized as quantum wire, quantum well and quantum dots respectively. Confined electrons exhibit discrete energy spectrum unlike continuous energy spectrum in bulk material as shown in Figure 7.

Example: Semiconductors such as ZnO, ZnSe, ZnS, GaN are wide bandgap materials show similar crystal structures. GaN-based technology made it possible to fabricate Blue, UV LEDs. Zinc-Oxide NPs have a wide bandgap of 3.4 eV hence used as an active layer in the heterostructured LEDs & Quantum well LASERs.

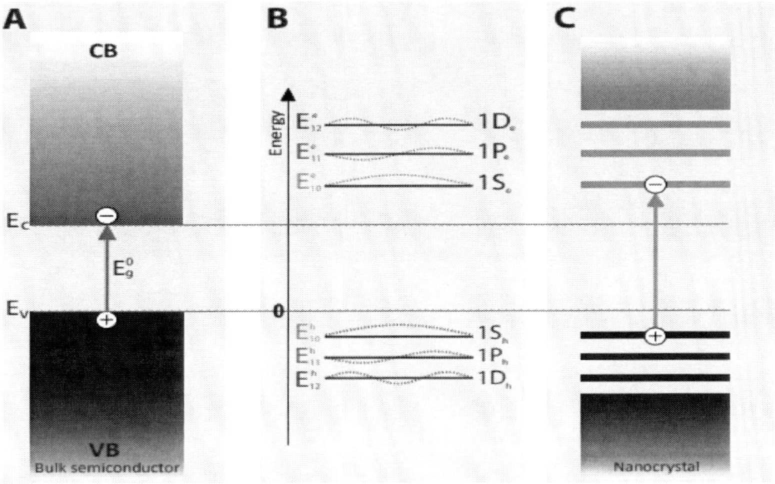

Figure 7. Quantum confinement of a semiconductor and its effect on the electronic structure. (a) Bulk semiconductor (b) Electron and hole energy levels (c) Nanocrystal quantum dot (Groeneveld 2012).

5.2. Absorption of Solar Radiations

Nanostructures offer much higher absorption of radiations in comparison to the identical bulk material. In Solar PV and Solar thermal applications, the efficiency and physical properties of the system depend on the shape and size of the material used. Hence, variation in the shape and size of NPs alter the efficiency of solar cells. ZnO based devices are widely used for high energy radiation space applications. TiO_2 and ZnO nanostructures are used in cosmetics as sunscreens because they are transparent to visible light and reflect the UV rays (Ealia et al. 2017).

5.3. The Larger Surface Area to Volume Ratio

In comparison to bulk material nanoparticles have a larger surface area to volume ratio as shown in Figure 8. This property increases the chemical reactivity of the material and NPs becomes more sensitive. The increased surface area is responsible for the variation in optical, mechanical and chemical properties of NPs as compared to bulk material (Mazur 2004). It is very difficult to synthesize the high purity nanomaterials as they are highly reactive and undergoes oxidation easily (Baladi et al. 2010). Let's take an example to show how this ratio increases with the decrease in the size of particles.

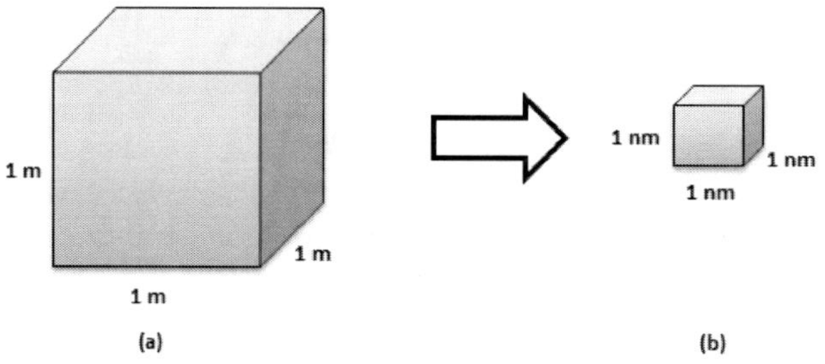

Figure 8. Comparison between the bulk material and nanoparticle based on the surface area to volume ratio.

Surface area of cube a = 6 × (1 m × 1 m) = 6 m^2
Volume of cube a = 1 m × 1 m × 1 m = 1 m^3
Surface area to volume ratio = $\frac{6\,m^2}{1\,m^3} = 6\,m^{-1}$
Similarly, for cube b
Surface area of cube b = 6 × (1 nm × 1 nm) = 6 nm^2
Volume of cube b = 1 nm × 1 nm × 1 nm = 1 nm^3
Surface area to volume ratio = $\frac{6\,nm^2}{1\,nm^3} = 6\,nm^{-1} = 6 \times 10^9\,m^{-1}$
The surface area to volume ratio for cube b is increased 10^9 time that of cube a.

5.4. Nano-Catalyst

Nanostructures have high selectivity and reactivity, depending on the shape, size and surface area to volume ratio as mentioned in the earlier sections. Catalytic reagents play an important in a chemical reaction as one can control the rate, temperature, reactivity, and unwanted side products accordingly. Catalysis is one of the prominent application of nanotechnology and NPs used is referred to as nanocatalysts. Nanocatalysts are energy efficient, reduce chemical waste and harmful byproducts of reactions. Nanocatalysts are found useful in water treatment and help to reduce global warming hence known as green catalysts (Singh and Tandon, 2014). ZnO nanostructures are non-toxic, low-cost and absorb the large fraction of the solar spectrum, hence used as a photocatalyst in the photodegradation of carbon-based organic pollutants (Ong et al. 2018). Au nanoparticles (>10 nm) show photocatalytic property and are used for water splitting into H_2 and O_2 (Tian et al. 2018).

6. Size and Shape Control of Nanostructures

6.1. Control of Size

The chemical and physical properties of nanostructures such as optical, electrical and reactivity are size dependent. (Ealia et al. 2017). Size depends on the process used for the synthesis of the nanostructure. The optical properties such as Absorbance, wavelength, and bandgap depends on the size of NPs. The variation of optical characteristics of Au nanospheres with the particle size has been reported and shown in Figure 9 (Liz-Marzán et al. 2006).

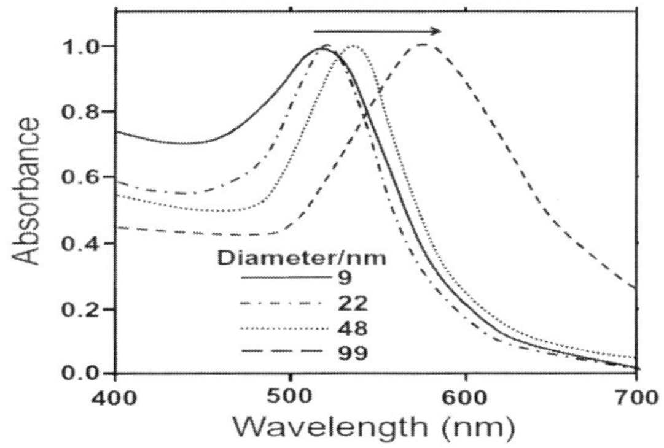

Figure 9. The effect of particle size on visible-light spectra of Au nanospheres (Liz-Marzán et al. 2006). Copyright 2006 by the American Chemical Society.

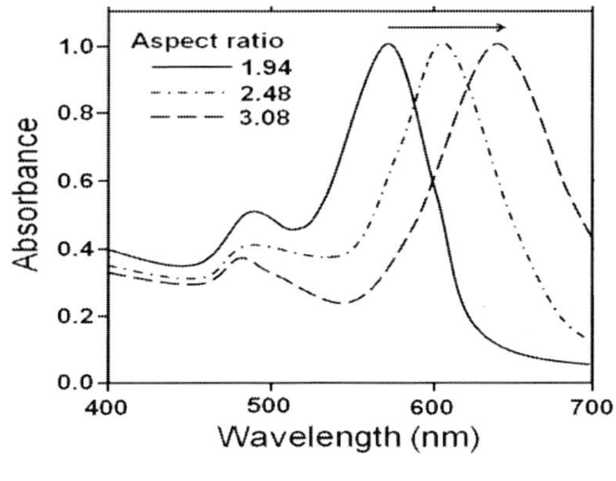

Figure 10. Spectra of Au rod-shaped NPs in visible light with a various aspect ratio (Liz-Marzán et al. 2006). Copyright 2006 by the American Chemical Society.

6.2. Control of Shape

The properties of NPs also depend on the shape of NPs. They are found in spherical, cylindrical rods, dots, etc. Gold NPs show different properties in spherical and cylindrical shaped nanorods. The variation of optical absorbance of Au rod-shaped NPs has been reported (Yu et al. 1997). In Figure 10, absorption spectra of Au nanorods vary with the change in the aspect ratio (ratio of the length of the longer side to the short side of the rod).

CONCLUSION

This chapter is focused on the classical approaches for the synthesis of nanomaterials and it further covers their different challenges, properties, and factors that affect the nanomaterials. Synthesis of nanostructures is now a mature process in the field of nanotechnology. Different research groups have studied and widely reported in the books and research papers. Different chemical, physical, and biological methods are available to synthesize a particular type of nanostructures. Generally, there are two classical synthesis approaches i.e., Top-down and Bottom-up approach, which are further divided into the chemical, physical and biological method. The chemical, photo-catalysis, electrical, optical and physical properties of nanostructures are shape and size dependent. Synthesis method and approach affect the size, shape, and purity of yield. There are some other factors that affect the properties of finally synthesized nanostructures are temperature, pressure, inert gas, time etc. maintained during the processing. All synthesis processes have their advantage as well as disadvantages over one another. So it is important to have an idea regarding the required properties of nanostructures before processing and synthesizing to ensure the synthesis process more efficiently. The synthesis process should opt according to the required characteristics of finally synthesized nanostructures.

ACKNOWLEDGMENTS

The authors acknowledge Prof. Sunita Rattan, Head of Institute, Amity Institute of Applied Sciences (AIAS), Amity University Uttar Pradesh, Noida. ST and RV acknowledge Prof. Sangeeta Tiwari and Prof. R. S. Pandey, Head, Dept. of Chemistry and Physics, AIAS, Amity University Uttar Pradesh, Noida. ST, RV and PK acknowledge authors for the kind permission to reproduce the Figures. ST would like to thanks Department of Science and Technology, Government of India, New Delhi, India.

REFERENCES

Asanithi P., Chaiyakun S., and Limsuwan P. 2012 "Growth of silver nanoparticles by DC magnetron sputtering." *Journal of Nanomaterials* 2012:1-9.

Azlina H. N., Hasnidawani J. N., Norita H. and Surip S. N. 2016 "Synthesis of SiO2 nanostructures using sol-gel method." *Acta Physica Polonica A* 129:842-844.

Baladi A., and Rasoul S. M. 2010 "Investigation of different liquid media and ablation times on pulsed laser ablation synthesis of aluminum nanoparticles." *Applied Surface Science* 256:7559-7564.

Bining T., Lei Q., Tian B., Zhang W., Cui Y., and Tian Y. 2018 "UV-driven overall water splitting using unsupported gold nanoparticles as photocatalysts." *Chemical Communications* 54: 1845-1848.

Choi H. S., Wenhao L., Misra P., Tanaka E., Zimmer P. J., Ipe I. B., Bawendi G. M., and Frangioni V. J. 2007 "Renal clearance of quantum dots." *Nature biotechnology* 25: 1165-1170.

Darezereshki E., Alizadeh M., Bakhtiari F., Schaffie M., and Ranjbar M. 2011 "A novel thermal decomposition method for the synthesis of ZnO nanoparticles from low concentration ZnSO4 solutions." *Applied Clay Science* 54: 107-111.

Ealia M. S. A., and Saravanakumar M. P. 2017 "A review on the classification, characterisation, synthesis of nanoparticles and their application." In *IOP Conference Series: Materials Science and Engineering* 263:1-12.

Groeneveld E. 2012 Eds. *"Synthesis and optical spectroscopy of (hetero)-nanocrystals."* Utrecht: Utrecht University, chapter 2:14-37.

Guzman M., Dille J. and Godet S. 2012 "Synthesis and antibacterial activity of silver nanoparticles against gram-positive and gram-negative bacteria." *Nanomedicine: Nanotechnology, biology and medicine* 8: 37-45.

Horikoshi S., and Serpone N., eds. 2013. "*Microwaves in nanoparticle synthesis: fundamentals and applications.*" John Wiley & Sons, ISBN: 978-3-527-33197-0:352.

Hur T B., Phuoc T. X., and Chyu M. K. 2009 "Synthesis of Mg-Al and Zn-Al-layered double hydroxide nanocrystals using laser ablation in water." *Optics and Lasers in Engineering* 47: 695-700.

Jiang Z. J., Liu C. Y., and Sun L. W. 2005 "Catalytic properties of silver nanoparticles supported on silica spheres." *The Journal of Physical Chemistry B* 109: 1730-1735.

Kalishwaralal K, Deepak V., Ramkumarpandian S., Nellaiah H. and Sangiliyandi G. 2008 "Extracellular biosynthesis of silver nanoparticles by the culture supernatant of Bacillus licheniformis." *Materials letters* 62: 4411-4413.

Kishore P. N. R. and Jeevanandam P. 2012 "A novel thermal decomposition approach for the synthesis of silica-iron oxide core–shell nanoparticles." *Journal of Alloys and Compounds* 522: 51-62.

Klaus T., Joerger R., Olsson E., and Granqvist C. G. 1999 "Silver-based crystalline nanoparticles, microbially fabricated." *Proceedings of the National Academy of Sciences* 96: 13611-13614.

Livage J., C. Sanchez, M. Henry M., and Doeuff S. 1989 "The chemistry of the sol-gel process." *Solid state ionics* 32: 633-638.

Marzán L., Luis M. 2006 "Tailoring surface plasmons through the morphology and assembly of metal nanoparticles." *Langmuir* 22: 32-41.

Masoud S. N., Javidi J., and Dadkhah M. 2013 "Ball milling synthesis of silica nanoparticle from rice husk ash for drug delivery application." *Combinatorial chemistry & high throughput screening* 16: 458-462.

Mazur M. 2004 "Electrochemically prepared silver nanoflakes and nanowires." *Electrochemistry Communications* 6: 400-403.

Mohammadi S., Harvey A., and Boodhoo V. K. 2014 "Synthesis of TiO2 nanoparticles in a spinning disc reactor" *Chemical Engineering Journal* 258: 171-184.

Nestor V., Alfredo R., Mendieta V. S., Marco A. C. L., Gómez-Espinosa R.M., Camacho-López M. A., and Arenas-Alatorre J. A. 2008 "Solventless synthesis and optical properties of Au and Ag nanoparticles using Camellia sinensis extract." *Materials Letters* 62: 3103-3105.

Ong C. B., Ng L. Y., and Abdul W. M. 2018 "A review of ZnO nanoparticles as solar photocatalysts: Synthesis, mechanisms and applications." *Renewable and Sustainable Energy Reviews* 81: 536-551.

Rajesh K. M., Ajitha B., Reddy A. K., Suneetha Y., Reddy P. S., and Ahn C. W. 2018 "A facile bio-synthesis of copper nanoparticles using Cuminum cyminum seed extract: antimicrobial studies." *Advances in Natural Sciences: Nanoscience and Nanotechnology* 9: 035005.

Reza G. H. 2014 "A review of methods for synthesis of Al nanoparticles." *Oriental Journal of Chemistry* 30: 1941-1949.

Rosaria D., Falconieri M., Gagliardi S., Popovici E., Serra E., Terranova G., and Borsella E.. 2013 "Synthesis of ceramic nanoparticles by laser pyrolysis: From research to applications." *Journal of analytical and applied pyrolysis* 104: 461-469.

Ruohong S., and Charpentier P. 2012 "Synthesis of metal oxide nanostructures by direct sol–gel chemistry in supercritical fluids." *Chemical reviews* 112: 3057-3082.

Salah N., Habib S. S., Khan Z. H., Memic A.,Azam A., Alarfaj E., Zahed N., and Al-Hamedi S.. 2011 "High-energy ball milling technique for ZnO nanoparticles as antibacterial material." *International journal of nanomedicine* 6: 863-869.

Scott C. D., Arepalli S., Nikolaev P., and Smalley R. E. 2002 "Growth mechanisms for single-wall carbon nanotubes in a laser ablation process." *Applied Physics A Materials Science and Processing* 74: 11-12.

Shahjahan M., Rahman M. H., Hossain M. S., Khatun M. A., Aminul Islam, and Begum M. H. A. 2017 "Synthesis and Characterization of Silver Nanoparticles by Sol-Gel Technique." *Nanoscience and Nanometrology* 3: 34-39.

Shankar S. S., Rai A., Ankamwar B.,Singh A., Ahmad A., and Sastry M. 2004 "Biological synthesis of triangular gold nanoprisms." *Nature materials* 3: 482-488.

SharifiI., Zamanian A. and Behnamghader A.. 2016 "A Simple Thermal Decomposition Method for Synthesis of Co0. 6Zn0. 4Fe2O4 Magnetic Nanoparticles." *Journal of Ultrafine Grained and Nanostructured Materials* 49: 87-91.

Singh S. B. and Tandon P. K. 2014 "Catalysis: a brief review on nano-catalyst." *J Energy Chem Eng,* 2:106-115.

Thakkar K. N., Mhatre S. S., and Parikh R. Y. 2010 "Biological synthesis of metallic nanoparticles." *Nanomedicine: Nanotechnology, Biology and Medicine* 6: 257-262.

Vishwakarma V., Subhranshu S. S., and Manoharan N. 2010 "Safety and risk associated with nanoparticles-a review." *Journal of minerals and materials characterization and engineering* 9: 455-459.

Vithiya K., and Sen. S. 2011 "Biosynthesis of nanoparticles." *International Journal of Pharmaceutical Sciences and Research* 2: 2781-2785.

Willner I., Baron R., and Willner B. 2006 "Growing metal nanoparticles by enzymes." *Advanced Materials* 18: 1109-1120.

Xu H., and Kenneth S. S. 2010 "Water Soluble fluorescent silver nanoclusters." *Advanced Materials* 22: 1078-1082.

Yoffe A. D. 1993 "Low-dimensional systems: quantum size effects and electronic properties of semiconductor microcrystallites (zero-dimensional systems) and some quasi-two-dimensional systems." *Advances in Physics* 42: 173-262.

Yu, Y. Y., Chang S. S., Lee C. L., and Wang C. R. C. 1997 "Gold nanorods: electrochemical synthesis and optical properties." *The Journal of Physical Chemistry B* 101: 6661-6664.

In: Photocatalysis
Editors: P. Singh, M. M. Abdullah, M. Ahmad et al.
ISBN: 978-1-53616-044-4

Chapter 7

CONTROLLED CHEMICAL SYNTHESIS OF NANOMATERIALS: A FUNDAMENTAL NECESSITY FOR PHOTOCATALYSIS

Suresh Sagadevan[1,*]***, K. Pradeev Raj***[2]***, Zaira Zaman Chowdhury***[1]***, Mohd. Rafie Johan***[1]***, Fauziah Abdul Aziz***[3]***, Preeti Singh***[4]***, J. Anita Lett***[5] ***and Jiban Podder***[6]

[1]Nanotechnology & Catalysis Research Centre (NANOCAT),
University of Malaya, Kuala Lumpur, Malaysia
[2]Department of Physics, CSI College of Engineering, Ooty, India.
[3]Department of Physics, Center for Defence Foundation Studies,
National Defence University of Malaysia,
Kem Sg. Besi, Kuala Lumpur, Malaysia
[4]Bio/Polymers Research Laboratory, Department of Chemistry,
Jamia Millia Islamia, New Delhi, India
[5]Department of Physics, Sathyabama Institute of Science and Technology,
Chennai, India
[6]Department of Physics, Bangladesh University
of Engineering & Technology, Dhaka, Bangladesh

[*] Corresponding Author's E-mail: sureshsagadevan@gmail.com.

ABSTRACT

Materials scientists and engineers have made a significant improvement in the development of the synthesis methodologies of nanomaterial solids. Generally, nanomaterials synthesis can be classified as bottom-up manufacturing, which involves building up of the atom or molecular constituents as against the top method which involves making smaller and smaller structures through etching from the bulk material as exemplified by the semiconductor industry. Synthesis includes materials of nanometre scale ranging from sub nanometers to several hundred nanometers mainly deals with single nano-objects, materials, and devices subsist. Nanotechnology- a rapidly expanding field- has gained significant attention among the numerous researchers during the past decades in the context of science, engineering, and technology and earned the public and the media interest worldwide. At present, nanoscience and technology represent the most active discipline around the globe and is considered to be the fastest growing technology revolution, human history has ever seen. This immense interest in the science of materials, confined within the atomic scales; fundamentally exhibits the unique properties with great potential of next-generation technologies in electronics, computing, optics, biotechnology, medical imaging, medicine, drug delivery, structural materials, aerospace, energy and lead to enhanced properties such as catalysts, tunable photoactivity, expanded quality, and numerous other intriguing attributes. It also plays a major role in the advancement of creative techniques to produce new products; to substitute existing production equipment and to reformulate new materials and synthetic concoctions with enhanced execution at the same time conservation of energy and materials and thereby aiding in diminished mischief to the earth just as natural remediation. Thus, the control of nanoscale morphology and microstructural properties of the nonmaterial has been the basis for photocatalytic performance. This chapter deals with the advanced chemical synthesis routes for nanostructures as a fundamental, essential necessity for the photocatalysis have been discussed.

Keywords: nanomaterials, photocatalysis, top-down, bottom-up, and sol-gel method

1. INTRODUCTION

Nanotechnology refers to the technology of manipulation of material targeted at the atomic level. Nowadays, nanotechnology has made a remarkable contribution in the field of technology due to its vital applications. Nanotechnology involves the synthesis and characterization of nanomaterials. Nanomaterials have been proven to be more effective in many fields as compared to their bulk forms such as nanometals, nano semiconductors, nano inorganic, and organic materials are among the few materials to be mentioned. The bottom-up and top-down approach of synthesis of nanomaterials provides diversity synthesis methods, which leads to a variety of products. It is well accepted that the properties of nanomaterials are size and shape dependent which keeps an ever-increasing demand in researchers in finding new protocols/methods to achieve the precise control of the oversize and shape of the nanomaterials (Pradeep 2012). Nanotechnology employs the use of physical, chemical and biological systems to produce individual particles or atoms to submicron measurements and likewise the fusion of the subsequent nanoparticles into bigger structures. It is a technology which promises numerous breakthroughs that can alter the course of innovative advances in an extensive way to suit diverse applications. The term nanoparticles signify a collection of atoms bonded together with a dimension between 1 to 100 nm. When the nanoparticle size is reduced, the properties remain the same initially; then for small changes in dimensions feeble changes in properties occur, until finally when the size drops below 100 nm, spectacular changes in properties are observed (Tjong and Chen 2004). Vast numbers of techniques are available to synthesize different types of nanomaterials in the form of particles, colloids, thin film, tubes, rods, etc. These technique being used depends upon the materials of interest and on the type of nanostructures viz., quantum dots, nanowires, nanorods, and nanoplates as per requirement (Pradeep 2012). In nanotechnology, rapid and vast advancement have acquired the critical interest of researchers for the natural and organic uses of nanostructures. Nanostructured materials are astounding adsorbents, catalysts, and sensors because of their substantial surface area, unrivaled toughness, selectivity, and high reactivity (Murphy 2002). To synthesize different types of nanomaterials in the form of clusters, colloids, rods, thin films, tubes, wires, etc. there is a large number of techniques are readily available. Some of the already existing conventional techniques to synthesize different types of materials are optimized to get nanomaterials and some new advanced techniques are also developed for the interdisciplinary field of nanotechnology. Therefore, to synthesize nanomaterials the various chemical, physical, hybrid, biological techniques are also available. This influences the properties of particles in separation and their interaction with other materials. Utilization of the metal oxide nanoparticles in photocatalytic strategy is the most excellent remedy for the natural and inorganic contaminants in wastewater. Controlling the size, shape, and structure of nanoparticles is innovatively critical, due to the strong relationship between these parameters to alter its optical, electrical, magnetic and catalytic properties (Zhang et al. 2005; Martucci et al. 2004). These nanomaterials are differentiated into compact materials and nanodispersions. The basic physical properties of nanomaterials are stated by the nano-objects they possess. The first type of nanomaterials comprises of nanostructured materials (Moriarty 2001), i.e., the materials which are isotropic in the macroscopic composition are those that include nanometer-sized units as repeating structural elements (Gusev et al. 2004). Quasi-zero-dimensional mesoscopic systems, quantum dots, quantized or Q-particles, etc. include the particles with low dimensions in the range of few to several tens of nanometers (Khairutdinov et al. 1996). Owing to size limitations there exist a qualitatively new behavior, thus making the nanoscale materials, an area of significant importance in the current century. Nanoparticles exhibit different properties when compared to larger particles, such as long and well-known ultra-dispersed powders with a grain size above 0.5 μm thus causing the nanoparticles to be spheroid in shape. Nanocrystallites are those nanoparticles that exhibit an ordered arrangement of atoms (or ions). "Quantum dots" or "artificial atoms" are the nanoparticles having clear-cut discrete electronic energy levels, which often possess compositions of typical semiconductor materials. The same set of electronic levels are represented by the magnetic nanoparticles. Nanoparticles can also be symbolized as a bridging path between bulk materials to molecules and structures at an atomic level that attracts scientific interest. Likewise in the chemical literature, the term "cluster" was used widely, but it is utilized to designate small nanoparticles with a size lower than 1 nm. Magnetic polynuclear coordinate compounds (magnetic molecular clusters) belong to a special type of magnetic materials. Molecular magnetic clusters are the entirely identical small magnetic nanoparticles, unlike nanoparticles. Exchange-modified paramagnetism is used to describe their magnetism (Alivisatos 1996; Gubin 2009). Sizes, shape, quantum confinement, response to external electrical and optical excitations of individual and coupled finite systems are some of the well defined major issues in nanoscience (Sathish et al. 2005; Balaji et al. 2006). The key objects of nanoscaled materials are entirely related to their size dependency. The evolution of structural, thermodynamic, electrical, magnetic, spectroscopic and catalytic features of the finite systems, towards science and technology can significantly exhibit a potential impact which can be analyzed by the size appropriating to the nanoscale (Fang et al. 2006; Park et al. 2007).

The natural use of nanotechnology address the headway of answers for the current biological issues and the preventive measures for future issues coming about in view of the cooperations of vitality and materials with the earth and any possible threats that may be displayed by nanotechnology itself. Environmental utilizations of nanotechnology address the advancement of answers for the current ecological issues, the preventive measures for future issues coming about in view of cooperation of energy and materials with the environment and thereby any conceivable dangers that might be presented by nanotechnology itself (Singh P. et al. 2016). Researchers are still looking for novel routes to prepare solid photocatalysts that would be able to transform solar into chemical energy more efficiently. Principally, photocatalysis can be separated from ordinary catalysis through the activated action of the catalytic solid, which occurs due to the assimilation of photon rather by thermal activation. This photonic activation subsequently requires the utilization of nanomaterial as a catalyst, provided that the radiation wavelengths are greater than its band gap that corresponds to the energy gap between both conduction and valence bands of the semiconductor. Activating a nanomaterial prompts the advancement of an electron from the valence to the conduction band with the concurrent production of a photogenerated hole inside the valence band.

Further, the transfer of photogenerated charge carriers to the photocatalyst surface enables the redox reactions to occur with adsorbed reactants, originating from gas or fluid phase contingent upon the application. At the catalyst surface, the redox reactions are separated into reduction and oxidative steps, involving on one hand, conduction band electrons and adsorbed electron acceptors for example, e.g., oxygen molecules (eCB−+A→A•−), and then again, valence band holes and adsorbed electron-electron contributors, for example, natural atoms or all the more, by and large, the focused on the targeted pollutant (hVB++ D→D•+). Indirect oxidation reactions also occur through the formation of highly oxidative hydroxyl radicals produced by the oxidation of water by holes. Therefore, the photocatalysis discipline exists through the capacity of a material (Lacombe S. and Keller N., 2012). For the reason that the photocatalytic reactions, which is of interest occur at the surface of nanomaterials used, thus a small molecular size is desirable to maximize surface region. Hence, an enhancement in the controlled synthesis of nanoparticles is required in order to utilize their properties in applications, by just not simply by the chemical composition and the physical arrangement of the atoms, but also by the molecular size that influences the behavior of these materials (Prada Cerro E. et al., 2019; Colmenares Carlos J. et al. 2009). In this chapter, by considering the nanomaterial as a key fundamental need for the photocatalysis, our aim is to discuss the various controlled chemical routes utilized for the synthesis of nanomaterials.

2. Synthesis Methods

Synthesis of nanomaterials with strict control over shape, dimension and crystalline structure making it significant for the solicitations of nanotechnology, which can be used in various multiple fields such as nano-medicine, catalytic process, and nano-electronic devices. Nano dimensional particles can be synthesized by two classical methods: namely the "top-down method" and "bottom-up" methods (Sagadevan Suresh 2013). Based on the principal of these two methods, the nanomaterials can be synthesized by various methods such as milling, sol-gel, etc.

The top-down includes the splitting up of a huge solid into smaller fragments, consecutively, which will ultimately reach to the nanometer range. This process is a combination of milling or attrition. The next, "bottom-up", a system of nanoparticles synthesis requires the preparation of smaller atoms or molecular units inside a vapour phase or in solution. This will produce the material on the nano-dimensional scale fabrication of nanomaterials due to several advantages. Figure 1 illustrates the general scheme for the two methods. There are several bottom-up techniques of preparing metal oxide nanomaterials, such as the solvothermal method, auto combustion technique, sonochemical method, gas-phase system, microwave synthesis, sol-gel processing, and wet chemical methods. The bottom-up approach is rather more established for the synthesis of nanomaterials.

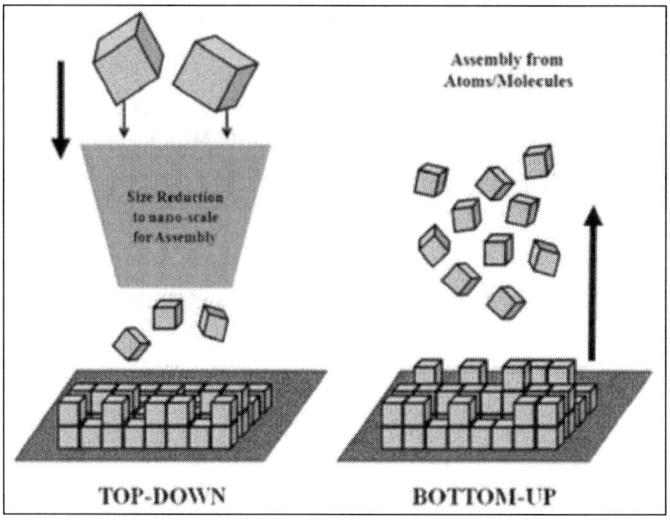

Figure 1. Schematic representation of bottom-up and top-down approaches.

2.1. Wet Chemical Methods

The initiative to develop smaller machines and components utilizing fewer resources and energy has been taking place for the last two decades. The expansion of novel strategies for the synthesis of advanced materials with improved properties at the nanometer level is therefore essential. There are two main general approaches: "bottom-up" and "top-down" for the synthesis of nano-materials. The earlier one tries to create organic and inorganic structures, atom-by-atom or molecule-by-molecule. This uses the atoms or molecules as building blocks for the basic design. After that, nano arrangements of atoms take place to provide macroscopic systems. 'Organic-inorganic hybrid' materials are recently produced widely using the solvothermal/hydrothermal and sol-gel process. This can allow a tailored chemical design. The process requires low temperature, which enables the properties of organic materials to retain their properly inside the hybrid materials and the materials are not destroyed. For the design and application of novel advanced functional nanoparticles, the uniformity of the shape and size is a significant issue. The nano-particles samples must be monodispersed inside the solution without agglomeration. It can be monodispersed when their size distribution is $< 5\%$. The morphology and size of the nanoparticles can be controlled by careful optimization of the synthesis process. The synthesis process can be undertaken through several approaches including chemical or physical techniques. The techniques can use gaseous, liquid and solid media. The physical methods usually synthesize the nanostructures top-down approach where the size of constituent materials is reduced in bulk materials. In the Bottom-up approach, the chemical methods are used to control the agglomeration of atoms/molecules. Wet chemical methods are most frequently used. They have several major advantages. Thus, this method is highly reliable and cost-effective, which enables one to control the shape and size of the nanoparticles. It can also prevent agglomeration of the particles as well as enhance the functionality of the particles using different capping ligands (Cushing et al. 2004). The method is reproducible to give colloidal solutions having a large range of particle size. The following wet chemical methods have been successfully employed for developing different types of nanoparticles:

1. Co-precipitation;
2. Sol-Gel;
3. Solvothermal/Hydrothermal.

2.1.1. Co-Precipitation Method

Co-precipitation is one of the main successful practices for synthesizing ZnO nanopowders having fine particle size distribution. In the co-precipitation process, partial precipitation of a particular ion inside the solution takes place. This method does not only co-precipitate the target ion but also other prevailing ions simultaneously in the solution. This process could prevent complicated actions such as refluxing of alkoxides, ensuing in quicker usage in contrast to other techniques. The flow chart of the co-precipitation procedure is presented in Figure 2. Co-precipitation response requires the instantaneous happening of nucleation, growth, together with the agglomeration process. There are three main processes of co-precipitation: i) inclusion, ii) occlusion and iii) adsorption.

If the impurity molecule resides inside the lattice site in the crystal arrangement of the carrier, a crystallographic flaw can be caused. This is the combined effect of the ionic radius and charges present in the impurity atoms. An adhesion or adsorption of atoms of impurity has been weakly bound (adsorbed) to the surface of the precipitate. A constriction materializes as the adsorbed impurity gets physically trapped within the crystal during the growth.

2.1.2. Sol-Gel Method

The sol-gel method finds considerable interest in the scientific and industrial field because of the several advantages as compared to other techniques (Hench and West 1992; Dimitriev et al. 2008). Sol-gel is a useful self-assembly process for nanomaterial formation. Solid particles that are suspended in a liquid are called a sol. The sol-gel method was used to synthesize ceramics, metal oxides, sulfides, borides, and nitrides. It is a wet-chemical method used to produce high purity oxide nanoparticles such as TiO_2, ZnO, etc. The Sol-gel process of synthesizing nanomaterials have been widely accepted amongst the researchers and is frequently used to prepare oxide materials. The sol-gel process was developed in the 1960s, mostly owing to the need for novel synthesis techniques in the nuclear industry. The sol-gel procedure ensures the progress of inorganic networks side by side from the colloidal suspension (sol). After that, the solidification takes place by freezing of the sol to obtain a liquid phase (gel). The precursors for preparing these colloids consist usually metal or metalloid elements. These metals or metalloids are surrounded by a number of reactive ligands. The original substance has been refined to acquire a dispersible oxide. This forms a sol when it reacts with water or dilutes acid. Elimination of the aqueous phase from the sol gives the gel. The particle dimension and shape are changed accordingly. The calcination of the gel makes the oxide. To make a uniform coating, the pH, temperature, and viscosity should be controlled. Instead of forming a thin film coating, crystalline powders are produced by suitable calcining the gel. The sequence of steps involved in a sol-gel process is displayed in Figure 3.

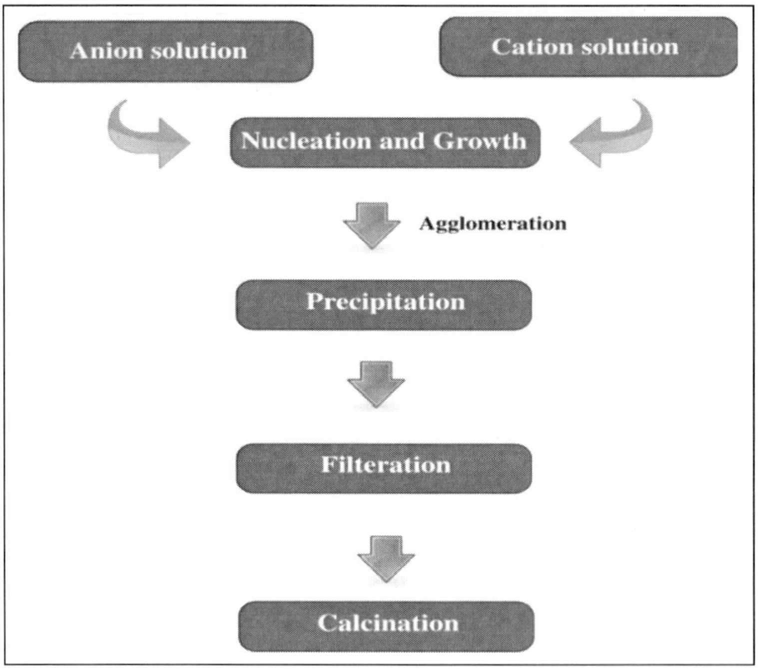

Figure 2. Flowchart of the co-precipitation method.

Advantages

The following are some the benefits of the sol-gel process:

- It was an economical route.
- It generates highly pure, well-controlled ceramics compared to other technologies. The sol-gel can be done at a low temperature.
- Instrument processing has been easy and synthesizes almost all the materials.
- It was possible to get unique materials like xerogels and zeolites.
- This method has been used to synthesize nanoparticle of different size.

Applications

The following are some of the applications of the sol-gel process.

- The nanomaterials produced by this technique are suitable for microelectronics and optoelectronics.
- Produces lightest materials.
- Optical coating, protective and decorative coatings can be done by this method.
- Sol-gel route ceramic fibers are used for fiber optic sensors and thermal insulation.
- The ultra-fine particle prepared by this technique is used for dental and biomedical applications.

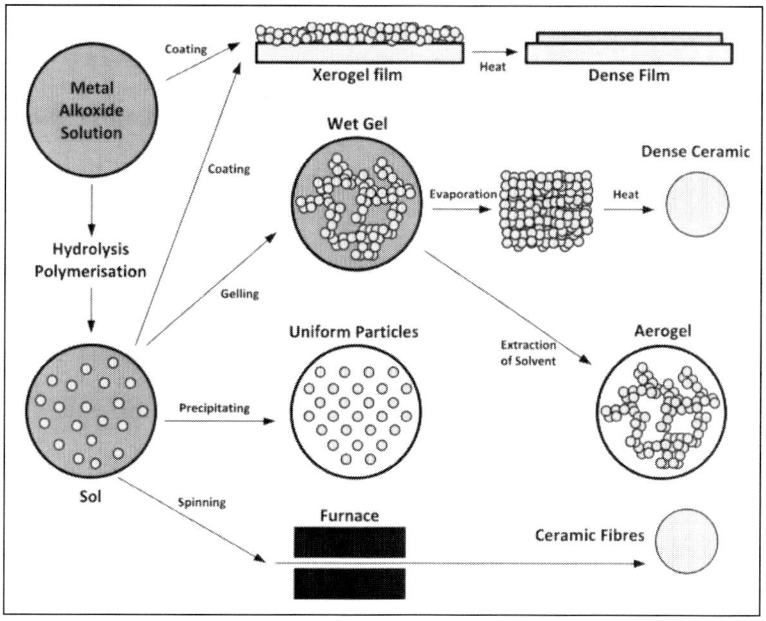

Figure 3. Flowchart of the sol-gel method.

2.1.3. Hydrothermal Method

Figure 4 illustrates the process of the hydrothermal method where the crystallization of the substances takes place at high-temperature from aqueous solutions at high vapour pressures. It is able to synthesize single crystals and it depends on the solubility of minerals in hot water under high pressure (Demazeau 2008). The growth of the crystal takes place inside a steel pressure vessel called autoclave, where a nutrient is added to water. A thermal gradient was maintained. The hotter end will melt the nutrient and the cooler end will trigger the seeds for further growth.

Several researchers have applied the hydrothermal approach to preparing undoped and doped nanoparticles. The crystallization boats used are the autoclaves. They are generally thick-walled steel cylinders with a hermetic seal that ought to resist high temperatures conditions and pressures for long-lasting intervals of time. Moreover, the autoclaved substance should be inert regarding the solvent. The closing is an essential constituent of the autoclave. Several patterns have already been produced for seals. The Bridgman seal was used most commonly. The corrosion of the internal cavity inside the autoclave has been avoided by a self-protective layer inserted inside. These inserts could have precisely the same shape of the autoclave and with no trouble fitting in the inner cavity (contact-type insert) or be a "floating" type insert that occupies only a little portion interior of an autoclave interior. The insertion was completed using carbon-free copper, iron, gold, silver, platinum, glass, titanium, or Teflon but it depends on the temperature and the solution used.

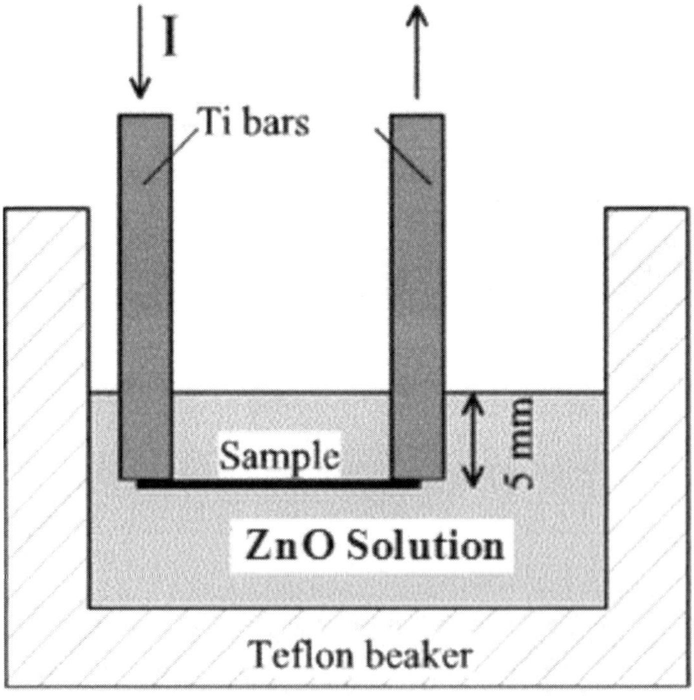

Figure 4. Experimental set up for hydrothermal synthesis.

Advantages

The following are some advantages of the hydrothermal process:

- Different types of substances are synthesized under hydrothermal conditions: elements, simple and complex oxides, carbonates, silicates, germanates, etc.

- The technique has been commonly applied for developing quartz, gems and other single crystals with industrial value.
- Some crystals namely emeralds, rubies, quartz, alexandrite, are made by this method.
- The process has been extremely competent to develop novel compounds with accurate physical properties. Even the methodical physicochemical approach can provide intricate multi-component systems suitable for elevated temperatures and pressures.

2.2. Chemical Vapour Deposition Technique

Chemical Vapour Deposition is an effective tool to synthesize nanopowder of different dimensions, nano-thin films, nanopowders, carbon nanotubes, etc. Experimental arrangement of Chemical Vapour Deposition is shown in Figure 5.

Various Types

- (APCVD): Atmospheric Pressure Chemical Vapour Deposition
- (LPCVD: Low-pressure Chemical Vapour Deposition
- (MGCVD): Metal Organic Chemical Vapour Deposition
- (PECVD): Plasma Enhanced Chemical Vapour Deposition
- (LCVD): Laser Chemical Vapour Deposition
- (PCVD): Photo Chemical Vapour Deposition

The source material was kept inside the reaction chamber. The best vacuum is maintained inside the chamber with the help of a diffusion pump. The source material is heated by resistive heating or inductive coil heating. The evacuated chamber is filled with argon gas at low pressure. There is a copper cooled target at the top of the chamber in which the metal vapour will condense into nano-thin films. The temperature of the source material was higher than its melting temperature, numerous numbers of metal ions are ejected from the source. These ions made a collision with inert gas atoms and lose their kinetic energy. Thus, the metal ions are nucleates homogeneously and diffuse towards the target. On the target surface, we develop thin films of dimension varying from nano size to micro size. The particle size inside the thin film was controlled by controlling the deposition time and the inert gas pressure.

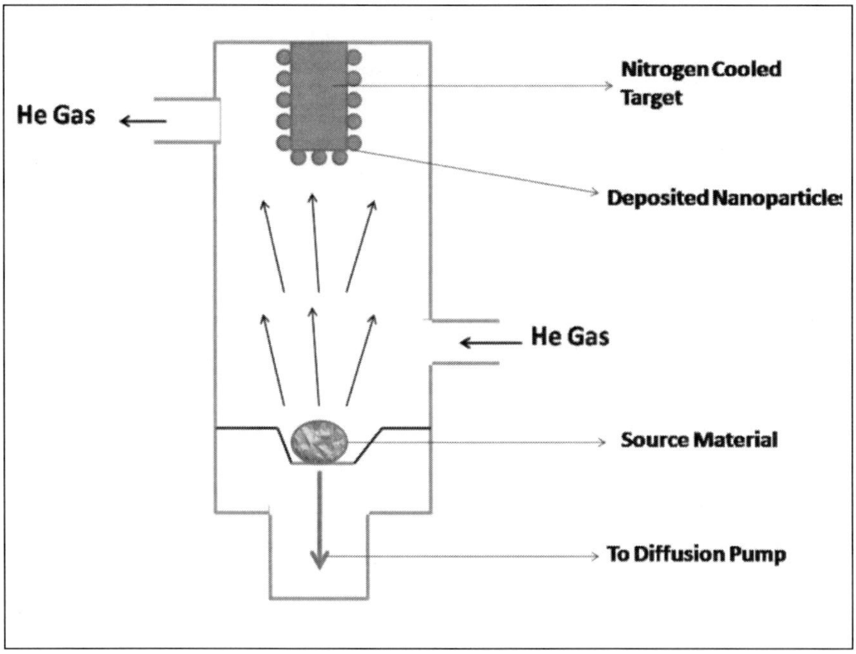

Figure 5. Experimental set up for chemical vapour deposition.

Applications

Some of the applications of CVD are as follows

- It was used to fabricate the nanomaterials, nano-thin films, and nanotubes.
- It can be used to produce the integrated circuits, sensors and optoelectronics devices.
- It was used in optical fibers which are used for telecommunications.
- It was used in the production of novel powders, fibers, catalysts, and nanomachines.

2.3. Electrodeposition Techniques

The electrochemical reaction is the basic principle behind the electrodeposition process. Experimental arrangement of electrodeposition is shown in Figure 6. In the electrodeposition method, three electrodes are used, namely, a working electrode, a counter electrode, and a reference electrode. A platinum electrode was usually used as a counter electrode. The working electrode consists of conducting materials such as Ti or Ni. The standard calomel electrode has been used as a reference electrode.

The electrolyte has been an aqueous solution. The temperature of the electrolyte has been kept constant. The power supply was used to apply a controlled current at a certain voltage. When the current is allowed to pass through the three electrodes electrochemical cell, a certain amount of atoms are liberated from an anode and is deposited on the surface of the cathode. The deposition of an atom can be made either on the cathode or anode. In the cathodic deposition, the deposition was made on the surface of the cathode. If the deposition has been made on the surface of the anode, then it is said to be an anodic deposition. In the galvanostatic deposition, the current flow through the circuit was made constant.

Advantages

- It was one of the simplest and inexpensive methods.
- This technique was used to produce nano-indented holes, mesoporous, silica, nanoporous polymers, and porous alumina.
- This method was a versatile technique to synthesize a variety of kinds of nanomaterials with required surface morphologies (nanorods, nanoparticles, and nanowires).
- The film thickness can be easily controlled by adjusting the rate of deposition.

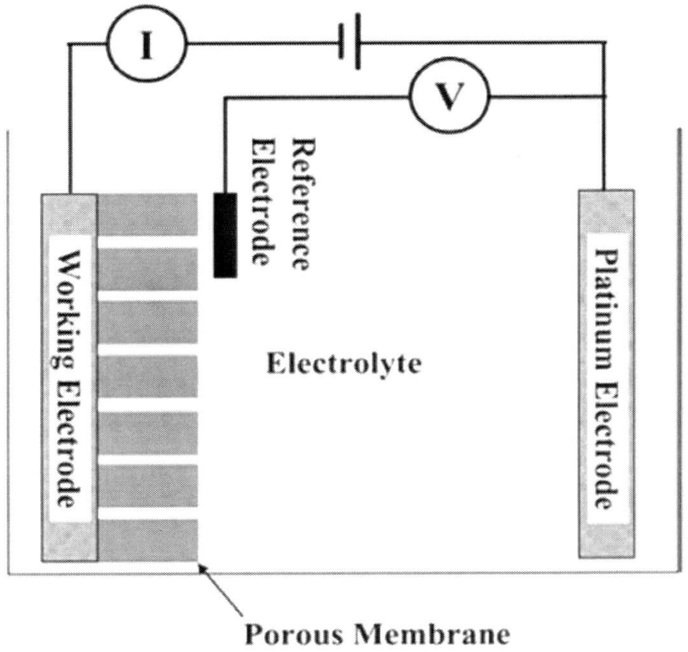

Figure 6. Experimental arrangement- Electro Deposition Method.

2.4. Microwave Synthesis

In the case of the Microwave synthesis process, microwave radiation has been applied to the chemical reactions. Under high-frequency electric fields, the microwave will usually heat any material containing movable electric charges, such as polar molecules inside the solvent or conducting ions inside the solid. Experimental arrangement of electrodeposition is shown in Figure 7.

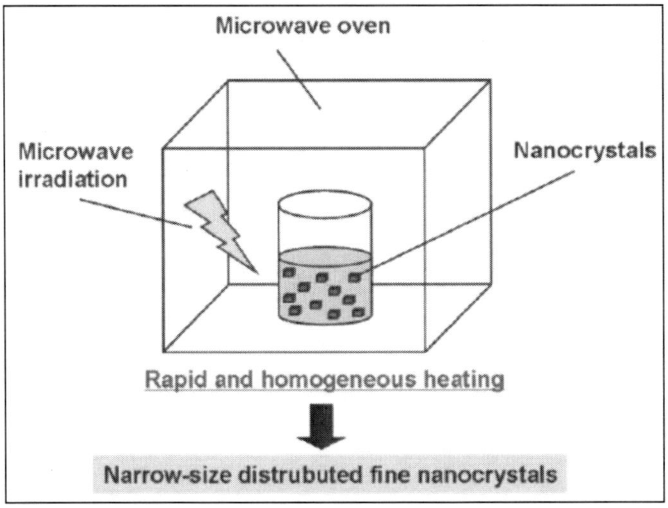

Figure 7. Experimental arrangements- Microwave Method.

Polar solvents are usually heated when their component molecules are imposed to rotate with the applied field. However, its' energy drop after the collisions. Semiconductors and other conductors become thermally excited. At that moment, ions or electrons inside them cause the current to flow, after that, the energy was lost due to the electrical resistance of the substances. Microwave-assisted preparation of nano-materials was a comparatively novel. It was an efficient method for the synthesis of an oxide containing materials. A variety of nanomaterials has been synthesized in an extraordinarily small span of time for irradiation under the microwave. Microwave techniques eliminate the use of high-temperature calcination for entire periods of time. Thus the process was fast and reproducible for the synthesis of crystalline metal oxide nanomaterials. For microwave energy, very fine grains in the nanocrystalline scale are formed within less time.

Advantages of the Microwave Method

- Reaction rate acceleration
- Milder reaction conditions

- Higher chemical yield
- Lower energy usage
- Different reaction selectivities

Conclusion

Finally, we give a point of view on the respective field of nanomaterial and its application in photocatalysis, in which the comprehension of the enhancement mechanisms combined with the genuinely controlled synthesis of nanomaterial. Nanomaterials act as an essentially incredible tool for the establishment of the design principles needed to take the field of photocatalysis more effective and successful. As the ideal properties of nanomaterials and their performances can be arranged ahead of time and the target material can be synthesized accordingly. This chapter describes in detail the various synthesis techniques employed for the preparation of nanostructures. The detailed description of the co-precipitation method and its advantages have been presented in this chapter. The synthesis of nano-materials was performed by utilizing different physical and chemical methods. The structure can be well defined with better morphological features based on the purification strategy and appropriate size selection. There are a large number of techniques available to synthesize different types of nanomaterials in the form of colloids, clusters, powders, tubes, rods, wires, thin films, etc. Some of the already existing conventional techniques to synthesize different types of materials are optimized to get novel nanomaterials and some new techniques are developed. Nanotechnology is an interdisciplinary subject. There are therefore various physical, chemical, biological and hybrid techniques available to synthesize nanomaterials. The technique being used depends upon the material of interest, type of nanomaterial viz., zero-dimensional (0-D), one dimensional (1-D) or two dimensional (2-D), their sizes and quantity. Considering the topics covered in this chapter, it is evident that the emergence of nanotechnology is bound to have significant implications in the research field of photocatalysis.

References

Alivisatos, A. P. (1996). "Perspectives on the Physical Chemistry of Semiconductor Nanocrystals" *Journal of Physics and Chemistry*, *100*, 13226-13239.

Balaji, S., Djaoued, Y. & Robichaud, J. (2006). "Phonon confinement studies in nanocrystalline anatase–TiO_2 thin films by micro Raman spectroscopy" *Journal of Raman Spectroscopy*, *37*, 1416-1422.

Carlos, C. J., Luque, R., Manuel, C. J., Colmenares, F., Karpiński, Z. & Angel, R. A. (2009). "Nanostructured Photocatalysts and Their Applications in the Photocatalytic Transformation of Lignocellulosic Biomass: An Overview" *Materials*, *2*, 2228-2258.

Cerro-Prada, E., García-Salgado, S., Ángeles, Q. M. & Varela, F. (2019). "Controlled Synthesis and Microstructural Properties of Sol-Gel TiO_2 Nanoparticles for Photocatalytic Cement Composites" *Nanomaterials*, *26*, 1-16.

Cushing, B. L., Kolesnichenko, V. L. & O Connor, C. J. (2004). "Recent Advances in the Liquid-Phase Syntheses of Inorganic Nanoparticles" *Chemical Reviews*, *104*, 3893-3946.

Demazeau, G. (2008). "Solvothermal reactions: an original route for the synthesis of novel materials" *Journal of Material Science*, *43*, 2104-2114.

Dimitriev, Y., Ivanova, Y. & Iordanova, R. (2008). "History Of Sol-Gel Science And Technology (Review)." *Journal of the University of Chemical Technology and Metallurgy*, *43*, 181-192.

Fang, K. C., Weng, C. I. & Ju, S. P. (2006). "An investigation into the structural features and thermal conductivity of silicon nanoparticles using molecular dynamics simulations" *Nanotechnology*, *17*, 3909-3914.

Gubin, S. P. (2009). *"Magnetic Nanoparticles"* Wiley-VCH Verlag Gmbh & Co. Kgaa, Weinheim.

Gusev, A. I. & Rampel, A. A. (2004). *"Nanocrystalline Materials"* Cambridge International Science Publishing, Cambridge.

Hench, L. L. & West, J. K. (1992). "Chemical processing of advanced materials" Wiley, New York.

Khairutdinov, R. F., Rubtsora, N. A. & Costa, S. M. B. (1996). "Size effect in steady-state and time-resolved luminescence of quantized MoS_2 particle colloidal solutions" *Journal of Luminescence*, *68*, 299-311.

Lacombe, S. & Keller, N. (2011). "Photocatalysis: fundamentals and applications" *Environmental Science and Pollution Research*, *19*, 3651–3654.

Martucci, A., Fick, J., LeBlanc, S. E., LoCascio, M. & Hache, A. (2004). "Optical properties of PbS quantum dot doped sol-gel films" *Journal of Non-Crystalline Solids*, *346*, 639-642.

Moriarty, P. (2001). "Nanostructured materials" *Reports on Progress in Physics*, *64*, 297-381.

Murphy, C. J. (2002). Materials science. Nanocubes and nanoboxes. *Science*, *13*, 2139-41.

Park, T. J., Papaefthymiou, G. C., Viescas, A. J., Moodenbaugh, A. R. & Wong, S. S. (2007). Size-Dependent Magnetic Properties of Single-Crystalline Multiferroic BiFeO3 Nanoparticles. *Nano letters*, *7*, 766-772.

Pradeep, T. (2012). *"Nano: The expanding Horizon, in A textbook of Nanoscience and nanotechnology"* New Delhi: The MacGraw Hill, 3–21.

Sagadevan, S. (2013). "Semiconductor Nanomaterials, Methods, and Applications: A Review" *Nanoscience and Nanotechnology*, *3*, 62-74.

Sathish, M., Viswanathan, B., Viswanath, R. P. & Gopinath, C. S. (2005). "Synthesis, Characterization, Electronic Structure, and Photocatalytic Activity of Nitrogen-Doped TiO_2 Nanocatalyst" *Chemistry of Materials*, *17*, 6349-6353.

Singh, P., Abdullah, M. M. & Ikram, S. (2012). "Role of Nanomaterials and their Applications as Photo-catalyst and Sensors: A Review" *Nano Research and Application*, *2*, 1-10.

Tjong, S. C. & Chen, H. (2004). "Nanocrystalline materials and coatings" *Material Science and Engineering Reports*, *45*, 1-88.

Zhang, H., Yang, D., Ma, X., Ji, Y., Li, S. Z. & Que, D. (2005). "Self-assembly of CdS: from nanoparticles to nanorods and arrayed nanorod bundles" *Material Chemistry and Physics*, *93*, 65-69.

In: Photocatalysis
Editors: P. Singh, M. M. Abdullah, M. Ahmad et al.
ISBN: 978-1-53616-044-4

Chapter 8

EFFECTIVE REMOVAL OF "NON-BIODEGRADABLE" POLLUTANTS FROM CONTAMINATED WATER

Divyanshi Mangla, Arshiya Abbasi, Shalu Aggarwal, Kaiser Manzoor, Suhail Ahmad and Saiqa Ikram*
Bio/Polymers Research Laboratory, Department of Chemistry,
Jamia Millia Islamia, New Delhi, India

ABSTRACT

Untreated domestic and industrial wastes are the two major sources of pollution worldwide. Inefficient waste disposal system causes long term issues to living bodies. Millions of tons of waste products are being disposed into rivers whose numbers are continuously increasing day by day. Depending upon the type, pollutant can be distinguished into two forms such as Bio-degradable and Non-biodegradable pollutants. The presented chapter represents the various removal techniques of non-biodegradable pollutants from waste water exploited so far by the various research scientists, such as adsorption, selectivity, disinfections and photocatalysis. In the current scenario, adsorption being traditional technique for water purification has many benefits over present techniques due to the enhancement in the field of nanotechnology.Also, it is very convenient& economical. Finally, Photocatalysis technique is mostly practicedbecause of its advantage of degrading pollutants into simpler compounds instead of their transformation under ambient conditions.

Keywords: pollutants, adsorption, nanoadsorbents, membrane process, disinfectants, photocatalysis

* Corresponding Author's E-mail: sikram@jmi.ac.in (Corresponding author).

1. INTRODUCTION

'Contaminated water', does not have any exact definition.Every person has a different idea for the 'contamination'. For instance, presence of even natural minerals canalso be a source of contamination for the chemists. But in a general sense "Contaminated water is the water which is contaminated or polluted with substances instead of natural minerals". Polluted water is a complex mixture of inorganic and organic materials. Pollution is introduced by humans through natural and industrial activities. These activities have a direct impact on the natural environment either waythe discharge istreated or untreated. Pollutant can be described as unwanted resources which enter into water bodies, making it dirty or polluted and consequently, unfit for drinking and other purposes. Unwanted resources can be of any form such as toxic gas, harmful chemical which enters in the environment and degrades the quality of land alsoaffecting air or water quality. A pollutant can cause long-term or short-term effects on human beings and animals directly or indirectly through bio-magnification.Heavy metals in sewage sludge have been found and studied of urban run-off water hasalreadybeen conducted in the past by many scientists (Abaychi et al. 1987; Wright J. et al. 2004).

Contaminated water consists of various settable solid particles that dispersed as colloids which do not get easily dissolved in solid. Polluted water-mixture contains enormous number of microscopic organisms such as bacteria and virus. Polluted water can be any 'used' water from different sources such as domestic, agricultural, commercial or industrial activities, sewer infiltration, surface run-off or storm water. Since the water can have many sources of contamination and the inputs are highly mutable.There can be the presence of an active microbial component also in contaminated water (JoAo P. S. C. 2010).

Depending upon the type of pollutant they can be distinguished as Bio-degradable and Non-biodegradable pollutants. Pollutants that can be easily converted into simpler and non-toxic substances naturally in due course of time with the help of certain micro-organisms to degrade them completely in land are referred to as biodegeradable. The biodegradable pollutants can be domestic wet wastes (kitchen waste), paper, wood, agriculture residues, cattle waste, wool, vegetable stuff or plants. The pollutants which cannot be converted into harmless substances by natural process are called non-biodegradable pollutants. Plastics, insecticides, pesticides, polythene-bags, mercury, lead, arsenic, metal articles like aluminum cans, iron products, silver foils, synthetic fibers, demolition wastes, etc.are non-biodegradable pollutants (Abdel-Raouf N. et al. 2012).These non-biodegradable pollutants harm aquatic as well as non-aquatic life. Species consuming the plastic experiencesa mild to severe digestive problems.Human wastes openly disposed onland resource takes very long period of time to get decomposed.The quality of air also gets depleted near the disposal sites due to breeding of mosquitos, flies, other insects and microorganisms.

Inorganic waste includes many pesticides and heavy metals such as mercury, chromium, arsenic etc. Such toxic metals when dumped into rivers are extremely harmful for both living and non-living beings (Haseena M. et al. 2017). Hence, the removal ofthese non-biodegradable pollutants from the environment completely should be an urgent concern for humans. In the present scenario, all the researchers and scientist are getting more and more attracted towards the metal oxide (MO) nanostructures due to theirvast applications in electronic as well as optoelectronic devices, sensors, medicines and renewable energy sources (Shipway et al. 2000).

2. Techniques to Remove Non-Biodegradable Pollutants from Waste Water

Due to active growth of civilization and mechanization, numerous pollutants are being discharged from industries and domestic holds into aquatic bodies. Some pollutants also consist of carcinogenic substances such as heavy metals, organic hetero compounds, inorganic dyes, pesticides and inorganic fertilizers which have caused adverse impact on bothplant as well as animal domain. Therefore, elimination of these carcinogenic substances from waste water is a serious matter of concern. For this reason, a number of traditional methods like precipitation, coagulation and other physicochemical technologies are being commonly used. But these methods have failed for many reasons such as still appearance of toxic substances, less discharge efficiency, complicated and harsh stimulating conditions, and cost-effectiveness (Das, S.K. et al. 2012).Currently, advancement in nanoscience and nanotechnology have recommended that environmental issues like waste water purification could be figured out through implementation of numerous nanomaterials as nano-adsorbents, nano-catalysts and nano-structured membranes. Nanomaterials are endorsed as profitable, methodical and eco-friendly substitutes for traditional waste water treatment agents and techniques (S J Shan et al. 2017).

3. Nanoadsorption

Adsorption is a surface phenomenon where toxic substances get adsorbed on solid substrates. Generally, this phenomenon is initiated by physical forces but sometimes formation of chemical bonds also takes place(Faust. et al. 1983).The ability of regular or traditional adsorbents are confined by their lack of selectivity and surface area (Qu, X. et al. 2013). Nanoadsorbents are used to remove biodegradable and non-biodegradable pollutants from polluted water. The special properties of nanoadsorbent is due to their very small size, high reactivity and large surface area. The ease of separation of nanoadsorbents from the solution makes them optimal adsorbent materials for treatment of polluted water and as large number of active sites are available for the interaction of different pollutantsit makes nanomaterial a potential candidate for Catalytic processes. Different types of Nanoadsorbentsuch as carbon based Nanoadsorbents, (Chowdhury S. Balasubramanian. 2014)metal based Nanoadsorbents (Das, S.K. et al. 2012; Zhang, W.-X., 2003), polymeric, magnetic or nonmagnetic oxide composite (Arshadi. M, et al. 2015; Chen L. et al. 2014; Rafiq. Z. et al. 2014) and zeolites are used for wastewater treatment (Gehrke, I. et al. 2015). Adsorption efficiency of targeted compounds depends on the nature of contaminated water.Higher concentrations of any substances like natural or anthropogenic, the capacity of a specific compound gets reduced because of competitive binding with the surface sites (Ayangbenro S. A. et al. 2017).

3.1. Carbon-Based Nanoadsorbents

3.1.1. Removal of Natural or Organic Contaminants

Carbon nanotubes (CNT's) are carbon based cylindrical nature nanoadsorbents. They are used as substitutes for activated carbon. They are classified as single-walled or multi-walled carbon nanotubes. These adsorbents have the unique quality of highly perceptible adsorption sites and large surface area. Hydrophobic surface of CNT's forms the loose bundles and aggregates which diminishes active surface area. This makes the adsorption of bulky organic contaminants in water easier due to high energy sites, larger pores in bundles and accessible sorption sites in aggregates (Pan, B. et al. 2008; Ji L. et al. 2009). These carbon nanotubes can also adsorb polar organic molecules via hydrogen bonding, π- π interactions, hydrophobic interactions, covalent binding, electrostatic interactions (Yang, K., Xing, B., 2010; Ji, L. et al. 2009; Chen, W. et al. 2007; Lin, D., Xing, B. 2008).

3.1.2. Removal of Heavy Metal Ions

The surface of CNT's can be oxidized or modified by using oxidizing agents like hydrogen peroxide (H_2O_2), $KMnO_4$and nitric acid for the removal of Cd^{2+} from aqueous solution (Li, Y.-H. 2003). Upon oxidation CNT's attains high adsorption sites for metal ions with faster kinetics. The surface of CNT's contain functional groups such ascarboxyl, sulphonyl, nitrile, hydroxyls, carbonyls etc. (Vukovic et al. 2010).These functional groups have good retention capacity for heavy metal ions, when pH is above the isoelectric point of oxidized CNT's. (Datsyuketal. 2008; Li, Y.-H et al. 2003; Musameh, M. et al. 2011; Peng, Y. et al. 2006; Lau, W.J. et al. 2015; Liu, L et al. 2012). These are also fit for the removal of Pb^{2+}, Cd^{2+}, Zn^{2+}, Cu^{2+}etc. (Li, Y.-H. 2003;Vukovic et al. 2010, Datsyuk et al. 2008; Li, Y.-H et al. 2003; Musameh, M. et al. 2011; Peng, Y. et al. 2006; Lau, W.J. et al. 2015; Liu, L et al. 2012; Lu, C. et al. 2006). Sponge made up of CNT's with a dash of boron has a good adsorbing capacity of oil from water. These sponge are also called reusable sponge because they are so efficient for removal of oil spills (oil remediation) evenafter many cycles of removal of oil from water (Hashim et al. 2012). Despite, these convincing advantage CNT's have certain limitations also, such as their high production costs, toxicity, losing of nano identity due to coagulation with some contaminants and algae (DeVolder et al. 2013; Iijima, S., 1991; Bottini. et al. 2006; Muller et al. 2006; Ali, I., 2012).

3.2. Metal-Based Nanoadsorbents

Effective and low cost materials like iron oxide, titanium dioxide, zinc oxide and aluminaare called metalnanoadsorbents. They are used for removalof heavy metal ions during water purification process. The mechanism of metal nanoadsorbents is based on the oxide group present in metal oxides complexes with heavy metal ions which readily gets dissolved in the contaminated water (Trivedi et al. 2000).With the decrease in particle size adsorption capacity and the specific area of nanoparticles also increases.This effect is called "nanoscale effect" in which surface structure creates new adsorption sites for metalions (Auffan, M. et al. 2008; Auffan, M. et al. 2009). Magnetic nanoadsorbents (MNP's) such as hematite (αFe_2O_3), maghemite (γFe_2O_3), and spinel ferrites are the best adsorbing materials for removing toxic elements from contaminated water. They are harmless to the environment in context of their magnetic nature, as they can be easily separated from reaction media by the application of external magnetic field. Variety of elements such as arsenic, chromium,cobalt, lead, copper, nickel in their ionic form can be removed by using metal nanoparticles (MNPs) (Badruddoza. et al. 2013; Lei, Y. et al. 2014; Ngomsik et al. 2012; Tan, L. et al. 2014; Tu, Y. J., et al. 2012). Also, undesired biopolymers can be adsorbed on these MNP's which can be altered by using their functional groups such as ($\equiv Si(CH_2)_3\ NH_2$ / $\equiv SiCH_3$) (Melnyk et al. 2012).Similarly, heavy metal ions like Zn^{2+},Cu^{2+},Pb^{2+} can be removed using maghnetite nanotubes and Fe^{2+},Zn^{2+},Cd^{2+},Cu^{2+},Pb^{2+} metal ions have been reported (Chowdhury et al. 2012;Roy, A., Bhattacharya, J., 2012;Karami, H., 2013). Compared with metal nanotubes, nanorods has higher adsorption capacity for $Zn^{2+}$and Pb^{2+} but the adsorption capacity for Cu^{2+} is minimum(Qi, X. et al. 2014).Rapid and selective adsorption of Hg^{2+} can be done by water soluble superparamagnetic magnetite nanoparticles (Qi, X. et al. 2014).Nano hematite particle model allows the adsorption of specific heavy metal ions andmanifest the adsorption of Pb^{2+}, Cd^{2+}, Cu^{2+} ions which show endothermic adsorption and adsorption of Zn^{2+} which show exothermic adsorption (Shipley, H et al. 2013). Paramagnetic ferrite-based nano particles ($CuFe_2O_4$) makes adsorptionattainable of oxidized forms of Arsenic. (As(V))as arsenate salts from contaminated ground water (Tu,

Y.-J. et al. 2012).The mechanism involved in adsorption of contaminants by magnetic nano-adsorbents is shown in Figure 1.

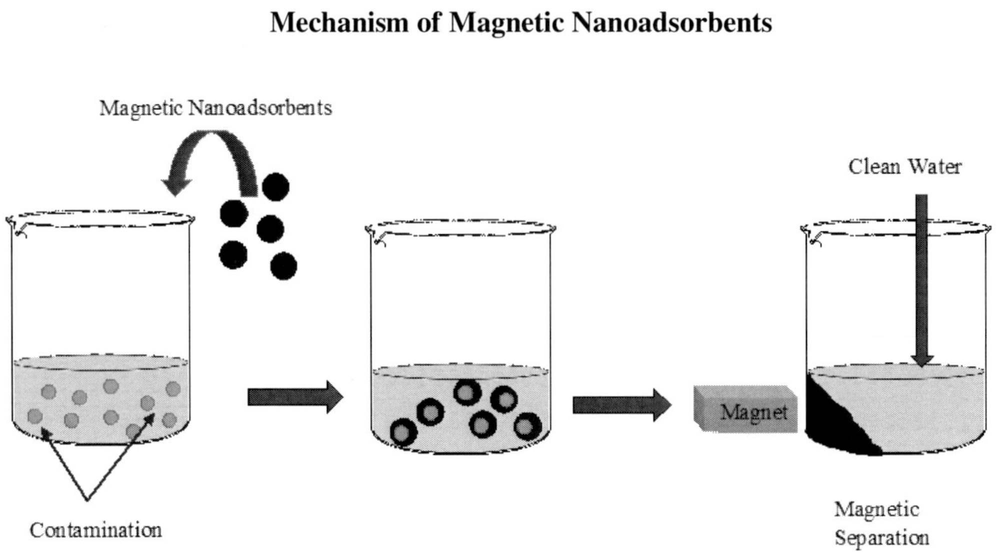

Figure 1. Adsorption of contaminants by magnetic nano-adsorbents.

4. Membrane and Membrane Process

A membrane is a penetrable, porous, thin-layered material that inhibits the entry of bacteria, viruses, metals, salts but synchronously allows water molecules to pass through it. Pressure-driven forces or Electrical forces are used on membranes in which pressure driven technology is best for water purification to obtain water of desired grade (Kumar, S. et al. 2014). Membrane separation process depends on pore and molecule size of material used. It is a reputable and mechanized process for waste water treatment (Yang, K., Xing, and B. 2010). This technology is accompanied with its disadvantages also, the choking of membranes makes the process complex which require high energy consumption due to pressure-driven process which minimizes the life time of membrane and membrane modules (Qu, X. et al. 2013).In order to boost the efficiency of membrane technology, the best suitably advantage method is to incorporate the functional nanomaterial into the membrane. This incorporation improves the membrane permeability, fouling resistance, mechanical and thermal stability. Figure 2 shows the mechanism involved in the removal of adsorption through semi-permeable membrane.

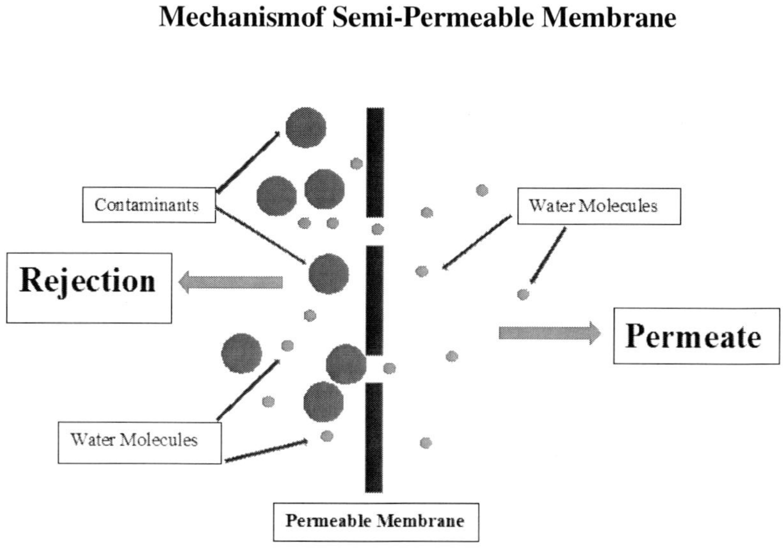

Figure 2. Removal of adsorption through semi-permeable membrane.

4.1. Nanofiber Membranes

Electrospinning is the clean, cheap and efficient technique to fabricate nanofibers. Electro spun nanofibers can be easily modified for different applications as these fibers consists of high surface area and porosity. Nanofiber membranes have wide applications like removal of micron-sized particle from water without any significant fouling (Ramakrishna et al. 2006), reverse osmosis, eradicate bacteria or viruses by size exclusion. But difficulties also arises by the violation of these membranes due to smaller pore sizes for relevant discharge of viral agents. Nanofibers can also be functionalized by organic molecules such as proteins, which make these fibers slightly bigger in diameter and bear conformational change upon wettingand is termed as "Bio-functionalization". During filtration and increase of pH above the iso-electric point of protein leads to evolution of hidden functional groups and swelling of protein which in turn enhances the capturing power of metal ions (Elbahri et al. 2012).For the separation of organic solvents, delicate and appropriate membranes are used which are made up of hydrophobic nanofiber materials(Feng et al. 2013).

4.2. Nanocomposite Membranes

Nano composite membranes are the surface modified membranes. They can be altered from mixed matrices. Inorganic Nano fillers are used by mixed matrix membranes. These membranes have considerable surface area and get easily blended with polymeric or inorganic oxide matrix. Variety of nanomaterials can be used as membrane such as hydrophilic metal oxide nanomaterials (Al_2O_3,TiO_2,zeolites, etc.), antimicrobial nanoparticles (nano-Ag, CNTs) and photocatalytic nanomaterials (bio-nanomaterials, TiO_2) (Qu, X., Alvarez, P.J.J., Li, Q., 2013). With the inclusion of hydrophilic metal oxide nano particles into the membranes, enhances mechanical and thermal stability of polymeric membranes. This inclusion also increases the water permeability, membrane surface hydrophilicity and restrict fouling of nanoparticle (Bae et al. 2005; Bottino et al. 2001; Maximous et al. 2010; Pendergast et al. 2011; Pendergast et al. 2010).These inorganic materials also compresses the negative impact of compaction and heat of membrane permeability. Nanocomposite membranes are semipermeable membranes whose top surface is applied in reverse osmosis as they are made up of ordered mesoporous carbons as Nano fillers formulated as thin-film polymeric matrices. Hydrophobic mesoporous carbon can be converted into hydrophilic carbon which in turn increases water permeability by atmosphere pressure plasma (Kim, E.-S., Deng, B., 2011). Studies revealed that various nanocomposites made from polyamide and nano-NaX zeolite (40-150nm) which were coated by interfacial polymerization using trimesoyl chloride and m-phenylenediamine monomers over porous polyethersulfone resulted in efficient permeability to pure water, leaving contaminants behind the membranes (Fathizadeh et al. 2011).

5. Use of Disinfectants

Organic pollutants can be classified into three divisions namely microorganisms, natural organic matter (NOM) and biological toxins. Microbial adulterants consist of human excreta and free living microbes (I. C. Escobar et al. 2001, I. Majsterek et al. 2004; D. Berry. et al. 2006; N. R. Dugan and D. J. Williams.2006; S. Srinivasan et al. 2008). Conventional water treatment systems are unable to eliminate cyanobacterial toxins. (W.-J. Huang et al. 2007; F. Al. Momani. et al. 2008).Ground and surface water both get polluted by biodegradable contaminants like bacteria, pathogens and viruses. Inorganic chemicals like chlorine dioxide, standard chlorine or ozone, etc. are used as disinfectants. But they are carcinogenic and produce harmful by-products like chlorite and chlorate (by chlorine dioxide) molecules, while ozone produces unknown organic products which are hazardous to both plant as well as animal life. Although, these disinfection technologies are traditional but still waterborne infections are occurring. So, there is an urgent need to resolve this issue by adopting such methods which are efficient to eliminate water-borne infections as well as produce harmless impacts on the environment. By keeping these points in the mind, there are several types of nanomaterials in the nature which sterilize the water and cure borne diseasescaused by microbes, such as silver, titanium and metal-oxide nanomaterials like TiO_2 and ZnO nanomaterials. They acquire antibacterial activities due to their charge capacity andmakes them most favorable and promising nanoparticles. Antimicrobial properties of silver and titanium oxide have been discussed below.

5.1. Silver Nanoparticles

Salts like silver nitrate and silver chloride are the main source of silver (Ag) nanoparticles which are most widely accepted due to their low toxicity and microbial inactivation fromwater (P. Jain and T. Pradeep 2005; J. A. Spadaro et al. 1974; G. Zhao and S. E. Stevens Jr. 1998; Y. Inoue et al. 2004). They bear antibacterial properties which are size as well as shape dependent (V.S.Kumar et al. 2004). Smaller Ag nanoparticles (8nm) are more competent than bigger sized particles while truncated triangular silver sized nanoparticles displays better antibacterial effects than spherical or rod-shaped nanoparticles, this indicates their shape dependentantibacterial property (V.S. Kumar. et al. 2004; J. R. Morones. et al. 2005; S. Makhluf. et al. 2005).The antibacterial effects of Ag nanoparticles include formation of free radicals which damage the bacterial cell membrane (S. Pal. et al. 2007; Z.-M. et al. 2011; Z.-M. et al. 2012; S. Y. Liau. et al. 1997; M. Danilczuk. et al. 2006). Immense antimicrobial activity is shown by immobilized nanomaterialsin which embedded Ag nanomaterials are adequate for both gram positive and gram negative bacteria (M. Danilczuk. et al. 2006). Incorporation of Ag nanoparticles in different types of polymers are used for the production of antimicrobial nanofibers and nanocomposites (A. Esteban-Cubillo. et al. 2006; L. Balogh et al. 2001; Y. Chen. et al. 2003). Advanced techniques are also used which are effective forantibacterial activitieslike water filters prepared from polyurethane's foam coated by Ag nanofibers against Escherichia coli (E. coli) (M. Botes et al. 2010) and pseudomonas. In remote areas of developing countries, Ag nanoparticles are used in terms of low cost potable microfilters (P. Jain et al. 2005).Lastly, incorporation of Ag nanocatalyst with carbon covered in alumina has been exhibited as appropriate microbial contaminant degradant in water (K. Zodrow et al. 2009). Despite Ag nanoparticles being used adequately for inactivating bacteria and viruses as well as reducing membrane biofouling, these particles have not been reported for long term competence because with the passage of time silver ions are lost (S. Chaturvedi et al. 2012; D.-G. Yu et al. 2003). So, there is urgent need for the researchers to do work in direction to reduce this loss of silver ions for long term control of membrane biofouling. As an alternative Ag nanoparticles can be doped with metallic nanoparticles or its composites with metal oxide nanoparticles would resolve this issue and most probably this would lead to the parallel removal of inorganic/organic compounds from waste water.

5.2. TiO_2 Nanoparticles

TiO_2 nanoparticles are economically beneficial and possess good photosensitive and nontoxic properties which makes them a promising photocatalyst used for water purification (J. S. Taurozzi et al. 2008). The basic action of these semiconductor based nanoparticles in photocatalysis process involves the formation of reactive oxidants such as OH radicals for eliminating microorganisms, bacteria, viruses and forth on (K. Hashimoto et al. 2012; H. Einaga et al. 1999; A. Fujishima et al. 2000; G. S. Shephard et al. 2002; J. A. et al. 2003; H.-L. et al. 2003; M. Cho et al. 2004; M. Cho et al. 2005; P. Hajkova et al. 2007). After conducting a phtocatalyses experiment, a simulated exposure of contaminated water with TiO_2 for about 8 hours, it has been reported that the viability of several water borne pathogens such as protozoa, fungi, E.coli and pseudomonas aeruginosa get reduced in water (L. Zan et al. 2007). Facial colifirms get completely inactivated under TiO_2, which proves that it has a better photocatalytic disinfection efficiency as compared to sunlight (J. Lonnen et al. 2005). Earlier TiO_2 has shown its photocatalytic capability only under UV light but nowadays its optical absorbance has been extended to visible light region by doping transition metals and anionic nonmetals such as nitrogen, carbon, sulfur or fluorine into TiO_2 (S. Gelover et al. 2006). Bacterial inactivation has been improved by doping Ag into TiO_2 nanoparticles, thereby enhancing the disinfection under UV radiations and solar radiations (M. Ni et al. 2007; L. Lin. et al. 2007; M. E. Rinc´on et al. 2007; H.Wang and J. P. Lewis 2005; D. Mitoraj, A. Ja'nczyk, M. Strusetal. 2007; T. Umebayashi et al. 2002). Strong bactericidal activity against E.coli was shown by visible light activated TiO_2 nanoparticles (T. Ohno et al. 2003; J. C. et al. 2002; M. Sökmen et al. 2001; H. M. Sung-Suh et al. 2004). TiO_2 photocatalysts can be further modified by doping metals like iron or coupling with transition metal oxides with promising results for water disinfection and in antibacterial effects against E. coli. Recentlt, results have been reported in which nitrogen and Pd doped with TiO_2 nanoparticles have shown better efficiency against E.coli bacteria's, Pseudomonas aeruginosa and Bacillus subtilis (V. Vamathevan et al. 2004; X. Zhang, M. Zhou, and L. Lei 2005).

6. Desalination

An important substitute for obtaining fresh water is desalination. Recent development in membrane technology that is Reverse Osmosis (RO), though expensive but covers 41% of desalination capability (K. Page et al. 2007; T. Humplik et al. 2011). Nanomaterial plays a crucial role in developing most reliable and low cost reactive membranes for the treatment of waste water (C. Fritz Mann et al. 2007). CNT's, zeolites and graphene are some nanomaterials which offer opportunities to limit the cost of desalination and enhance the energy efficacy(A. Srivastava et al. 2004).Good salt rejection is exhibited by thin film of nanocomposite membranes which consists of Ag and TiO_2 nanoparticles and composite ceramic membranes coated by iron oxide nanoparticles (Fe_2O_3) (J. K. Holt et al. 2006).Alumina ceramic membranes modified by silica nanoparticles have also shown high rejection of sodium chloride (C. Song et al. 2009).Similarly, RO have zeolite based membranes which exhibited high flux with better ion rejection characteristics (M. M. Cortalezzi et al. 2002).Recent studies revealed that the graphene membranes have higher potential than polymeric RO membranes (B.-H. Jeong et al. 2007).Lipotropic liquid crystals and aquaporins are some other nanostructures which also exhibited high flux and selective water transportation (M. E. Suk et al., 2010).Water permeability and high salt rejection capabilities have been increased in RO and NF membranes by designing new ways offered by zeolite polyamide thin film nanocomposite membranes (X. Gong, J. Li, H. Lu et al. 2007). At the opening of CNT's, functional groups are grafted such as carboxyl groups have shown better selective rejection of some components simultaneously.This would lead to reduce permeability thereby rendering CNT's incapable for desalination(E. M. V. Hoek and A.K. Ghosh 2009; M. S. Mauter and M. Elimelech, 2008).In order to achieve economically feasible and commercially acceptable desalination membranes, nanocomposite membranes may serve ideally but basic understanding of transport mechanism along with proper pore size selection by keeping the uniformity is required. For long term operation and maintaining membrane performance, effects of real seawater feed on the efficiency of different nanomaterials need to be investigated. For this purpose, advanced oxidation process such as photocatalysis has been introduced.

7. Photocatalytic Removal of Pollutants (Dyes/Metals) from Waste Water

There is a great need for the development of clean and renewable energy sources to meet the increasing global energy demands and resolve the environmental issues caused by the overuse of fossil fuels. One of the best options is the conversion of solar energy into hydrogen through water splitting process, with the help of semiconductor-based photocatalysts (Tachibana et al. 2012; Tran et al. 2012; Xiang et al. 2012). The semiconductor-based photocatalytic systems must be designed by keeping these 3 key points in psyche:

(1) The semiconductor should have a narrow band gap to absorb as much light as possible; meanwhile, the bottom of its conduction band has to be more negative than the reduction potential of water to produce H_2. For overall water splitting,

the top of its valence band must be more positive than the oxidation potential of water to produce O_2.

(2) There should be efficient charge separation and fast charge transport simultaneously avoiding the bulk.

(3) Kinetically feasible surface chemical reactions must take place between these carriers and water or other molecules; meanwhile, the surface backward reaction of H_2 and O_2 to water can be successfully suppressed.

Therefore, various scientists and researchers are trying to develop photocatalysts with specific bulk and surface properties as well as energy band structures to satisfy the above requirements (Kudo et al. 2009; Chen et al. 2010).

Metal oxides such as TiO_2, $SrTiO_3$, etc. have been widely investigated for photocatalytic hydrogen generation, butare not ideal due to the poor solar energy utilization (Tong et al. 2012; Wang et al. 2013). Their wide energy band gaps are only active in UV region. Some other oxides including WO_3, $BiVO_4$, and so forth are visible-light-responsive but cannot conduct water reduction to produce H_2 because their conduction bands are lower than the reduction potential of water (Yerga et al. 2009). CdS possesses a narrow band gap with appropriate band levels for water splitting but it is toxic and usually unstable due to self-oxidation (Abe et al. 2010).g-C_3N_4 is considered to be the most stable allotrope among various carbon nitrides under ambient conditions as it undergoesphotocatalytic water splitting in UV-visible light. There is an urgent need of developing efficient photocatalytic systems for water splitting as it can lead to theproduction of solar fuels (Yerga et al. 2009; Gust et al. 2009; Kudo et al. 2009; Kamat et al. 2007). In the field of photocatalysis the number of semiconductors showing highly advancedphotocatalytic activity under visible-light illumination is still limited. A vast majority of the study of photocatalytic activities are carried out with modified TiO_2 (Yerga et al. 2009; Zhu et al. 2009; Hernadez et al. 2009).

In the overall water splitting process, the water is reduced to hydrogen and simultaneously oxidized to oxygen. From these two semi-reactions, the generation of hydrogen is considered simpler. In the principle process, hydrogen consists of protons present in the water which accepts theelectrons and noble metals acting as hydrogen evolution centers. In contrast to the simplicity of hydrogen generation, formation of oxygen from water is conceptually more challenging. Since, mechanically it has to occur through several steps requiring four positive holes and the formation of O-O bonds (Siilva et al. 2009). Therefore, the oxygen evolution is significant factor in determining the overall efficiency in the photocatalytic water splitting. The semi-reactions ofhydrogen and oxygen generation from water, can be independently studied by adding sacrificial electron donors (for hydrogen generation) and electron acceptors (for oxygen formation) during the photocatalytic process. Therefore, each reaction can be studied independently by decoupling them. Thus, there is an urgent requirement for the development of novel efficient photocatalysts with visible-light activity with comparable or higher efficiency than those currently known (Maeda et al. 2007; Maeda et al. 2005). Researchers should take interest in finding the efficient semiconductor for oxygen evolution. The nanoparticulated ceria prepared by a novel biopolymer-templating methodology and containing appropriate gold loadings is a stable and efficient photocatalyst for oxygen evolution (Primo et al. 2011).In addition, it is also reported that the deposition of gold nanoparticles at low loading increases the photocatalytic activity for visible-light-producing samples that exhibit higher photocatalytic activity than the same material upon irradiation at its band gap (Alvaro et al. 2010).Moreover, the ceria samples containing gold under visible light irradiation outperform the photocatalytic activity ofWO_3 under UV irradiation (Primo et al. 2011).

7.1. What Is a Photocatalyst?

A photocatalyst is defined as a substance which is activated by adsorption of a photon and is capable of accelerating a reaction without being consumed. These substances are invariably semiconductors. Semiconducting oxide photocatalysts arein great demand due to their potential applications in solar energy conversion and environmental purification. Semiconductor heterogeneous photocatalysis has enormous potential to treat organic contaminants in water and air that pollute the environment. This process is known as advanced oxidation process (AOP) and is a suitable process for the oxidation of a wide range of organic compounds (Yang Deng et al. 2015).

Photocatalysis is a science of applying catalyst that speeds up the chemical reactions that requires light. A photocatalyst is defined as a material that absorbs light, produces electron–hole pairs that enable chemical transformations of the reaction participants and regenerate its chemical composition after each cycle of such interactions (Chan et at. 2011; Djurisic et al. 2014; Pelizetti et at. 1994; Hisatomi et al. 2014; Hoffmann et al. 1995; Racele et al. 2005). The two types of photocatalytic reactions are homogeneous and heterogeneous photocatalysis. Among these two, heterogeneous photocatalysis is in more demand due to its greater efficiency in degrading recalcitrant organic compounds (Vincenzo Augugliaro et al. 2006).The significant features of the photocatalytic system are listed below:

- desired band gap
- suitable morphology
- high surface area
- stability and reusability (Pelizetti et at. 1994; Hisatomi et al. 2014; Hoffmann et al. 1995; Aracely et al. 2015)

Metal oxides such as oxides of vanadium, chromium, titanium, zinc, tin, and cerium having the above characteristics follow similar primary photocatalytic processes. In this process, a metaloxide is activated with either UV light, visible light or a combination of both, and the photoexcited electrons are promoted from the valence band to the conduction band, forming an electron-hole pair. The photogenerated electron–hole pair is able to reduce/oxidize a compound adsorbed on the photocatalyst surface. The photocatalytic activity of metal oxide comes from two sources, firstly by the generation of OH radicals by oxidation of OH^- anions and then by generation of O_2 radicals by reduction of O_2. Both the radicals and anions degrade the pollutants either by reacting with them or otherwise transform them to less harmful byproducts (Hoffmann et al. 1995; Aracely et al. 2015). Among these metal oxides TiO_2, ZnO, SnO2 and CeO_2 are abundant in nature, and have also been extensively used as heterogeneous photocatalyst (Aracely et al. 2015; Fujishima et al. 1972). Heterogeneous photocatalyst have biocompatibility, favorable electronic structure, light absorption properties, exceptional stability in various conditions and the capability to generate charge carriers when stimulated with required amount of light energy which makes them an efficient photocatalyst (Pelizetti et at. 1994; Hisatomi et al. 2014; Hoffmann et al. 1995; Aracely et al. 2015; Wang et at.2014). Heterogeneous photocatalysis metal oxides such as TiO_2 and CeO_2 has proved their efficiency in degrading a wide range of distinct pollutants into biodegradable compounds and their bio-mineralization into harmless products such as carbon dioxide and water. This photocatalytic process needed for the purification of waste water, by removal of bacteria and other pollutants, as this can make water reusable. Among oxidation reactions, most of the photocatalytic processes focuses on the conversion of highly toxic chemicals to either less toxic chemicals or CO_2 and H_2O (Wang et al. 2014; Chen et al. 2012; Pelaez et al. 2012). Metal oxides are the important component of photocatalyst and their physical along with the chemical properties are important for the photocatalytic performance, which are basically size, shape, morphology and composition dependent. Their synthesis procedure employed can control the size, shape and morphology of the materials prepared which can facilitate the formation of powders or thin films with the required characteristics that enhance the performance of the catalyst (Yu et al. 2014; Zhou et al. 2014; Wang et al. 2015). It is also the source and type of light used that can affect the performance of the material being used as a photocatalyst (Khan, M.M. et al. 2015).

7.2. Types of Photocatalysts and Their Characteristics

Many metal oxide semiconductors are considered to be the most suitable photocatalysts due to their photo-corrosive resistance and wide band gap energies (Fox et al. 1993). $TiO_2$is one of the most effective and extensively used photocatalystin wastewater treatment studies. The advantages ofTiO_2 over other photocatalyst is that it is cost effective, thermally stable, non-toxic, chemically and biologically inert and is capable of promoting oxidation of organic compounds (Mandelbaum et al. 1993). The photocatalytic activity of TiO_2 depends on its surface and structural properties, basically the surface area, band gap energy, crystal composition, particle size distribution and porosity (Ahmed et al. 2011). TiO_2 is also known as titania, titanic oxide, titanium white, titanic anhydride, or titanic acid anhydride.Following are the three types of reactors used in various methodologies fordegradation of organic pollutants in water:

7.2.1. Photocatalytic Reactors

Photocatalytic reactors can be classified based on the suspension state of the photocatalyst. Photocatalytic reactors work by using UV or solar radiations. Solar photocatalytic reactors are widely used for the photoxidation of organic contaminants in water. Such kind of reactors can be divided into concentrating or non-concentrating reactors and have their own certain advantages and disadvantages. The advantages of non-concentrating reactors have negligible optical losses and therefore can be used directly. They also diffuses sun irradiation but are larger in size as compared tothe concentrating reactors and have high frictional pressure losses (de Lasa et al. 2005). However, the use of solar radiated photoreactors is limited due to the intrinsic nature of TiO_2 particles.

7.2.2. Slurry Reactors

Among photocatalytic reactors, TiO_2 slurry reactors are most commonly used in water treatment. Their photocatalytic activity is highly significant and more efficient as compared with the immobilized photocatalyst which provides large total surface area of photocatalyst per unit volume and it is one of the most important factor in determining efficiency of a photocatalytic reactor (Chog et al. 2010). However, there is a complication with such reactors that requires separation of the sub-micron TiO_2 particles from the treated water.

7.2.3. Immobilized TiO_2 Reactors

Photocatalytic reactors with immobilized TiO_2 are those in which catalyst is fixed to support via physical surface forces or chemical bonds. These reactors have special benefits of not requiring catalyst recovery and allow the continuous use of the photocatalyst (de Lass et al. 2005). Hybrid photocatalytic membrane reactors have been developed to achieve the purpose of downstream separation of photocatalysts. The photocatalytic membrane reactors can be generalized into two categories (1) irradiation of the membrane module and (2) irradiation of feed tank containing photocatalyst in suspension (Molinari et al. 2002). Various membranes such as microfiltration, ultrafiltration, and nano filtration membranes may be used for this purpose depending on the requirements of the treated water quality (Chong et al. 2010). Photocatalytic membrane reactors have been successfully used for the degradation of trichloroethylene and 4-nitrophenol (Artale et al. 2000; Tsuru et al. 2003).

CONCLUSION

In this chapter we made an effort to cover a wide range of wastewater effluent removal techniques for water treatment so that the reader can easily get an idea about various types of removal methods and basic principle involved behind each technique. Among these techniques, one is the carbonaceous nanomaterials that have attracted a wide attention to remove pollutants fromwater. The exclusive properties of these nanomaterials brought about a great chance to recast wastewater treatment. Further wastewater treatment was also carried out using silver and titanium based nanoparticles which proved marvelous results in this area and presently, scientists are working on the modification of these nanomaterials to stimulate their applications by improving inherent properties, exhibiting a brighter future in wastewater treatment. From the above discussed techniques for wastewater treatment photocatalyst, an advanced oxidation technique is the most important one used for the removal of pollutants from wastewater. This technique has excellent properties in removing organic pollutants like herbicides, pesticides and micro-pollutants from water. In order to destroy recalcitrant organic compounds from wastewater, advanced oxidation process (AOP) is the most robust and promising technique which results in the enhancement of oxidation rate with the formation of active radicals. Previous studiesrevealed that 99% organic pollutants were removed by TiO_2nanocatalysts within 2 hours. Despite some expenditures being involved in photocatalysis, it has considerable ecological and economic advantages over the conventional methods for long term application.

REFERENCES

Abaychi J. K. 1987 "Concentrations of trace elements in aquatic vascular plants from shatt al. Arab river" *Iraqian Journal of Biological Sciences Research* 18:123-129.

Abdel-RaoufN., Al-HomaidanA.A. andIbraheemI.B.M.2012 "Microalgae and wastewater treatment" *Saudi Journal of Biological Sciences* 19: 257–275.

Abdullah M., Low G., Mathews R.W. 1990 "Effects of common inorganic ions on rates of photocatalytic oxidation of organic carbon over illuminated titanium dioxide" *Journal of Physical Chemistry* 94: 6820.

Abe R. 2010 "Recent Progress on Photocatalytic and Photoelectrochemical Water Splitting under Visible Light Irradiation" *Journal of Photochemistry and Photobiol*ogy C 11: 179–209.

Ahmed S., Rasul M.G., Brown R., Hashib M.A. 2011 "Influence of Parameters on the Heterogeneous Photocatalytic Degradation of Pesticides and Phenolic Contaminants in Wastewater: A Short Review" *Journal of Environmental Management* 92: 311-330.

Ali I., 2012 "New generation adsorbents for water treatment" *Chemical Reviews* 112: 5073–5091.

Al Momani F., Smith D.W. and Gamal El-Din M. 2008 "Degradation of cyanobacteria toxin by advanced oxidation processes" *Journal of Hazardous Materials* 150:238–249.

Alvaro M.,Cojocaru B., Ismail A., Petrea N., Ferrer B., Harraz F., VasileP.I. and Hermenegildo G. 2010 "Visible-light photocatalytic activity of gold nanoparticles supported on template-synthesized mesoporous titania for the decontamination of the chemical warfare agent Soman" *Applied Catalysis B: Environmental* 99:191-197.

Arshadi M., Faraji, A.R. and Amiri, M.J. 2015 "Modification of aluminum-silicate nanoparticles by melamine-based dendrimer l-cysteine methyl esters for adsorptive characteristic of Hg(II) ions from the synthetic and Persian Gulf water" *Chemical Engineering Journal* 266:345–355.

Artale M.A., Augugliaro V., Drioli E., Golemme G., Grande C., Loddo V., MolinariR., Palmisano L., Schiavello M. 2000 "Preparation and Characterisation of Membranes with Entrapped TiO2 and Preliminary Photocatalytic Tests." *Annali di Chimica* 91: 127–136.

Auffan M., Rose J., Bottero J.-Y., Lowry G.V., Jolivet J.-P., Wiesner M.R. 2009 "Towards a definition of inorganic nanoparticles from an environmental, health and safety perspective" *Nature Nanotechnology.* 4: 634–641.

Auffan M., Rose J., Proux O., Borschneck D., Masion A., Chaurand P., Hazemann J.-L., Chaneac C., Jolivet J.-P., Wiesner M.R., VanGeen A. and BotteroJ.-Y. 2008 "Enhanced adsorption of arsenic onto maghemites nanoparticles: As(III) as a probe of the surface structure and heterogeneity" *Langmuir* 24: 3215–3222.

Ayangbenro S.A. and Babalola O.O. 2017 "A New Strategy for Heavy Metal Polluted Environments: A Review of Microbial Biosorbents" *International Journal Environmental Research and Public Health* 14: 94.

Badruddoza A.Z.M., Shawon Z.B.Z., Rahman M.T., Hao K.W.,Hidajat K., Uddin M.S. 2013 "Ionically modified magnetic nanomaterials for arsenic and chromium removal from water" *Chemical Engineering Journal* 225: 607–615.

Bae T.-H., Tak T.-M. 2005 "Effect of TiO2 nanoparticles on fouling mitigation of ultrafiltration membranes for activated sludge filtration" *Journal of Membrane Science* 249: 1–8.

Balogh L., Swanson D.R., Tomalia D.A., Hagnauer G.L., and McManus A.T. 2001 "Dendrimer-Silver Complexes and Nanocomposites as Antimicrobial Agents" *Nano Letters* 1:18–21.

Berry D., Xi C., and Raskin L. 2006 "Microbial ecology of drinking water distribution systems" *Current Opinion in Biotechnology* 17:297–302.

Botes M. and Cloete T.E. 2010 "The potential of nanofibers and nanobiocides in water purification" *Critical Reviews in Microbiology* 36:68–81.

Bottini M., Bruckner S., Nika K., Bottini N., Bellucci S., Magrini A., Bergamaschi A. andMustelin T. 2006 "Multi-walled carbon nanotubes induce T lymphocyte apoptosis" *Toxicology Letters* 160: 121–126.

Bottino A., Capannelli G., Dasti V. and Piaggio P.2001 "Preparation and properties of novel organic-inorganic porous membranes" *Separation and Purification Technology* 22–23:269–275.

Chan S.H.S., Wu T.Y. and Juan J.C 2011 "Recent developments of metal oxide semiconductors as photocatalysts in advanced oxidation processes (AOPs) for treatment of dye waste-water" *Journal of Chemical Technology and Biotechnology* 86:1130–1158.

Chaturvedi S., Dave P.N. and Shah N.K.2012 "Applications of nano-catalyst in new era," *Journal of Saudi Chemical Society*16:307–325.

Chen H., Nanayakkara C.E. and Grassian V.H. 2012 "Titanium dioxide photocatalysis in atmospheric chemistry" *Chemical Reviews* 112:5919–5948.

Chen L., Xin H., Fang Y., Zhang C., Zhang F., Cao X., Zhang C. and Li X. 2014 "Application of metal oxide heterostructures in arsenic removal from contaminated water" *Journal of Nanomaterials* 2014:1-10.

Chen W., Duan L. andZhu D. 2007 "Adsorption of polar and nonpolar organic chemicals to carbon nanotubes" *Environmental Science and Technology* 41: 8295–8300.

Chen X. B., Shen S. H., Guo L. J., Mao S. S. 2010 "Semiconductor-Based Photocatalytic Hydrogen Generation" *Chemical Review* 110: 6503–6570.

Chen Y., Wang L., Jiang S., and Yu H. 2003 "Study on Novel Antibacterial Polymer Materials (I) Preparation of Zeolite Antibacterial Agents and Antibacterial Polymer Composite and Their Antibacterial Properties" *Journal of Polymer Materials* 20:279–284.

Cho M., Chung H., Choi W. and Yoon J.2004 "Linear correlation between inactivation of *E. coli* and OH radical concentration in TiO2 photocatalytic disinfection" *Water Research* 38:1069–1077.

Cho M., Chung H., Choi W. and Yoon J.2005 "Different inactivation behaviors of MS-2 phage and *Escherichia coli* in TiO_2 photocatalytic disinfection" *Applied and Environmental Microbiology*71:270–275.

Chong M.N., Jin B., Chow C.W.K. and Saint C.2010 "Recent Developments in Photocatalytic Water Treatment Technology: A Review" *Water Resources* 44: 2997-3027.

Chowdhury S. and Balasubramanian R. 2014 "Recent advances in the use of graphene-family nanoadsorbents for removal of toxic pollutants from wastewater" *Advances in Colloid and Interface Sciences* 204: 35–56.

Chowdhury S.R., Yanful E.K., Pratt A.R. 2012 "Chemical states in XPS and Raman analysis during removal of Cr(VI) from contaminated water by mixed maghemite–magnetite nanoparticles" *Journal of Hazardous Material* 235-236:246–256.

Crittenden J.C., Zhang Y., Hand D.W., Perram D.L. and Marchand E.G. 1996 "Solar detoxificationof fuel-contaminated groundwater using fixed-bed photocatalysts" *Water Environment Research* 68: 270-277.

Das S.K., Khan M.M.R., Guha A.K., Das A.R., Mandal A.B. 2012 "Silver-nanobiohybride material: synthesis, characterization and application in water purification" *Bioresource Technology* 124:495–499.

Datsyuk V., Kalyva M., Papagelis K., Parthenios J., Tasis D., Siokou A., Kallitsis I. and Galiotis C. 2008 "Chemical oxidation of multiwalled carbon nanotubes" *Carbon* 46:833–840.

De Lasa H., Serrano B. and Salaices M. 2005 *"Photocatalytic Reaction Engineering"* Springer Science: USA, ISBN 978-0-387-27591-8.

De Volder M.F., Tawfick S.H., Baughman R.H. and Hart, J. 2013 "Carbon nanotubes: present and future commercial applications"*Science* 339:535–539.

Deng Y. and Zhao R. 2015 "Advanced Oxidation Processes (AOPs) in Wastewater Treatment" *Current Pollution Reports* 1:167-176.

Djurisic A.B., Leung Y.H., A.M.C. 2014 "Strategies for improving the efficiency of semiconductor metal oxide photocatalysis" *Materials Horizons* 1: 400-410.

Dugan N. R. and D. J. Williams 2006 "Cyanobacteria passage through drinking water filters during perturbation episodes as a function of cell morphology, coagulant and initial filter loading rate" *Harmful Algae* 5:26–35.

Einaga H., Futamura S. and Ibusuki T. 1999 "Photocatalytic decomposition of benzene over TiO2 in a humidified air stream" *Physical Chemistry Chemical Physics* 1:4903–4908.

Elbahri M., Homaeigohar S., Abdelaziz R., Dai T., Khalil R. and Zillohu A.U., 2012 "Smart metal–polymer bionanocomposites as omnidirectional plasmonic black absorber formed by nanofluid filtration" *Advance Functional Materials* 22:4771–4777.

Escobar I.C, Randall A.A. and Taylor J.S., 2001 "Bacterial growth in distribution systems: effect of assimilable organic carbon and biodegradable dissolved organic carbon" *Environmental Science and Technology* 35:3442–3447.

Esteban-Cubillo A., Pecharroman C., Aguilar E., J. Santaren J., and Moya J.S., 2006. "Antibacterial activity of copper monodispersed nanoparticles into sepiolite" *Journal of Materials Science* 41:5208–5212.

Fathizadeh M., Aroujalian A. and Raisi A. 2011 "Effect of added NaXnano-zeolite into polyamide as a top thin layer of membrane on water flux and salt rejection in a reverse osmosis process" *Journal of Membrane Science* 375:88–95.

Faust S.D. and Aly O.M. 1983 Ed.1 *"Chemistry of Water Treatment"* CRC Press, Taylor and Francios Groups, ISBN 9781575040110 CAT# LA0115.

Feng C., Khulbe K.C., Matsuur T., Tabe S. and Ismail A.F. 2013 "Preparation and characterization of electro-spun nanofiber membranes and their possible applications in water treatment" *Separation andPurification Technology* 102:118–135.

Fox M.A. and Dulay M.T. 1993 "Heterogeneous Photocatalysis" *Chemical Reviews* 93:341–357.

Fritzmann C., Lowenberg J., Wintgens T. and Melin T. 2007 "State of- the-art of reverse osmosis desalination" *Desalination* 216:1–76.

Fujishima A. and Honda K. 1972 "Electrochemical photolysis of water at a semiconductor electrode" *Nature* 238: 37– 38.

Fujishima A., Rao T.N. and Tryk D.A. 2000 "Titanium dioxide photocatalysis," *Journal of Photochemistry and Photobiology C:Photochemistry Reviews* 1:1–21.

Gaya U.I. and Abdullah A.H. 2008 "Heterogeneous Photocatalytic Degradation of Organic Contaminants over Titanium Dioxide: A Review of Fundamentals, Progress and Problems" *Journal of Photochemistry and Photobiology C Photochemistry Reviews* 9:1-12.

Gehrke I., Geiser, A. and Somborn-Schulz A. 2015 "Innovations in nanotechnology for water treatment" *Nanotechnology Science andAppl*ications 8:1–17.

Gelover S., Gomez L.A., Reyes K. and Teresa Leal M. 2006 "A demonstration of water disinfection using TiO2 films and sunlight" *Water Research* 40:3274–3280.

Gong X., Li J. and Lu H. 2007 "A charge-driven molecular water pump," *Nature Nanotechnology 2:709–712.*

Gust D., Moore T. A. and Moore A. L. 2009 "Solar Fuels via Artificial Photosynthesis" *Account of Chemical Research* 42:1890-8.

Hajkova P., Spatenka P., Horsky J., Horska I., and Kolouch A. 2007 "Photocatalytic effect of TiO2 films on viruses and bacteria" *Plasma Processes and Polymers* 4:397–401.

Hashim D.P., Narayanan N.T., Romo-Herrera J.M., Cullen D.A., Hahm M.G., Lezzi P., Suttle J.R., KelkhofD., Munoz-Sandoval E., Ganguli S., Roy A.K., Smith D.J., Vajtai R., Sumpter B.G., Meunier V., Terrones H., Terrones M. and Ajayan P.M. 2012 "Covalently bonded three-dimensional carbon nanotube solids via boron induced nanojunctions" *Scientific Rep*orts 2: 363-9.

Hashimoto K., Irie H., and Fujishima A.2005 "TiO2 photocatalysis: a historical overview and future prospects" *Japanese Journalof Applied Physics, Part 1: Regular Papers and Short Notes and Review Papers* 44:8269–8285.

Hernandez R.A.and MedinaR.I. 2015 "Photocatalytic Semiconductors" *Springer* ISBN 978-3-319-10998-5.

Hernandez-Alonso M. D., Fresno F., Suarez S. and Coronado J. M. 2009 "Development of alternative Photocatalysts to TiO_2: Challenges and opportunities" *Energy Environmental Science* 2: 1231–1257.

Hisatomi T., Kubota J. and Domen K.2014 "Recent advances insemiconductors for photocatalytic and photoelectron chemical water splitting" *Chemical Society of Reviews* 43: 7520–7535.

Hoek E. M. V.and Ghosh A.K. 2009 "Nanotechnology-Based Membranes for Water Purification," in *Nanotechnology Applications Advances in Materials Science and Engineering for Clean Water* chapter 4:47–58.

Hoffmann M.R., Martin S.T., Choi W. and Bahnemann D.W.1995 "Environmental applications of semiconductor photocatalysis" *Chemical Reviews* 95:69–96.

Holt J.K., Park H.G. and Wang Y.2006 "Fast mass transport through sub-2-nanometer carbon nanotubes" *Science* 312:1034–1037.

Huang W.J., Cheng B.L. and Cheng Y.L. 2007 "Adsorption of microcystin-LR by three types of activated carbon" *Journal of Hazardous Materials* 141:115–122.

Humplik T., Lee J., OHern S.C., Fellman B.A., Baig M.A., Hassan S.F., Atieh M.A., Rahman F., Laoui T., Karnik R. and Wang E.N. 2011"Nanostructured materials for water desalination," *Nanotechnology* 22:1-19.

Ibanez J.A., Litter M.I. and Pizarro R.A. 2003 "Photocatalytic bactericidal effect of TiO2 on Enterobacter cloacae. Comparative study with other Gram (-) bacteria" *Journal of Photochemistry and Photobiology A: Chemistry* 157:81–85.

Iijima S. 1991 "Helical microtubules of graphitic carbon" *Nature* 354: 56–58.

Inoue Y., Hoshino M. and Takahashi H.2002 "Bactericidal activity of Ag-zeolite mediated by reactive oxygen species under aerated conditions" *Journal of Inorganic Biochemistry* 92:37–42.

Jain P.and Pradeep T. 2005 "Potential of silver nanoparticle-coated polyurethane foam as an antibacterial water filter," *Biotechnology and Bioengineering* 90:59–63.

Jeong B.H., E. M. V. Hoek E.M.V., Yan Y.2007 "Interfacial polymerization of thin film nanocomposites: a new concept for reverse osmosis membranes," *Journal of Membrane Science* 294:1–7.

Ji, L., Chen, W., Duan, L. and Zhu, D. 2009 "Mechanisms for strong adsorption of tetracycline to carbon nanotubes: a comparative study using activated carbon and graphite as adsorbents" *Environmental Science Technology* 43:s2322–2327.

Joao P. S. C. 2010 "Water Microbiology Bacterial Pathogens and Water" *Int. J. Environ. Res. Public Health* 7:3657-3703.

Kamat, P. V. 2007 "Meeting the Clean Energy Demand: Nanostructure Architectures for Solar Energy Conversion" *Journal of Physi*cal *Chemistry C* 111:2834-39.

Karami, H. 2013. "Heavy metal removal from water by magnetite nanorods" *Chemical Engineering Journal* 219:209–216.

Khan M.M., Adil S.F. and Al-Mayouf A.2015 "Metal oxides as photocatalysts" *Journal of Saudi Chemical Society*19:462-464.

Kim E.-S. and Deng B. 2011 "Fabrication of polyamide thin-film nano-composite (PA-TFN) membrane with hydrophilized ordered mesoporous carbon (H-OMC) for water purifications". *Journal of Membrane Science* 375:46–54.

Kudo A., Miseki Y. 2009 "Heterogeneous Photocatalyst Materials for Water Splitting" *Chemical Society of Review* 38:253−278.

Kumar S., Ahlawat W., Bhanjana G., Heydarifard S., Nazhad M.M., Dilbaghi N. 2014. "Nanotechnology-based water treatment strategies" *Journal of Nanosciences and Nanotechnology*14:1838–1858.

Kumar V.S., Nagaraja B.M., Shashikala V. 2004 "Highly efficient Ag/C catalyst prepared by electro-chemical deposition method in controlling microorganisms in water," *Journal of Molecular Catalysis A: Chemical* 223:313–319.

Lau W.J., Gray S., Matsuura T., Emadzadeh D., Paul Chen J. and Ismail A.F., 2015 "A review on polyamide thin film nanocomposite (TFN) membranes: history, applications, challenges and approaches" *Water Research* 80:306–324.

Lei Y., Chen F., Luo Y. and Zhang L., 2014 "Three-dimensional magnetic graphene oxide foam/Fe3O4 nanocomposite as an efficient absorbent for Cr(VI) removal" *Journal of Material Science* 49:4236–4245.

Li Y.H., Ding J., Luan Z., Di Z., Zhu Y., Xu C., Wu D. and Wei B. 2003 "Competitive adsorption of Pb2+, Cu2+ and Cd2+ ions from aqueous solutions by multi-walled carbon nanotubes" *Carbon* 41: 2787–2792.

Li Y.H., Wang S., Luan Z., Ding J., Xu C. andWu D. 2003"Adsorption of Cadmium(II) from aqueous solution by surface oxidized carbon nanotubes" *Carbon* 41:1057–1062.

Liau S.Y., Read D. C., Pugh W. J., Furr J. R. and Russell A.D., 1997 "Interaction of silver nitrate with readily identifiable groups relationship to the antibacterial action of silver ions" *Letters in Applied Microbiology* 25:279–283.

Lin C. and Lin K. 2007 "Photocatalytic Oxidation of Toxic Organohalides with TiO2/UV: The effects of Humic Substances and Organic Mixtures" *Chemosphere* 66: 1872–1877.

Lin D. and Xing B. 2008. "Adsorption of phenolic compounds by carbon nanotubes: role of aromaticity and substitution of hydroxyl groups" *Environmental Science and Technology*42: 7254–7259.

Lin L., Lin W. and Zhu Y.X. 2005 "Uniformcarbon-covered titaniaand its photocatalytic property" *Journal of Molecular Catalysis A: Chemical* 236:46–53.

Liu H.L. and Yang T.C.K. 2003 "Photocatalytic inactivation of *Escherichia coli* and Lactobacillus helveticus by ZnO and TiO2 activatedwithultraviolet light" *Process Biochemistry* 39: 475–481.

Liu L., Liu J. and Sun D.D. 2012 "Graphene oxide enwrapped Ag_3PO_4 composite: towards a highly efficient and stable visible-light-induced photocatalyst for water purification" *Catalyst Science and Technology* 2:2525–2532.

Lonnen J., Kilvington S., Kehoe S.C., Al-Touati F. and McGuigan K.G.2005 "Solar and photocatalytic disinfection of protozoan, fungal and bacterial microbes in drinking water" *Water Research* 39:877–883.

Lu C., Chiu H. and Liu C. 2006 "Removal of zinc (II) from aqueous solution by purified carbon nanotubes. Kinetics and equilibrium studies" *Industrial and Engineering Chemistry Research* 45:2850–2855.

Maeda K. and Domen K. 2007"New Non-Oxide Photocatalysts Designed for Overall Water Splitting under Visible Light" *Journal of Physics Chemisrty C*111: 7851–7861.

Maeda K.; Takata T.; Har M., Saito N., Inoue Y., Kobayashi H. and Domen K.2005 "GaN:ZnO Solid Solution as a Photocatalyst for Visible-Light-Driven Overall Water Splitting" *Journal of American Chemical Soc*iety 127: 8286–8287.

Majsterek I., Sicinska P., Tarczynska M., Zalewski M. and Walter Z. 2004 "Toxicity of microcystin from cyanobacteria growing in a source of drinking water" *Comparative Biochemistry and Physiology-C Toxicology and Pharmacology* 139:175–179.

Makhluf S., Dror R., Nitzan Y., Abramovich Y., Jelinek R., and Gedanken A. 2005. "Microwave-assisted synthesis of nanocrystallineMgO and its use as a bacteriocide" *Advanced Functional Materials* 15:1708–1715.

Mandelbaum P., Regazzoni A., Belsa M., Bilme S.1999 "Photo-electron-oxidation of Alcohol on Titanium Dioxide Thin Film Electrodes" *Journal of Physics and Chemistry B* 103: 5505-5511.

Maximous N., Nakhla G., Wong K., Wan W., 2010 "Optimization of Al2O3/PES membranes for wastewater filtration" *Separation and Purification Technology* 73:294–301.

Melnyk I.V. and Zub Y.L. 2012 "Preparation and characterization of magnetic nanoparticles with bifunctional surface layer Si(CH2)3NH2/SiCH3 (or SiC3H7–n)"*Microporous Mesoporous Materials* 154:196–199.

Mitoraj D., Janczyk A. and Strus M. 2007 "Visible light inactivation of bacteria and fungi bymodifiedtitaniumdioxide" *Photochemical and Photobiological Sciences* 6:642–648.

Molinari R., Palmisano L., Drioli E., Schiavello M. 2002 "Studies on Various Reactor Configurations for Coupling Photocatalysis and Membrane Processes in Water Purification" *Journal of Membrane Science* 206:399–415.

Morones J.R., Elechiguerra J.L., Camacho A. 2005 "The bactericidal effect of silver nanoparticles" *Nanotechnology* 16:2346–2353.

Muller J., Huaux F. and Lison D. 2006 "Respiratory toxicity of carbon nanotubes: how worried should we be?" *Carbon* 44:1048–1056.

Musameh M., Hickey, M. and Kyratzis, I. 2011 "Carbon nanotube-based extraction and electrochemical detection of heavy metals" *Research on Chemical Intermediates* 37:675–689.

Nakamura, R., Okamura, T., Ohashi, N., Imanishi, A., Nakato, Y. 2005 "Molecular Mechanisms of Photo-induced Oxygen Evolution, PL Emission, and Surface Roughening at Atomically Smooth (110) and (100) *n*-TiO_2(Rutile) Surfaces in Aqueous Acidic Solutions". *Energy Environmental Sciences* 4:1946–1971.

Navarro Yerga R. M., Alvarez Galvan M. C., del Valle F., Villoriade la Mano J. A. and Fierro J. L. 2009 "Water Splitting on SemiconductorCatalysts under Visible-Light Irradiation" *Chem Sus Chem* 2: 471−485.

Ngomsik, A.-F., Bee, A., Talbot, D., Cote, G., 2012. "Magnetic solid-liquid extraction of Eu(III), La(III), Ni(II) and Co(II) with maghemite nanoparticles" *Separation and Purification Technology* 86:1–8.

Ohno T., Mitsui T. andMatsumura M. 2003. "Photocatalytic activity of S-doped TiO2 photocatalyst under visible light," *Chemistry Letters* 32:364–365.

Page K., Palgrave R.G., Parkin I.P., Wilson M., Savin S.L.P. and Chadwick A.V. 2007 "Titania and silver-titania composite films on glass—potent antimicrobial coatings" *Journal of Materials Chemistry* 17:95–104.

Pal S., Tak Y.K. and Song J.M. 2007 "Does the antibacterial activity of silver nanoparticles depend on the shape of the nanoparticle? A study of the gram-negative bacterium *Escherichia coli*" *Applied and Environmental Microbiology* 73:1712–1720.

Pan, B., Lin, D., Mashayekhi, H. and Xing, B., 2008 "Adsorption and hysteresis of bisphenol A and 17O ± -ethinyl estradiol on carbon nanomaterials" *Environmental Science and Technology* 42:5480–5485.

Parent Y., Blake D., Magrini-Bair K., Lyons C., Turchi C., Watt A., Wolfrum E., Praire M. 1996. "Solar Photocatalytic Process for the Purification of Water: State of Development and Barriers to Commercialization" *Solar Energy* 56: 429 437.

Pelizzetti E. and Minero C. 1994 "Metal oxides as photocatalysts for environmental detoxification" *Comments Inorganic Chemistry* 15: 297–337.

Pendergast, M.T.M., Nygaard, J.M., Ghosh, A.K., Hoek, E.M.V., 2010. "Using nanocomposite materials technology to understand and control reverse osmosis membrane compaction". *Desalination* 261: 255–263.

Peng, Y., Liu, H., 2006 "Effects of oxidation by hydrogen peroxide on the structures of multiwalled carbon nanotubes". *Industrial Engineering and Chemical Research* 45:6483–6488.

Primo A., Marino T., Corma, A.Molinari, Raffaele,Hermenegildo G. 2011 "Efficient Visible-Light Photocatalytic Water Splitting by Minute Amounts of Gold Supported on Nanoparticulate CeO_2 Obtained by a Biopolymer Templating Method" *Journal of the American Chemical Society* 133: 6930-6933.

Qi X., Li N., Xu Q., Chen D., Li H., Lu J. 2014 "Water-soluble Fe3O4 superparamagnetic nanocomposites for the removal of low concentration mercury(II) ions from water" *Royal Society of Chemistry Advances* 4:47643–47648.

Qu X., Alvarez P.J.J. and Li Q., 2013 "Applications of nanotechnology in water and wastewater treatment" *Water Res.* 47:3931–3946.

Rafiq Z., Nazir, R., Shah, M.R. and Ali, S., 2014 "Utilization of magnesium and zinc oxide nano-adsorbents as potential materials for treatment of copper electroplating industry wastewater" *Journal of Environmental Chemistry Engineering* 2:642–651.

Ramakrishna, S., Fujihara, K., Teo, W.-E., Yong, T., Ma, Z. and Ramaseshan, R., 2006. "Electrospun nanofibers: solving global issues"*Materials Today* 9:40–50

Rincon M.E., Trujillo-Camacho M.E. and Cuentas-Gallegos A.K 2005 "Sol-Gel Titanium Oxides Sensitized By Nanometric Carbon Blacks: Comparison with the Optoelectronic and Photocatalytic Properties of Physical Mixtures" *Catalysis Today* 107-108:606-611.

Roy A. and BhattacharyaJ. 2012 "Removal of Cu(II), Zn(II) and Pb(II) from water using microwave-assisted synthesized maghemite nanotubes" *Chemical Engineering Journal* 211–212:493–500.

Saquib M. and MuneerM. 2003 "TiO2-mediated Photocatalytic Degradation of a Triphenyl Methane Dye (Gentian Violet), in Aqueous Suspensions" *Dyes and Pigments* 56:37-49.

Shan S.J., Zhao Y., Tang H. and Cui F.Y. 2017 "A Mini-review of Carbonaceous Nanomaterials for Removal of Contaminants from Wastewater" *IOP Confenece of Series: Earth and Environmental Science* 68012003.

Shephard G.S., Stockenstrom S., Villiers D.D., Engelbrecht W.J. and Wessels G.F.S. 2002"Degradation of microcystin toxins in a falling film photocatalytic reactor with immobilized titanium dioxide catalyst" *Water Research* 36:140–146.

Shipley, H., Engates, K., Grover, V., 2013. "Removal of Pb(II), Cd(II), Cu(II), and Zn(II) by hematite nanoparticles: effect of sorbent concentration, pH, temperature, and exhaustion". *Environmental Science and Pollution Research* 20:1727–1736.

Shipway A. N., Katz E. and Willner I. 2000 "Nanoparticle Arrays on Surfaces for Electronic, Optical, and Sensor Applications" *ChemPhysChem*1:18–52.

Silva, C. G., Bouizi, Y., Fornes, V., Garcia, H. 2009 "Layered double hydroxides as highly efficient photocatalysts for visible light oxygen generation from water." *Journal of American Chemical Soc*iety 131:13833–13839.

Song C. and Corry B.2009 "Intrinsic ion selectivity of narrow hydrophobic pores" *Journal of Physical Chemistry B* 113:7642–7649.

Spadaro J.A., Berger T.J., Barranco S.D., Chapin S.E. and Becker R.O. 1974 "Antibacterial effects of silver electrodes with weak direct current" *Antimicrobial agents and chemotherapy* 6: 637–642.

Srinivasan S., Harrington J.W., Xagoraraki I. and Goel R. 2008 "Factors affecting bulk to total bacteria ratio in drinking water distribution systems" *Water Research* 42: 3393-3404.

Srivastava A.,Srivastava O.N., Talapatra S., R. Vajtai R. and Ajayan P.M., 2004 "Carbon nanotube filters" *Nature Materials* 3:610–614.

Sung-Suh H.M., Choi J.R., Hah H.J., Koo S.M. and Bae Y.C. 2004 "Comparison of Ag deposition effects on the photocatalytic activity of nanoparticulate TiO2 under visible and UV light irradiation" *Journal of Photochemistry and Photobiology A:Chemistry*163:37–44.

Tachibana, Y., Vayssieres, L., Durrant, J. R.2012 "Artificial Photosynthesisfor Solar Water-Splitting" *Natural Photonics* 6:511−518.

Tan, L., Xu, J., Xue, X., Lou, Z., Zhu, J., Baig, S.A., Xu, X., 2014. "Multifunctional nanocomposite Fe3O4@SiO2-mPD/SP for selective removal of Pb(II) and Cr(VI) from aqueous solutions" *Royal Society of Chemistry Advances* 4:45920–45929.

Taurozzi J.S., Arul H., Bosak V.Z. 2008 "Effect of filler incorporation route on the properties of polysulfone-silver nanocomposite membranes of different porosities" *Journal of Membrane Science* 325:58-68.

Tong H., Ouyang S.; Bi Y., Umezawa N., Oshikiri, M. and Ye J. 2012 "Nano-Photocatalytic Materials: Possibilities and Challenges" *Advanced Materials* 24:229−251.

Tran P. D., Wong L. H., Barber J. and Loo J. S. C. 2012 "Recent Advancesin Hybrid Photocatalysts for Solar Fuel Production" *Energy Environmental Science* 5:5902−5918.

Trivedi, P. and Axe, L. 2000 "Modeling Cd and Zn sorption to hydrous metal oxides"*Environmental Science and Technology* 34:2215–2223.

Tsuru T.,Toyosada T., Yoshioka T., Asaeda M. 2003 "Photocatalytic Membrane Reactor using Porous Titanium Dioxide Membranes"*Journal of Chemical Engineering Japan* 36: 1063–1069.

Tu Y.J., You C.F., Chang C.-K., Wang S.L. and Chan T.-S. 2012 "Arsenate adsorption from water using a novel fabricated copper ferrite" *Chemical Engineering Journal* 198–199:440–448.

Umebayashi T., Yamaki T., Itoh H. and Asai K. 2002 "Band gap narrowing of titanium dioxide by sulfur doping" *Applied Physics Letters*81:454–456.

Vamathevan V., R. Amal R., BeydounD., Low G., and McEvoy S.2004 "Silvermetallisation of titania particles: effects on photoactivity for the oxidation of organics" *Chemical Engineering Journal* 98:127–139.

Vukovic G.D., Marinkovic A.D., Colic M., Ristic M.D.., Aleksic R., PericGrujic A.A. andUskokovic P.S. 2010 "Removal of cadmium from aqueous solutions by oxidized and ethylenediamine-functionalized multi-walled carbon nanotubes" *Chemical Engineering Journal* 157:238–248.

Wang B., Zhang G., Leng X., Sun Z., Zheng S. 2015 "Characterization and improved solar light activity of vanadium doped TiO2/diatomite hybrid catalysts" *Journal of Hazardous Material* 285:212-220.

Wang H. and Lewis J.P. 2005 "Effects of dopant states on photoactivity in carbon-doped TiO2" *Journal of Physics Condensed Matter* 17:209–213.

Wang H., L. Zhang L., Chen Z., J. Hu J., Li S., Wang Z., Liu J. and Wang X. 2014 "Semiconductor heterojunction photocatalysts: design, construction, and photocatalytic performances" *Chemical Society ofRev*iews 43:5234–5244.

Wang, Y., Wang, Q., Zhan, X., Wang, F., Safdar, M. and He, J. 2013 "Visible Light Driven Type II Heterostructures and Their Enhanced Photocatalysis Properties: A Review"*Nanoscale* 5:8326−8339.

Wright J., Gundry S. and Conroy R. 2004"Household drinking water in developing countries: a systematic review of microbiological contamination between source and point of use" *Tropical Medicine and International Health* 9:106–117.

Xiang Q. and Yu J. 2013 "Graphene-Based Photocatalysts for HydrogenGeneration" *Journal of Physics Chemisrty and Letters 4*:753−759.

Xiang Q., Yu J. and Jaroniec M. 2012 "Graphene-Based Semiconductor Photocatalysts" *Chemical Society ofReviews* 41:782−796.

Xiu Z.M., Ma J. and Alvarez P.J.J. 2011 "Differential effect of common ligands and molecular oxygen on antimicrobial activity of silver nanoparticles versus silver ions" *Environmental Science and Technology* 45:9003–9008.

Xiu Z.M., Zhang Q.B., Puppala H.L., Colvin V.L. and Alvarez P.J.J. 2012 "Negligible particle-specific antibacterial activity of silver nanoparticles" *Nano Letters* 12:4271–4275.

Yang K. and Xing B. 2010 "Adsorption of organic compounds by carbon nanomaterials in aqueous phase: polanyi theory and its application" *Chemical Reviews* 110:5989–6008.

Yerga, R. M. N., Galvan, M. C. A., del Valle, F., de la Mano,J. A. V., Fierro, J. L. G. 2009 "Water splitting on semiconductor catalysts under visiblelight irradiation" *Chemsumchem* 2:417-425.

Yu D.G., Teng M.Y., Chou W.L. and Yang M.C. 2003 "Characterization and inhibitory effect of antibacterial PAN-based hollow fiber loaded with silver nitrate" *Journal of Membrane Science* 225:115–123.

Yu J., Low J., Xiao W., Zhou P., Jaroniec M., 2014 "Enhanced photocatalytic CO_2-reduction activity of anatase TiO_2 by coexposed {001} and {101} facets" *Journal of American Chemical Soc*iety 136: 8839–8842.

Yu J.C., Yu J., Ho W., Jiang Z. and Zhang L. 2002 "Effects of Doping on the photocatalytic activity and microstructures of nanocrystalline TiO2 powders" *Chemistry of Materials*14:3808–3816.

Zan L., Fa W., Peng T. and Gong Z.K. 2007 "Photocatalysis effect of nanometer TiO2 and TiO2-coated ceramic plate on HepatitisB virus" *Journal of Photochemistry and Photobiology B: Biology* 86:165–169.

Zhang W.X. 2003 "Nanoscale iron particles for environmental remediation: an overview"*Journal of Nanoparticles Research* 5:323–332.

Zhang X., Zhou M. and Lei L. 2005 "Preparation of an Ag-TiO2 photocatalyst coated on activated carbon by MOCVD" *Materials Chemistry and Physics* 91:73–79.

Zhao G. and Stevens S.E. 1998 "Multiple parameters for the comprehensive evaluation of the susceptibility of *Escherichiacoli*to the silver ion" *BioMetals* 11: 27–32.

Zhou P., Yu J. and Jaroniec M. 2014 "All-solid-state Z-scheme photocatalytic systems" *Advance Mater*ials 26:4920–4935.

Zhu, J. F., Zach and M. Cur. 2009 "Nanostructured materials for photocatalytic hydrogen production" *Opinion on Colloid and Interface Sciences* 14:260–269.

Zodrow K., Brunet L., Mahendra S. 2009 "Polysulfone ultrafiltration membranes impregnated with silver nanoparticles show improved biofouling resistance and virus removal" *Water Research* 43:715–723.

In: Photocatalysis
Editors: P. Singh, M. M. Abdullah, M. Ahmad et al.
ISBN: 978-1-53616-044-4

Chapter 9

METAL OXIDES BASED PHOTOCATALYST FOR THE DEGRADATION OF ORGANIC POLLUTANTS IN WATER

Azar Ullah Mirza, Abdul Kareem, Shahnawaz Ahmad Bhat, Fahmina Zafar and Nahid Nishat*
Inorganic Material Research Laboratory, Department of Chemistry,
Jamia Millia Islamia, New Delhi, India

ABSTRACT

Due to environmental importance, effective measures are required for the removal of organic contaminants from waste water. Traditional methods of removal include coagulation, adsorption, membrane separation and secondary pollutant generation. In the past, a class of metal oxide catalyst has been synthesized due to a wide range of applications such as antimicrobial, anticancer and sensor etc. Today use of metal oxide based photocatalyst has gained much attention because of its unique characteristic properties such as optical, electronic, electrical, thermal and emerged out as a most efficient degradation technique for the removal of organic pollutants. Recently, major problems of water pollutants aredyes from various sources such as paper, craft, tannery, pulp and pharmaceutical textile industries. Although the removal of organic pollutant prior to discharge into the environment is important. Organic pollutants (lindane, dichlorodiphenyltrichloroethane, industrial chemicals, fertilizers, pesticides and dyes) have been introduced into water resources. It is hazardous to the biotic organism. Photocatalyst has emerged a promising technology for waste water treatment and it becomes an important part of the degradation process. Photocatalytic degradation is the

*Corresponding Author's E-mail: nishat_nchem08@yahoo.com.

best protocol for waste water treatment. There are various techniques used for the degradation of organic contaminants such as solvent extraction, adsorption, photodegradation and the most advanced technology is photocatalytic degradation of organic pollutant. This chapter represents the history, principle and photocatalytic degradation of organic pollutants by using different metal oxide based materials.

Keywords: metal oxide, photocatalytic degradation, organic pollutants, waste water

1. Introduction

The harmful industrial liquid or gaseous effluents cause health problems in human beings. Some classes of dyes are toxic and carcinogenic that are disastrous and pose several threats to the biotic life.Wastewaters from laboratories, industries, factories, etc. creates problems to the environment. The discharged waste of water contains harmful dyes which are toxic to living organisms (Borker and Salker AV. 2006).Tin,zinc,cerium,titanium,vanadium and chromium are transition metal oxide having photocatalytic characteristics which induces a charge separation process (positive holes) are due to oxidation of organic substrates (Hisatomiet al. 2014; Hoffmann et al. 1995;Fu et al. 1996).They are important class of semiconducting materials which have application in solar energy transformation, gas, electronic, catalysis, sensors and magnetic storage devices. Semiconductor materials which have unique magnetic, optical and electronic properties (Ramgiret al. 2013; Jani 2013; Shalanet al. 2013; De Montferrand et al. 2013;Ahmadi 2011). Although, several methods were used for the preparation of suitable nanostructured catalysts and optimize the wavelength of light exposure depend on the selection of semiconducting material with suitable band gap energy, enhance the pollutant degradation from waste water (Huet al. 1999; Suriet al. 1993; Liu et al. 2012). On the other hand, photocatalysts have vast potential for the removal of pollutants from waste water (Kanakarajuet al. 2014; Chong et al. 2010; Pincella et al. 2014).Photocatalysis have gained considerable attention from the science community as the important way to solve the environment problemsmainly dispose of residual pollutants from waste water. The photocatalysts involved in a photochemical reactions at the surface of metal oxide whichconsists of at least two main types of reactions simultaneously, the first step is oxidation occur from photo-induction of positive holes and second step is reduction from negative photo induced electrons (Fujishima et al. 2008).

Prior treatment of disposal is an important for discharged ofindustrial effluents contains toxic organic chemicals.A toxic non-biodegradable 4-chlorophenol is a known endocrine disruptor, which present in waste as a byproduct of paper, pulp, dyestuff, agrochemical and pharmaceutical industries (Keith 1998; Theurich et al. 1996).The characteristic features of metal oxide such as chemical stability, redox potential, compatibility of growth over various supports, suitability of electronic band structure and non-toxicity, boost the photocatalytic degradation process to remove such chemicals. The metal oxides obtained from different precursors (such as metal salts, metal acetates, metal foils and metal alkoxides) have diversity of morphologies that effect the photocatalytic degradation of dyes.

Effective removal of pollutants from waste water by conventional treatment methods such as activated charcoal adsorption, chemical precipitation and ion-exchange method.Biodegradation of 4-chlorophenol is not a complete and slow but also byproducts formed are much more toxic than the contaminated product. However, treatment is required because of the transfer of contaminants from one media to another media (Jain et al. 2004; Chen D and Ray AK. 1999; Sherrardet al. 1996).Water pollutants are most dangerous for aquatic life and also creates a bad impact on the ecosystem. Although, the degradation of pollutants from water are essentially importantto prevent the negative effects on the environment and human health. Dyes are one of the dangerous class of pollutants in water. Water is not safe for drinking and also it is difficult for the treatment of aquatic pollutants (Forgacset al. 2004; Rai et al. 2005).

2. Historical Background of Photocatalysis

Fujishima and Honda in 1972 discovered the photocatalysis of water on titanium oxide electrode.Therefore, the history of photocatalyzed materials by photocatalytic degradation methods can be traced back in 1960 and is considered as a landmark event that stimulated the evaluation of conversion of photonic energy (Fujishima and Honda 1972; Tong et al. 2012; Muller and Steinbach 1970; Steinbach 1969; Doerffler and Hauffe 1964). Teoh and Co-workers have carried out the investigation of photocatalytic degradation processes of model compounds and their mechanism. The degradation process can strongly affect the photocatalytic performance of materials because of direct effect on structure of compound. Studies show that the degradation of dyes in solution using various analytical techniques (Ultraviolet-Visible spectroscopy, Fourier transform infrared spectroscopy, Liquid chromatography-mass spectrometry etc.) is convenient and simple. Using an organic dye as a model compound can give unreliable results as this may not shows the intrinsic photocatalysis of semiconducting material, mainly for materials in visible light activated materialare good examples of such dyes (cango red, alizarin yellow and methyl orange) (Teoh et al. 2012).Based on the literature, a photocatalytic application of titanium oxide nanoparticles have been achieved since the end of the 1990s and development in the 21st century. Titanium oxide nanoparticles have gainedmuch attention in comparison with the bulk materials. Due to large surface-area, titanium oxide nanostructures during the photo-induced reactions, enhanced the light absorption rate, increase the photo-induced carrier density and leads to higher surface photoactivity. Nanomaterialsincreasesphotocatalytic reaction rate, because of high surface-volume ratio, increases the surface absorption of OH^- and H_2O (Lan et al. 2013).

3. Principle behind Photocatalysis

Photocatalytic degradation is a field which consists of number of reactions such as synthesis of organic compounds, photoreduction, transference of hydrogen, splitting of water, deposition of metal, removal of gaseous pollutants, detoxification of water, anticancer therapy, disinfectionand isotopic exchange (Herrmann 1999; Carp et al. 2004). Metal such as gold, silver and iron were used for the photocatalytic degradation of pollutants. Among these metal oxides based photocatalytic degradation has an alternative for thephotocatalytic degradation of water.In Photocatalytic degradation, molecules are in direct contact with the surface of catalyst. Photo catalytic reaction is initiated when excitation of photoelectron takes place from valence band (filled) to conduction band (empty), the absorption of energyis equal to or more of semiconductor photocatalyst, leaving behind a hole in the valence band. There is a generation of hole and a pair of electron.

There are various reactions which shown below:

Photocatalyticexcitation: Metal oxide/Photocatalyst + hv$\rightarrow e^+ + h^+$ (1)

Ionosorption of oxygen: $O_2(ads) + e^- \rightarrow (O_2^{\cdot-})$ (2)

Water Ionization: $H_2O \rightarrow H^+ + OH^-$ (3)

Super oxide protonation: $(O_2^{\cdot-})$, + H$\rightarrow$ $HOO^{\cdot}$ (4)

The scavenging activity of hydroperoxyl radical in (IV) as O_2 has prolonged twice the photonhole life time:

$HOO^{\cdot} + e^- \rightarrow HO_2$

$HOO^- + H^+ \rightarrow HO_2$

At the surface of photo excited catalyst, the oxidation and reduction takes place is shown in Figure 1. The recombination of electron and hole occurs unless when the availability of oxygen to scavenge the electron to obtained a superoxide ($O_2^{\cdot-}$), the hydroperoxyl radical ($HO_2^{\cdot}$) and Hydrogen peroxide (H_2O_2), a protonated form (Dung 2005; Habibi and Vosooghian 2005).

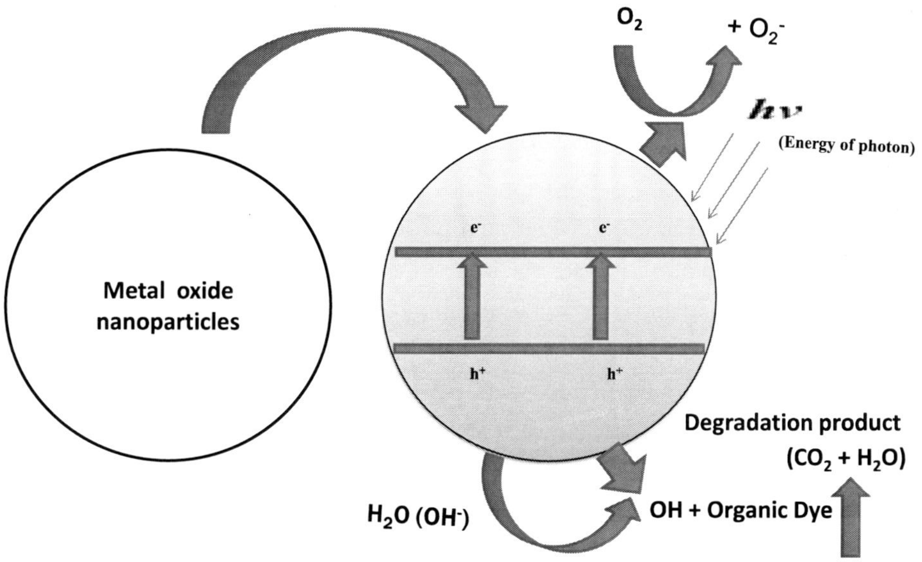

Figure 1. Schematic diagram of photodegradationofdyein solar irradiation illustrating principle of photocatalysis.

4. PHOTOCATALYTIC DEGRADATION OF ORGANIC POLLUTANTS

The photocatalytic degradation is a significant method in the area of waste water treatment containing organic contaminants. Thismethod has several importance such as cheap, free from waste disposal problem, necessity of mild and complete mineralization (Legrini et al. 1993; Mills et al. 1993). Ullahet al.(2008) synthesized manganese pure and doped photocatalyst via wet-chemical techniques. This photocatalyst can be used for the degradation of organic contaminants with irradiation of UV-Visible light. The evaluation of photocatalytic reduction using a basic dye (aniline and methylene blue).Tayadeet al.(2006) prepared mesoporous TiO_2 from titanium isopropoxide by wet impregnation method and modified the synthesis composite bydifferent metal ions (Fe, Co, Ni and Ag) having different importance. Later the synthesized composite were used for the photocatalyticdegradation of organic contaminants (nitrobenzene and acetophenone) present in the aqueous solution under ultraviolet light irradiation.Zang et al. (2014) prepared microparticles of $CuFeO_2$ and used them as a fenton like catalyst. The prepared nanoparticles is rhombohedral with size ranges from 2-3 μm and show photocatalytic degradation of Bisphenol A. Guo et al. (2013) synthesized heterogeneous catalyst graphene-Fe_2O_3 for photo degradation of organic pollutants via impregnation technique.Degradation of 4-nitrophenol and Rhodamine B takes place under ultraviolet light in the presence of hydrogen peroxide. It is best catalyst for the degradation of visible light and water treatment.Ahmed et al. (2013) successfully prepared Fe_2O_3/TiO_2 nanoparticles by sol-gel method. This catalyst is used for the degradation ofmethyleneblue.TheFe_2O_3/TiO_2nanoparticles can be used for photodegradation and adsorption process and is efficient photocatalyst for the degradation of organic pollutants. Asif et al. (2014) prepared CO_3O_4/Fe_2O_3nanofibers based solar photocatalyst and is used forthe degradation of brilliant cresol blue and acridine orange dyes under solar light. TheCO_3O_4/Fe_2O_3nanocompositewould be the best candidate and efficient in solar photocatalysis for the degradation of degradation of toxic contaminants. Liu et al. (2012) produced Fe/TiO_2nanocatalyst by sol-gel process and calcination followed by chemical reductive deposition of iron. TheFe/TiO_2catalyst were tested for the degradation of chlorophenols (2,4-dichlorophenol) and total organic carbon removal.Tadjarodi et al.(2015) synthesized copper oxide nanoparticles and explained theirmesoporous structure. Copper oxide nanoparticles were characterized by FT-IR, XRD, SEM and DRS and is applied for the photocatalytic degradation of Rhodamine B dye in waste water.Zhu et al. (2011) successfully synthesized cuprous micro/nanocrystals by microwave irradiation liquid phase reduction method. The materials were characterized by XRD, UV-Visible/DRS, SEM, TEM and HRTEM show various morphologiesincludes truncated, octahedral, truncated cubic and cubic microcrystalsand applied for the photocatalytic degradation against Rhodamine B dye. Khorshidi et al. (2016) produced Ag/ZnO/CuOnanocomposite and evaluation of photocatalytic activity from waste water against basic violet 16 and red 18 dyes as an organic contaminants. Motahari et al. (2014) synthesized nickle oxide nanostructured materials by thermal decomposition method. Studies show that nickle oxide nanomaterials act as a promising photocatalyst for the degradation of Rhodamine B dye in waste water.Gupta et al. (2017) synthesized iron oxide nanoparticles in two forms like goethite (α-FeOOH) and magnetite (Fe_3O_4). The particles were characterized and applied for the degradation of anthracene in presence of UV light.

Currently, researchinphotocatalytic degradation has evaluation of several nanophotocatalyst and shows their physiochemical properties. A photocatalyst has photocorrosion activity, chemical, biological inertness and low toxicity. Nanomaterials are used mostly in every sector of life such as energy production, electronics, environmental issues, industry, drugs, materials, modification and manufacturing. The nanomaterials applied inphotocatalysis have been used in the degradation of pollutants from water.Some metal oxide nanomaterials are SiO_2, MgO, TiO_2, CaO, ZnO and CeO_2 (Khan et al. 2017;Hirota et al. 2010;Wei et al. 2009).They show better properties and excellent photocatalytic degradation of organic pollutants.Many studies have been carried out to measurethephotocatalytic degradation of organic pollutants in water in the presence of metal oxide nanomaterials is given in Table 1.

Table 1. Shows that some nanomaterials used in photocatalytic degradation

S. NO.	Photocatalyst	Pollutant	Percentage of dye degradation	Reference
01	Cu_2O	Methyl orange	97%	(Yang et al. 2006.)
02	TiO_2	Acid Blue 80	-	(Su et al. 2008.)
03	TiO_2	Tetracycline	-	(Wang et al. 2017.)
04	WO_3	Methylene Blue	-	(Ofori et al. 2015.)
05	ZnO-Carbon	Rhodamine B	-	(Li et al2013.)
06	TiO_2	Methyl orange	-	(Ruan 2001.)
07	CuO	Crystal violet	97%	(Vaidehi 2018.)
08	NiO	Acid red 1	-	(Song X and Gao L. 2008.)
09	$CO_3O_4/BiVO_4$	Phenol	-	(Long et al. 2006.)
10	TiO_2/Graphene Oxide	Methylene Blue	-	(Zhang et al. 2011.)
11	α-Fe_2O_3	Congo red	95.4%	(Satheesh et al. 2016.)
12	γ-Fe_2O_3	Acridine orange	98%	(Qadri et al. 2009.)
13	Fe_3O_4	Crystal violet	-	(Muthukumaran et al. 2016.)
14	ZnO	Phenol	88%	(Parida KM and Parija S. 2006.)
15	CuO/ZnO	Acid red 88	-	(Sathishkumar et al. 2011.)
16	TiO_2	4-chlorophenol	-	(Venkatachalam et al. 2007.)
17	Pr-TiO_2	Phenol	50 and 90%	(Chiou CH and Juang RS. 2007.)
18	ZnO	4-nitrophenol	92 and 98%	(Parida KM, Dash SS and Das DP. 2006.)
19	Ag/TiO_2	2,4,6 Trichlorophenol	95%	(Rengaraj S and Li XZ. 2006.)
20	Au/TiO_2	Methylene Blue	-	(Thomas J and Yoon M. 2012.)
21	Au/TiO_2	Orange 16	-	(Hsiao et al. 2011.)
22	ZnO	Methyl orange	99.70%	(Chen et al. 2017.)
23	CuO	Methylene Blue	72.57%, 93.48%, 49.71and 95.71%.	(Meshram et al. 2012.)
24	Nickle dioxide	Crystal violet	97%	(He et al. 2010.)
25	ZnO-CeO_2	Acridine orange	84.55 and 48.66%	(Faisal et al. 2011.)
26	Sn-ZnO	Methylene blue	13 and 29-52%	(Sun et al. 2011.)
27	TiO_2	Methylene blue	90%	(Dariani et al. 2016.)

4.1. Iron Oxide Materials

Iron oxide nanomaterials are an n-type semiconducting material and has been used as good photocatalyst for the photocatalytic degradation under visible light. Comparison with titanium oxide absorbs ultraviolet light with wavelength less than 3380 nm covers 5 percent of the solar spectrum because of 3.2 eV band-gap and (Fe_2O_3) band-gap is 2.2 eV (Akhavan et al. 2009). Haung et al. (2015) synthesized nanomaterials (Fe_3O_4) phanerochaetechrysosporium with oxalate secretion was developed for degradation of phenol under light. The maximum degradation efficiency is 93.41% under sunlight while 40.36% under dark. It could be useful for the efficient treatment of phenol in waste water. Wang et al. (2010)synthesized magnetic nanoparticles (Fe_3O_4) with enhancement in peroxidase-activity by ultrasound assisted reverse co-precipitation method and characterize with BET,XRD and SEM. Production of nanomaterials possess small size and peroxidase activity of magnetic nanomaterials were evaluated and assessment of photocatalytic degradation of Rhodamine B. Chen et al. (2017) synthesized iron oxide magnetic (Fe_3O_4) nanomaterialsand is most promising heterogeneous catalyst for the oxidative degradation of organic pollutants by oxidative combined isothermal method for the degradation of Rhodamine B and other pollutants in water.

4.2. Titanium Oxide Materials

Titanium oxide nanoparticles are the emerging and most important photocatalysts for water purification (Adesina 2004; Li et al. 2008).The basic mechanism of a semiconductorbasedphotocatalyst having good photocatalytic activity, non-toxicity and low cost (Hashimoto et al. 2005). Titanium oxide involves the production of reactive oxidants (OH radicals) for disinfection of microorganisms, fungi, bacteria, viruses and algae etc. (Einaga1999; Fujishima et al. 2000; Shephard2002; Ibanez et al. 2003; Liu and Yang TC. 2003; Cho et al. 2004; Cho et al. 2005).The photocatalytic ability of titanium oxide is enhanced under ultraviolet and drastically increased in the visible-light region (Zan et al. 2007). Bhatkhande et al. (2003) studied titanium dioxide as a catalyst for the photocatalytic degradation of nitrobenzene. Degradation of nitrobenzene in different factors for ascertaining the penetration effect of photon depth in the degradation process. Ray et al. (2009) synthesized titanium oxide for the photocatalytic degradation of phenol in aqueous solution. Using Box-Benkhen technique four parameters were observed such as catalyst size, concentration of titanium dioxide phenol concentration and dissolved oxygen concentration. Titanium oxide is 10 nm and applied for the photocatalytic degradation of phenol with activation energy is 13.55kJ/mol and obeyed Arrhenius dependence.

4.3. Copper Oxide Materials

Cuprous oxide (Cu_2O) is a promising semiconducting material used for the conversion of solar energy into electrical and chemical energy. These materials have a direct band gap of 2-2.2 eV vacancies (0.4 eV) above the valence band (Nian et al. 2008; De Jongh et al. 1999; Fernando and Wetthasinghe 2000; Akimoto et al. 2006; Yang and Zhu 2003; De Jongh 1999). Zang et al. (2010) synthesized cuprous oxide (Cu_2O) microcubes and nanoparticles by reduction of copper sulphate using ascorbic acid at room temperature. The microscale materials show high photocatalytic degradation of methyl orange by visible light. Copper oxide microcubes have high photodegradation of methyl orange (98.1%) and photocatalytic activity. On the other hand Meshram et al. (2012) synthesized cupric oxide nanostructures by chemical precipitation and hydrothermal methods having different morphologies such as vesicular, platelet and nanosheet. The structures were characterized and applied for the degradation of methylene blue with different degradation rates (93.48%, 95.71%, 72.57% and 49.71%).

4.4. Nickle Oxide Materials

Nickle oxide (NiO) nanoparticles is a crucial inorganic material and have great attention because of potential properties and applications that can be grown and used in many fields such as lithium ion batteries, capacitors and solar cell (Wu et al. 2007; Rai et al. 2013; Zhang2004;Bandara and Weerasinghe 2005). Hayat et al. (2011) synthesized nickle oxide nanoparticle by sol-gel method for the photocatalytic oxidation of phenol under irradiation of UV laser source. The degradation of phenol show a pseudo first order rate kinetics. Ajoudanian et al. (2015) synthesized nickle oxide nanoparticle by incorporation of clinoptilolite by ion exchange and calcination. The synthesized catalysts were used for the photocatalytic degradation of cephalexin. The degradation efficiency is 73.5% and 89%.

4.5. Zinc Oxide Materials

A zinc oxide nanomaterial has been extensively used in photocatalysis (Yumoto et al. 1999), solar cells (Hara et al. 2000) and sensors (Sberveglieri et al. 1995; Rodriguez et al. 2000) and is semiconducting material with band gap of 3.7 eV and large excitation binding energy of 60 meV. Zinc oxide can be used for the degradation of organic pollutants in nearly neutral solution and sterilization of viruses and bacteria (Sato et al. (1998); Ye et al. 2006). Semiconductor photocatalysts on a nanometer-scale become attractive due to their unique physiochemical properties as compared to the bulk materials (Curri et al. 2003; Hariharan 2006). Shen et al. (2008) synthesized zinc oxide nanoparticles by deposition of zinc hydrate on the surface of silica nanoparticles and dispersion of zinc hydrate in starch gel. The materials were characterized and size is less than 50 nm and 60 nm. The 60 nm was produced from starch gel and show poor photocatalytic activity while the other produced from deposition of silica nanoparticle displayed higher photocatalytic activity. Pardeshi et al. (2008) examined the zinc oxide mediated solar photocatalytic degradation of phenol from waste water. Photocatalytic degradation of phenol were found to be more effective under sunlight as compared to visible light irradiation and occurs in neutral or weakly acidic solutions. It can be reused for photocorrosion.

4.6. Cobalt Oxide Materials

Cobalt oxide (CO_3O_4) is an important antiferromagnetic semiconductor material with utility in ceramic pigments, gas sensing, catalysts and lithium ion batteries (Warang et al. 2012; Casas-Cabanas et al. 2009; Chou et al. 2008; Li et al. 2005; Sugimoto and Matijevic et al. 1979). Warang et al. (2012) prepared cobalt oxide nanoparticles by reactive pulsed laser deposition (nanoparticle assembled thin coated catalyst) method. Photocatalytic degradation of methylene blue under visible light irradiation. The reusability of cobalt oxide catalyst were evaluated. Huang et al. (2014) synthesized micropancake-like and micro flower like cobalt oxide (Co_3O_4) nanoparticles via aqueous solution template free route combination with thermal treatment. The cobalt oxide nanoparticles were used as a photocatalyst for the degradation of methyl orange, p-nitrophenol, eosin B, Rhodamine B and methylene blue.

CONCLUSION

Development of metal oxide based material is most promising for photocatalytic degradation of phenolic compounds and organic pollutantsunder solar/ultraviolet visible lightirradiation in waste water resources.Therefore, this chapter shows the brief overview, history and mechanism of metal oxide based photocatalytic materials. Metal oxide based photocatalyst have some advantages for degradation of organic pollutants because of versatile synthesis methods (precipitation, sol-gel, solvothermal, hydrothermal and ultrasonication etc.), well defined structure and property, modular nature of metal oxide synthesis, structural features of active sites leads to more efficiency in harnessing of solar energy, porosity and high surface area can facilitate the diffusion of pollutant molecules through catalytic efficiency and better visible response of light.Variety of chemical additives and techniques has been used to reduce the charge carrier recombination loss and extended the spectral result of metal oxide using metal, non-metal doping, codoping and semiconductor coupling of metal oxide.Cheap starting material availability, feasibility, good stability, easy recycling, efficiency of formed products with high purity, yield and unique properties can be considered as best material for various applications in different fields. The use of solar radiations has to be the improvement by virtue of designing a photo reactor in order to minimize the cost treatment.These photocatalyst may be considered as an advancement in large-scale utilization of heterogenous catalyst under visible light to address contamination of water and environmental pollution.

ACKNOWLEDGMENTS

The Authors would like to express our thanks to head department of chemistry, JamiaMilliaIslamiafor providing necessary facilities. AzarUllahMirzawould like thanks to UGC-New Delhi forNon-net Fellowship.Dr. Fahmina Zafar is thankful to DST New Delhi for Postdoc Fellowship.

REFERENCES

Adesina AA. 2004. "Industrial exploitation of photocatalysis: progress, perspectives and prospects."*Catalysis Surveys from Asia*. 8(4):265-273.

Ahmadi SJ, Outokesh M, Hosseinpour M and Mousavand T. 2011. "A simple granulation technique for preparing high-porosity nano copper oxide (II) catalyst beads." *Particuology*. 9:480-485.

Ahmed MA, El-Katori EE and Gharni ZH. 2013. "Photocatalytic degradation of methylene blue dye using Fe_2O_3/TiO_2 nanoparticles prepared by sol-gel method." *Journal of Alloys and Compounds*. 553:19-29.

Ajoudanian N and Nezamzadeh-Ejhieh A. 2015. "Enhanced photocatalytic activity of nickel oxide supported on clinoptilolite nanoparticles for the photodegradation of aqueous cephalexin." *Materials Science in Semiconductor Processing*. 36:162-169.

Akhavan O and Azimirad R. 2009. "Photocatalytic property of Fe_2O_3nanograin chains coated by TiO_2nanolayer in visible light irradiation." *Applied Catalysis A: General*. 369(1-2):77-82.

Akimoto K, Ishizuka S, Yanagita M, Nawa Y, Paul GK and Sakurai T. 2006. "Thin film deposition of Cu_2O and application for solar cells." *Solar energy*. 80:715-722.

Asif SA, Khan SB and Asiri AM. 2014. "Efficient solar photocatalyst based on cobalt oxide/iron oxide composite nanofibers for the detoxification of organic pollutants." *Nanoscale research letters*. 9:510.

Bandara J and Weerasinghe H. 2005. Solid-state dye-sensitized solar cell with p-type NiO as a hole collector. *Sol Energy Mater Sol Cells*. 85:385-390.

Bhatkhande DS, Pangarkar VG and Beenackers AA. 2003. "Photocatalytic degradation of nitrobenzene using titanium dioxide and concentrated solar radiation: chemical effects and scaleup." *Water Research.* 37(6):1223-1230.

Borker P and Salker AV. 2006. "Photocatalytic degradation of textile azo dye over Ce1-xSnxO_2series."*Materials Science and Engineering: B.* 133:55-60.

Carp O, Huisman CL and Reller A. 2004. "Photoinduced reactivity of titanium dioxide." *Progress in solid state chemistry.* 32:33-177.

Casas-Cabanas M, Binotto G, Larcher D, Lecup A, Giordani V and Tarascon JM. 2009. "Defect chemistry and catalytic activity of nanosized Co_3O_4."*Chemistry of materials.*21(9):1939-1947.

Chen D and Ray AK. 1999. "Photocatalytic kinetics of phenol and its derivatives over UV irradiated TiO_2."*Applied Catalysis B: Environmental.* 23:143-157.

Chen F, Xie S, Huang X and Qiu X. 2017. "Ionothermal synthesis of Fe_3O_4 magnetic nanoparticles as efficient heterogeneous Fenton-like catalysts for degradation of organic pollutants with H_2O_2." *Journal of hazardous materials.* 322:152-162.

Chen X, Wu Z, Liu D and Gao Z. 2017. "Preparation of ZnOphotocatalyst for the efficient and rapid photocatalytic degradation of azo dyes."*Nanoscale research letters.* 12(1):143.

Chiou CH and Juang RS. 2007. "Photocatalytic degradation of phenol in aqueous solutions by Pr-doped TiO_2 nanoparticles."*Journal of Hazardous Materials.* 49:1-7.

Cho M, Chung H, Choi W and Yoon J. "Linear correlation between inactivation of E. coli and OH radical concentration in TiO_2photocatalytic disinfection." 2004. *Water research.* 38 (4):1069-1077.

Cho M, Chung H, Choi W and Yoon J. 2005. "Different inactivation behaviors of MS-2 phage and Escherichia coli in TiO_2photocatalytic disinfection."*Applied and environmental microbiology.* 71(1):270-275.

Chong MN, Jin B, Chow CW and Saint C. 2010. "Recent developments in photocatalytic water treatment technology: a review."*Water research.* 44:2997-3027.

Chou SL, Wang JZ, Liu HK and Dou SX. 2008. "Electrochemical deposition of porous Co_3O_4 nanostructured thin film for lithium-ion battery."*Journal of Power Sources.* 182:359-364.

Curri ML, Comparelli R, Cozzoli PD, Mascolo G and Agostiano A. 2003. "Colloidal oxide nanoparticles for the photocatalytic degradation of organic dye."*Materials Science and Engineering: C.* 23:285-289.

Dariani RS, Esmaeili A, Mortezaali A and Dehghanpour S. 2016. "Photocatalytic reaction and degradation of methylene blue on TiO_2nano-sized particles."*Optik-International Journal for Light and Electron Optics.* 127(18):7143-7154.

De Jongh PE, Vanmaekelbergh D and Kelly JJ. 1999. "Cu_2O: a catalyst for the photochemical decomposition of water."*Chemical Communications.* 12:1069-70.

De Jongh PE, Vanmaekelbergh D and Kelly JJ. 1999. "Cu_2O: electrodeposition and characterization."*Chemistry of materials*. 11:3512-3517.

De Montferrand C, Hu L, Milosevic I, Russier V, Bonnin D, Motte L, Brioude A and Lalatonne Y. 2013. "Iron oxide nanoparticles with sizes, shapes and compositions resulting in different magnetization signatures as potential labels for multiparametric detection." *Actabiomaterialia*. 9:6150-6157.

Doerffler W and Hauffe K. 1964. Heterogeneous photocatalysis II. "The mechanism of the carbon monoxide oxidation at dark and illuminated zinc oxide surfaces." *Journal of Catalysis*. 3(2):171-178.

Dung NT, Van Khoa N and Herrmann JM. 2005. "Photocatalytic degradation of reactive dye RED-3BA in aqueous TiO_2 suspension under UV-visible light." *International Journal of Photoenergy*.7:11-15.

Einaga H, Futamura S and Ibusuki T. 1999. "Photocatalytic decomposition of benzene over TiO_2 in a humidified airstream."*Physical Chemistry Chemical Physics*. 1(20):4903-4908.

Faisal M, Khan SB, Rahman MM, Jamal A, Akhtar K and Abdullah MM. 2011. "Role of ZnO-CeO_2 nanostructures as a photo-catalyst and chemi-sensor." *Journal of Materials Science & Technology*. 27(7):594-600.

Fernando CAN and Wetthasinghe SK. 2000. "Investigation of photoelectrochemical characteristics of n-type Cu_2O films." *Solar energy materials and solar cells*. 63(3):299-308.

Forgacs E, Cserhati T and Oros G. 2004. "Removal of synthetic dyes from wastewaters: a review." *Environment international*. 30:953-971.

Fu X, Clark LA, Yang Q and Anderson MA. 1996. "Enhanced photocatalytic performance of titania-based binary metal oxides: TiO_2/SiO_2 and TiO_2/ZrO_2." *Environmental Science & Technology*.30:647-653.

Fujishima A and Honda K. 1972. "Electrochemical photolysis of water at a semiconductor electrode." *Nature*. 238(5358):37.

Fujishima A, Rao TN and Tryk DA. 2000. "Titanium dioxide photocatalysis." *Journal of photochemistry and photobiology C: Photochemistry reviews*. 1(1):1-21.

Fujishima A, Zhang X and Tryk DA. 2008. "TiO_2photocatalysis and related surface phenomena." *Surface science reports*. 63:515-582.

Guo S, Zhang G, Guo Y and Jimmy CY. 2013. "Graphene oxide-Fe_2O_3 hybrid material as highly efficient heterogeneous catalyst for degradation of organic contaminants." *Carbon*. 60:437-44.

Gupta H, Kumar R., Park HS and Jeon BH. 2017. "Photocatalytic efficiency of iron oxide nanoparticles for the degradation of priority pollutant anthracene." *Geosystem Engineering*. 20:21-27.

Habibi MH and Vosooghian H. 2005. "Photocatalytic degradation of some organic sulfides as environmental pollutants using titanium dioxide suspension."*Journal of Photochemistry and Photobiology A: Chemistry.*174:45-52.

Hara K, Horiguchi T, Kinoshita T, Sayama K, Sugihara H and Arakawa H. 2000. "Highly efficient photon-to-electron conversion with mercurochrome-sensitized nanoporous oxide semiconductor solar cells."*Solar Energy Materials and Solar Cells.* 64(2):115-134.

Hariharan C. 2006. "Photocatalytic degradation of organic contaminants in water by ZnO nanoparticles: Revisited."*Applied Catalysis A: General.* 304:55-61.

Hashimoto K, Irie H and Fujishima A. 2005. "TiO_2photocatalysis: a historical overview and future prospects." *Japanese journal of applied physics.* 44(12R):8269.

Hayat K, Gondal MA, Khaled MM and Ahmed S. 2011. "Effect of operational key parameters on photocatalytic degradation of phenol using nano nickel oxide synthesized by sol-gel method." *Journal of Molecular Catalysis A: Chemical.* 336(1-2):64-71.

He H, Yang S, Yu K, Ju Y, Sun C and Wang L. 2010. "Microwave induced catalytic degradation of crystal violet in nano-nickel dioxide suspensions." *Journal of Hazardous Materials.* 173(1-3):393-400.

Herrmann JM. 1999. "Heterogeneous photocatalysis: fundamentals and applications to the removal of various types of aqueous pollutants."*Catalysis today.*15;53(1):115-29.

Hirota K, Sugimoto M, Kato M, Tsukagoshi K, Tanigawa T and Sugimoto H. 2010. "Preparation of zinc oxide ceramics with a sustainable antibacterial activity under dark conditions." *Ceramics International.*36:497-506.

Hisatomi T, Kubota J and Domen K. 2014. "Recent advances in semiconductors for photocatalytic and photoelectrochemical water splitting." *Chemical Society Reviews.* 43:7520-7535.

Hoffmann MR, Martin ST, Choi W and Bahnemann DW. 1995. "Environmental applications of semiconductor photocatalysis." *Chemical reviews.* 95:69-96.

Hsiao RC, Roselin LS, Hsu HL, Selvin R and Juang RS. 2011. "Photocatalytic degradation of reactive orange 16 dye over Au-doped TiO_2 in aqueous suspension." *International Journal of Materials Engineering Innovation.* 2(1):96-108.

Hu JY, Wang ZS, Ng WJ and Ong SL. 1999. "The effect of water treatment processes on the biological stability of potable water." *Water Research.* 33:2587-2592.

Huang DL, Wang C, Xu P, Zeng GM, Lu BA, Li NJ, Huang C, Lai C, Zhao M H, Xu J J and Luo XY. 2015. "A coupled photocatalytic-biological process for phenol degradation in the Phanerochaetechrysosporium-oxalate-Fe_3O_4 system."*International Biodeterioration & Biodegradation.* 97:115-123.

Huang J, Ren H, Chen K and Shim J J. 2014. Controlled synthesis of porous Co_3O_4 micro/nanostructures and their photocatalysis property. *Superlattices and Microstructures.* 75: 843-856.

Ibanez JA, Litter MI and Pizarro RA. 2003. "Photocatalytic bactericidal effect of TiO_2 on Enterobacter cloacae: Comparative study with other Gram (−) bacteria." *Journal of Photochemistry and photobiology A: Chemistry.* 157(1):81-85.

Jain AK, Gupta VK, Jain S and Suhas. 2004. "Removal of chlorophenols using industrial wastes." *Environmental science & technology*. 38:1195-200.

Jani AM, Losic D and Voelcker NH. 2013. "Nanoporous anodic aluminium oxide: advances in surface engineering and emerging applications." *Progress in Materials Science*.58:636-704.

Kanakaraju D, Motti CA, Glass BD and Oelgemoller M. 2014. "Solar photolysis versus TiO_2-mediated solar photocatalysis: a kinetic study of the degradation of naproxen and diclofenac in various water matrices." *Environmental Science and Pollution Research.*12:27-47.

Keith LH. 1998. "Environmental endocrine disruptors." *Pure and applied Chemistry.* 70: 2319-2319.

Khan I, Saeed K and Khan I. 2017. "Nanoparticles: Properties, applications and toxicities." *Arabian Journal of Chemistry.* 18 May doi:http://dx.doi.org/10.1016/ j.arabjc.2017.05.011.

Khorshidi N, Khorrami SA, Olya ME and Mottiee F. 2016. "Photodegradation of basic dyes using nanocomposite (Ag-zinc oxide-copper oxide) and kinetic studies." *Oriental Journal of Chemistry.* 32:1205-1214.

Lan Y, Lu Y and Ren Z. 2013. "Mini review on photocatalysis of titanium dioxide nanoparticles and their solar applications." *Nano Energy.* 1;2(5):1031-45.

Legrini O, Oliveros E and Braun AM. 1993. "Photochemical processes for water treatment." *Chemical reviews.* 93:671-98.

Li Q, Mahendra S, Lyon DY, Brunet L, Liga MV, Li D and Alvarez PJ. 2008."Antimicrobial nanomaterials for water disinfection and microbial control: potential applications and implications."*Water research.* 42(18):4591-4602.

Li WY, Xu LN and Chen J. 2005. "Co_3O_4nanomaterials in lithium-ion batteries and gas sensors. *Advanced Functional Materials."* 15:851-857.

Li Y, Zhang BP, Zhao JX, Ge ZH, Zhao XK and Zou L. 2013. "ZnO/carbon quantum dots heterostructure with enhanced photocatalytic properties." *Applied Surface Science*. 279:367-73.

Liu HL and Yang TC. 2003. "Photocatalytic inactivation of Escherichia coli and Lactobacillus helveticus by ZnO and TiO_2 activated with ultraviolet light." *Process Biochemistry.* 39(4):475-481.

Liu L, Chen F, Yang F, Chen Y and Crittenden J. 2012. "Photocatalytic degradation of 2,4-dichlorophenol using nanoscale Fe/TiO_2." *Chemical Engineering Journal.* 181:189-95.

Liu R, Wang P, Wang X, Yu H and Yu J. 2012. "UV and visible-light photocatalytic activity of simultaneously deposited and doped Ag/Ag (I)-TiO_2photocatalyst." *The Journal of Physical Chemistry C.* 116:17721-17728.

Long M, Cai W, Cai J, Zhou B, Chai X and Wu Y. 2006. Efficient photocatalytic degradation of phenol over $Co_3O_4/BiVO_4$ composite under visible light irradiation. *The Journal of Physical Chemistry B.*110:20211-20216.

Meshram SP, Adhyapak PV, Mulik UP and Amalnerkar DP. 2012. "Facile synthesis of CuOnanomorphs and their morphology dependent sunlight driven photocatalytic properties." *Chemical engineering journal.* 204:158-168.

Meshram SP, Adhyapak PV, Mulik UP and Amalnerkar DP. 2012. "Facile synthesis of CuOnanomorphs and their morphology dependent sunlight driven photocatalytic properties." *Chemical engineering journal.*204:158-168.

Mills A, Davies RH and Worsley D. 1993. "Water purification by semiconductor photocatalysis." *Chemical Society Reviews.* 22(6):417-25.

Motahari F, Mozdianfard MR, Soofivand F and Salavati-Niasari M. 2014. "NiO nanostructures: synthesis, characterization and photocatalyst application in dye wastewater treatment." *RSC Advances.*4:27654-27660.

Muller HD and Steinbach F. 1970. "Decomposition of isopropyl alcohol photosensitized by zinc oxide." *Nature.* 225(5234):728-729.

Muthukumaran C, Sivakumar VM and Thirumarimurugan M. 2016. "Adsorption isotherms and kinetic studies of crystal violet dye removal from aqueous solution using surfactant modified magnetic nanoadsorbent." *Journal of the Taiwan Institute of Chemical Engineers.* 63:354-362.

Nian JN, Hu CC and Teng H. 2008. "Electrodeposited p-type Cu_2O for H_2 evolution from photoelectrolysis of water under visible light illumination." *International Journal of Hydrogen Energy.* 33:2897-2903.

Ofori FA, Sheikh FA, Appiah-Ntiamoah R, Yang X and Kim H. 2015. "A Simple Method of Electrospun Tungsten Trioxide Nanofibers with Enhanced Visible-Light Photocatalytic Activity." *Nano-Micro Letters.*7:291-297.

Pardeshi SK and Patil AB. 2008. "A simple route for photocatalytic degradation of phenol in aqueous zinc oxide suspension using solar energy." *Solar Energy.* 82(8):700-705.

Parida KM and Parija S. 2006. "Photocatalytic degradation of phenol under solar radiation using microwave irradiated zinc oxide." *Solar Energy.*80:1048-1054.

Parida KM, Dash SS and Das DP. 2006. "Physico-chemical characterization and photocatalytic activity of zinc oxide presented by various methods." *Journal of Colloid and Interface Science.* 298:787-793.

Pincella F, Isozaki K and Miki K. 2014. "A visible light-driven plasmonicphotocatalyst." *Light: Science & Applications.* 3:e133.

Qadri S, Ganoe A and Haik Y. 2009. "Removal and recovery of acridine orange from solutions by use of magnetic nanoparticles." *Journal of hazardous materials.* 169:318-323.

Rai AK, Anh LT, Park CJ and Kim J. 2013. "Electrochemical study of NiO nanoparticles electrode for application in rechargeable lithium-ion batteries." *Ceramics International.* 39:6611-6618.

Rai HS, Bhattacharyya MS, Singh J, Bansal TK, Vats P and Banerjee UC. 2005. "Removal of dyes from the effluent of textile and dyestuff manufacturing industry: a review of emerging techniques with reference to biological treatment." *Critical reviews in environmental science and technology.* 35:219-238.

Ramgir N, Datta N, Kaur M, Kailasaganapathi S, Debnath AK, Aswal DK and Gupta SK. 2013. "Metal oxide nanowires for chemiresistive gas sensors: issues, challenges and prospects." *Colloids and Surfaces A: Physicochemical and Engineering Aspects.* 439:101-116.

Ray S, Lalman JA and Biswas N. 2009. "Using the Box-Benkhen technique to statistically model phenol photocatalytic degradation by titanium dioxide nanoparticles." *Chemical Engineering Journal.* 150(1):15-24.

Rengaraj S and Li XZ. 2006. Enhanced photocatalytic activity of TiO_2 by doping with Ag for degradation of 2, 4, 6-trichlorophenol in aqueous suspension. *Journal of molecular catalysis A: Chemical.* 243:60-67.

Rodriguez JA, Jirsak T, Dvorak J, Sambasivan S and Fischer D. 2000. "Reaction of NO_2 with Zn and ZnO: Photoemission, XANES, and density functional studies on the formation of NO_3." *The Journal of Physical Chemistry B.* 104(2):319-328.

Ruan S, Wu F, Zhang T, Gao W, Xu B and Zhao M. 2001. Surface state studies of TiO_2 nanoparticles and photocatalytic degradation of methyl orange in aqueous TiO_2 dispersions. *Materials Chemistry and Physics.* 69:7-9.

Satheesh R, Vignesh K, Rajarajan M, Suganthi A, Sreekantan S, Kang M and Kwak BS. 2016. "Removal of congo red from water using quercetin modified α-Fe_2O_3 nanoparticles as effective nanoadsorbent." *Materials Chemistry and Physics.* 180:53-65.

Sathishkumar P, Sweena R, Wu JJ and Anandan S. 2011. "Synthesis of CuO-ZnOnanophotocatalyst for visible light assisted degradation of a textile dye in aqueous solution." *Chemical Engineering Journal.* 171:136-140.

Sato K, Aoki M and Noyori R. 1998. "A green route to adipic acid: Direct oxidation of cyclohexenes with 30 percent hydrogen peroxide." *Science,* 281:1646-1647.

Sberveglieri G, Groppelli S, Nelli P, Tintinelli A and Giunta G. 1995. "A novel method for the preparation of NH_3 sensors based on ZnO-In thin films." *Sensors and Actuators B: Chemical.* 25:588-590.

Shalan AE, Rashad MM, Yu Y, Lira-Cantu M and Abdel-Mottaleb MS. 2013. "Controlling the microstructure and properties of titaniananopowders for high efficiency dye sensitized solar cells." *ElectrochimicaActa*. 89:469-478.

Shen W, Li Z, Wang H, Liu Y, Guo Q and Zhang Y. 2008. "Photocatalytic degradation for methylene blue using zinc oxide prepared by codeposition and sol-gel methods." *Journal of Hazardous Materials.* 152(1):172-175.

Shephard GS, Stockenstrom S, de Villiers D, Engelbrecht WJ and Wessels GF. 2002. "Degradation of microcystin toxins in a falling film photocatalytic reactor with immobilized titanium dioxide catalyst." *Water Research.* 36(1):140-146.

Sherrard KB, Marriott PJ, Amiet RG, McCormick, MJ, Colton R and Millington K. 1996. "Spectroscopic analysis of heterogeneous photocatalysis products of nonylphenol and primary alcohol ethoxylate nonionic surfactants." *Chemosphere.* 33:1921-1940.

Song X and Gao L. 2008. "Facile synthesis and hierarchical assembly of hollow nickel oxide architectures bearing enhanced photocatalytic properties." *The Journal of Physical Chemistry C.* 112:15299-15305.

Steinbach F. 1969. "Influence of metal support and ultraviolet irradiation on the catalytic activity of nickel oxide." *Nature.* 221(5181): 657.

Su Y, Deng L, Zhang N, Wang X and Zhu X. 2008. "Photocatalytic degradation of C.I. Acid Blue 80 in aqueous suspensions of titanium dioxide under sunlight." *Reaction Kinetics and Catalysis Letters.* 98:227.

Sugimoto T and Matijevic E. 1979. "Colloidal cobalt hydrous oxides, preparation and properties of monodispersed Co_3O_4." *Journal of Inorganic and Nuclear Chemistry.* 41:165-172.

Sun JH, Dong SY, Feng JL, Yin XJ and Zhao XC. 2011. "Enhanced sunlight photocatalytic performance of Sn-doped ZnO for Methylene Blue degradation." *Journal of Molecular Catalysis A: Chemical.* 335(1-2):145-150.

Suri RP, Liu J, Hand DW, Crittenden JC, Perram DL and Mullins ME. 1993. "Heterogeneous photocatalytic oxidation of hazardous organic contaminants in water." *Water Environment Research.* 65:665-673.

Tadjarodi A, Akhavan O and Bijanzad K. 2015. "Photocatalytic activity of CuO nanoparticles incorporated in mesoporous structure prepared from bis (2-aminonicotinato) copper (II) microflakes." *Transactions of Nonferrous Metals Society of China.* 25:3634-3642.

Tayade RJ, Kulkarni RG and Jasra RV. 2006. "Transition metal ion impregnated mesoporous TiO_2 for photocatalytic degradation of organic contaminants in water." *Industrial & engineering chemistry research.*45:5231-5238.

Teoh WY, Scott JA and Amal R. 2012. "Progress in heterogeneous photocatalysis: from classical radical chemistry to engineering nanomaterials and solar reactors." *The Journal of Physical Chemistry Letters.* 3(5):629-639.

Theurich J, Lindner M and Bahnemann DW. 1996. "Photocatalytic degradation of 4-chlorophenol in aerated aqueous titanium dioxide suspensions: a kinetic and mechanistic study." *Langmuir.* 12:6368-6376.

Thomas J and Yoon M. 2012. "Facile synthesis of pure TiO_2 (B) nanofibers doped with gold nanoparticles and solar photocatalytic activities." *Applied Catalysis B: Environmental.* 111-112:502-508.

Tong H, Ouyang S, Bi Y, Umezawa N, Oshikiri M and Ye J. 2012. "Nano-photocatalytic materials: possibilities and challenges." *Advanced materials.* 24(2):229-251.

Ullah R and Dutta J. 2008. "Photocatalytic degradation of organic dyes with manganese-doped ZnO nanoparticles." *Journal of Hazardous materials.*156:194-200.

Vaidehi D, Bhuvaneshwari V, Bharathi D and Sheetal BP. 2018. "Antibacterial and photocatalytic activity of copper oxide nanoparticles synthesized using Solanumlycopersicum leaf extract." *Materials Research Express.* 5:085403.

Venkatachalam N, Palanichamy M and Murugesan V. 2007. "Sol-gel preparation and characterization of alkaline earth metal doped nano TiO_2: efficient photocatalytic degradation of 4-chlorophenol." *Journal of Molecular Catalysis A: Chemical.* 273:177-185.

Wang N, Zhu L, Wang D, Wang M, Lin Z and Tang H. 2010. "Sono-assisted preparation of highly-efficient peroxidase-like Fe_3O_4 magnetic nanoparticles for catalytic removal of organic pollutants with H_2O_2." *UltrasonicsSonochemistry.* 17(3):526-533.

Wang X, Jia J and Wang Y. 2017. "Combination of photocatalysis with hydrodynamic cavitation for degradation of tetracycline." *Chemical Engineering Journal.* 315:274-282.

Warang T, Patel N, Santini A, Bazzanella N, Kale A and Miotello A. 2012. "Pulsed laser deposition of Co_3O_4 nanoparticles assembled coating: Role of substrate temperature to tailor disordered to crystalline phase and related photocatalytic activity in degradation of methylene blue." *Applied Catalysis A: General.* 423:21-27.

Warang T, Patel N, Santini A, Bazzanella N, Kale A and Miotello A. 2012. Pulsed laser deposition of Co_3O_4 nanoparticles assembled coating: Role of substrate temperature to tailor disordered to crystalline phase and related photocatalytic activity in degradation of methylene blue. *Applied Catalysis A: General.* 423:21-27.

Wei J, Mashayekhi H and Xing B. 2009. "Bacterial toxicity comparison between nano- and micro-scaled oxide particles." *Environmental Pollution.* 157:1619-1625.

Wu Y, He Y, Wu T, Chen T, Weng W and Wan H. 2007. Influence of some parameters on the synthesis of nanosizedNiO material by modified sol-gel method. *Materials Letters.* 61: 3174-3178.

Yang H, Ouyang J, Tang A, Xiao Y, Li X, Dong X and Yu Y. 2006. "Electrochemical synthesis and photocatalytic property of cuprous oxide nanoparticles." *Materials Research Bulletin.* 13;41(7):1310-8.

Yang M and Zhu JJ. 2003. "Spherical hollow assembly composed of Cu_2O nanoparticles." *Journal of Crystal growth.* 256:134-138.

Ye C, Bando Y, Shen G and Golberg D.2006. "Thickness-dependent photocatalytic performance of ZnOnanoplatelets." *The journal of physical chemistry B.* 110(31):15146-15151.

Yumoto H, Inoue T, Li SJ, Sako T and Nishiyama K. 1999. "Application of ITO films to photocatalysis." *Thin Solid Films.* 345:38-41.

Zan L, Fa W, Peng T and Gong ZK. 2007. "Photocatalysis effect of nanometer TiO_2 and TiO_2-coated ceramic plate on Hepatitis B virus." *Journal of Photochemistry and Photobiology B: Biology.* 86:165-169.

Zhang FB, Zhou YK and Li HL. 2004. "NanocrystallineNiO as an electrode material for electrochemical capacitor." *Materials Chemistry and Physics.* 83:260-264.

Zhang X, Ding Y, Tang H, Han X, Zhu L and Wang N. 2014. "Degradation of bisphenol A by hydrogen peroxide activated with $CuFeO_2$microparticles as a heterogeneous Fenton-like catalyst: efficiency, stability and mechanism." *Chemical Engineering Journal.* 236:251-262.

Zhang X, Song J, Jiao J and Mei X. 2010. "Preparation and photocatalytic activity of cuprous oxides." *Solid State Sciences.*12(7):1215-1219.

Zhang YP, Xu JJ, Sun ZH, Li CZ and Pan CX. 2011. "Preparation of graphene and TiO_2 layer by layer composite with highly photocatalytic efficiency." *Progress in Natural Science: Materials International.* 21:467-471.

Zhu Q, Zhang Y, Wang J, Zhou F and Chu PK. 2011. "Microwave synthesis of cuprous oxide micro/nanocrystals with different morphologies and photocatalytic activities." *Journal of Materials Science & Technology.* 27:289-95.

In: Photocatalysis
Editors: P. Singh, M. M. Abdullah, M. Ahmad et al.
ISBN: 978-1-53616-044-4

Chapter 10

PHOTOCATALYTIC DEGRADATION OF SYNTHETIC DYES USING NANO METAL OXIDES

***B. M. Nagabhushana*[1,*] *and M. N. Zulfiqar Ahmed*[2]**

[1]Department of Chemistry, M.S. Ramaiah Institute of Technology,Bangalore, India
[2]Jawaharlal Nehru Technological University Anantapur, Anantapuramu, India

ABSTRACT

This book chapter describes the environmental effects caused due to the presence of dyes in water and the various methods employed for their removal. It also illustrate the application of nano metal oxides (ZnO andFe_2O_3) and nano composite ($Zn_2Fe_2O_4$ –ZnO) in photocatalytic degradation of synthetic dyes like acid orange 7 dye. These ZnO andFe_2O_3nano materials and $Zn_2Fe_2O_4$ – ZnO composite were derived from the solution combustion method. In photocatalitic degradation studies the effect of various factors such as initial dye concentration, dosage of the photocatalyst and irradiation time on the rate of photocatalysis was investigated in detail and fond that, The photocatalytic process was found to be depending on the dosage of photo catalyst, irradiation time, initial dye concentration, It is observed that nano composite shows a better photocatalytic activity compared to individual oxide powder. So, nano structured materials could be as effective photocatalyst for removal of dyes from textile and paper effluents.

Keywords: photocatlytic activity, dyes, nano composites, dosage effect, stirring time, initial dye concentration

---

* Corresponding Author's E-mail: bmnshan@yahoo.com.

1. INTRODUCTION

Water pollution or contamination has become a universal problem due to the industrialization and populace explosion. Engineering (Industries) and domestic toxic wastewaters have several objectionable components. The toxic wastewater treatment and its proper management have become essential to conserve the water resources. The contamination problem in India is inferior to some of the highly industrialized developed countries. The marine environment has been affected harmfully by human behaviour, Unacceptable sanitation and by unregulated huge release of public and engineering wastes. A number of the techniques to minimize the water contaminated problem.

The aquatic water quality get polluted by municipal wastes, marine dumping, domestic activities, modern agricultural methods, mining activities, underground storage leakages, Industries oil spillage and radioactive wastes. But the real pollution is made by different industries or manufacturing units. The unsystematic or random discharge of deadly chemicals and dye stuffs through waste water or effluents from different types of industries (i.e., textile, fertilizers production process, steel, mines, oil, tanneries, sugar mills, detergent production process, canneries, refineries and electroplating units) enter into water system and pollutes the source and causes harmful effects on living organisms.

Dyes or colourants are extensively used in different types of manufacturing units like leather, plastic, textile, food and paper to develop or increase the visual values of their manufactured products. These many industries generate colored effluents or wastewater knowingly or unknowingly enter into natural source of water system leading to disagreeable consequences to the ecosystem or environment. The presence of color or dyes stuffs in the aquatic water can cause destructive damage to marine or aquatic life by hindering the chemical oxygen demand and photosynthetic processes.

Prior to 1857, the natural dyes were exclusively used by the dyeing industry. However, by 1900 nearly 90% of the industrial dyes were synthetic. The dye industry played a vital role in the development of structural organic chemistry, which in turn provided a sound scientific foundation for the dye industry. The synthetic dye industry originated directly from the studies of coal tar. It attracted the attention of chemists as a source of new organic compounds which could be easily isolated bydistillation. Rapid advances in the understanding of chemical structure together with strong industrial-academic interactions as well as favourable governmental practices resulted in a systematic development in the synthetic dye industry based on solid scientific foundations. Until the 1850s many dyes were obtained from natural sources such as vegetables, plants, trees, lichens and insects. Countless attempts were being made to extract dyes from brightly coloured plants and flowers; however, only a few dyes found widespread use. Most of the attempts failed because of the fact that natural dyes are not highly stable and occur as components ofcomplex mixtures, the successful separation of which was very difficult and unlikely by the crude methods available in the ancient times.

However, studies of these dyes in the 1800s provided a base for development ofthe so called synthetic dyes.

Dyes contain conjugated systems ofbenzenerings with simple unsaturated groups such as —NO_2, —N = N—, —C = O etc. calledchromophores, and polar groups such as —NH_2, —OH etc. called auxochromes. The chromophores and auxochromes are together referred as chromogens. The colour of a dye is due to the absorption of visiblelightby the chromophores present in it. Organic compoundsabsorb electromagnetic radiation, but only those containing several conjugated double bonds appear coloured by the absorption of visible light. Without substituents, the chromophores do not absorb visible light, but the presence of auxochromes shifts the absorption of these chromogens into the visible region of the electromagnetic spectrum. The auxochromes also extend theconjugated system.

The absorption spectrum is used for the characterization of specific compounds. In the visible spectrum, the absorption pattern exhibits a broad band with maximum at the longer wavelength side corresponding to more extended conjugation. The position and shape of the absorption band affect the appearance of the observed colour. Majority of the compounds absorb in the UV region with a few bsorptions extending into the violet (400–430 nm) region. Hence, these compounds appear yellowish to the eye, that is the the perceived colour is complementary to the absorbed colour. Progressive absorption into the visible region gives orange (430–480 nm), red (480–550 nm), violet (550–600 nm), and blue (600–700 nm); absorption at 400–450 and 580–700 nm gives green. The brilliance of a particular colour increases with decrease in bandwidth. Synthetic dyes tend to give more brilliant colours which undoubtedly led to their rapid development and dominance over the natural dyes which give rather dull and diffuse colorations. Additionally, the synthetic dyes are less expensive, brighter, more colour-fast and can be easily applied on the fabric.

The synthetic dyes have become indispensable tools for various types of industries. These dyes are common water pollutants and they may frequently be found in trace quantities in industrial wastewater.These are extensively used as colorants in plastic, textile dyeing and the highly sophisticated biotechnology industry. Dyes arealso employed by industries for inks and tinting. They also employed in other industries including paper and pulp, adhesives, ceramics, art supplies, beverages, food, soaps, cosmetics, construction, glass, paints, polymers, waxes, biomedicines etc. Presently, a large number of dyes are being manufactured to meet the needs of various types of industries.

Commercially, dyes are available in different forms such as dry powders, granules, pellets, pastes, liquids and chips. They are classified as cationic, anionic and non-ionic dyes. Basic dyes are cationic dyes while direct and reactive dyes are anionic dyes. The non-ionic dyes include the disperse dyes.

Anionic dyes contain the azo or anthraquinone groups as the chromophores. The presence of azo groups results in the formation of toxic amines. Sulphonated azo dyes are readily soluble in water and hence their removal is very difficult. Anthroquinone-based dyes are highly resistant to degradation due to the presence of fused aromatic structure in them. Hence, the effluents containing such dyes retain their colour for a longer period of time. Basic dyes possess a high brilliance and intensity and are therefore, highly visible even at very low concentrations. The metal-complexed dyes are mainly based on the metal chromium which is carcinogenic in nature. Among the several types of dyes, the azo dyes account for more than 60% of the total dyes manufactured (Fu and Viraraghavan, 2001; Wesenberg et al., 2003; Paszczynski et al., 1992).

2. Harmful Effects of Synthetic Dyes

More than 1,000,000 commercially available dyes exist and their annual production exceeds 7 x 10^5 tonnes. However, this brightly coloured, changed new world is not without a downside. The chemicals used in the preparation of these synthetic dyes are toxic and exhibit carcinogenic effects while some of them are even explosive. Aniline, for example, is the basis for the preparation of azo dyes which are regarded as highly poisonous and release carcinogenic amines in water. Aniline is dangerous to work with due to its toxic and highly inflammable properties (Ruh Ullah, and Joydeep Dutta, 2008; Chung Ro Lee et al., 2001). Apart from aniline, some of the other toxic chemicals used in the dyeing process include the following:

(i) Dioxin - a carcinogen and a possible hormone disrupter.
(ii) Heavy metals such as Cr, Cu and Zn - toxic and possess carcinogenic effects and other health hazards.
(iii) Formaldehyde - a suspected carcinogen.

The workers involved in the manufacture of dyes and dyeing of garments are obviously at a greater risk of tumours, cancers, and lung diseases. Almost every industrial dyeing process involves an aqueous solution of a dye in which the fabrics are immersed or washed. After dyeing a batch of fabrics, it is less expensive to release the used water (dye effluent) into nearby sources of water rather than to clean and reuse it.It is because of this reason that most of the dyeing industries throughout the world release millions of tonnes of dye effluents into the rivers. Most of the dyes are readily soluble in water and are also quiet stable. In view of their extra stability, these dyes ultimately find their way into the aquatic ecosystem and have adverse and sometimes irreversible effects on both animals and plants. Some of the dye effluents release potentially harmful and carcinogenic substances into the water which pose a serious threat to the aquatic living organisms. Apart from being carcinogenic, these dyes also have adverse effects on mankind including malfunctioning of kidneys, lungs, intestine, liver etc.

Dyes inhibit the penetration of light into the water which leads to the death of the phytoplankton. It is very difficult to treat the wastewater containing these dyes due to the fact that these dyes are recalcitrant organic molecules, resistant to aerobic digestion and are stable to heat, light and oxidizing agents (Daneshvar et al., 2007; Abbas et al., 2008).

3. Methods of Dye Removal

In view of the adverse effects of the synthetic dyes, it is important to treat the industrial effluents containing dyes. The various methods used for the removal of dyes from different industrial effluents are given in Table 1 (Anjaneyulu et al., 2009). Each method has its own advantages and disadvantages. The conventional biological methods used for the treatment of industrial waste water are ineffective which leads to the release of an intensely coloured discharge from the treatment facilities (Harrelkas et al., 2009; Devipriya et al., 2005; Mills et al., 2006; Yan et al., 2009).Recently, nanocatalysis (photocatalysis using nanomaterials) and adsorption using nano metal oxides have proved to be good techniques for dye removal.

Table 1. Different methods used for dye removal from industrial effluents

Method of dye removal	Advantages	Disadvantages
Fenton's reagent	Effective removal of both water soluble and water insoluble dyes.	Large scale production of sludge.
Ozonation	No change in the volume of the effluent since applied in the gaseous state.	Short half-life (about 20 min).
Photochemical	No sludge production.	Formation of by-products.
Sodium hypochlorite	Effective for azo dyes.	Release of aromatic amines into the water.
Electrochemical destruction	Break-down compounds are not harmful.	High electricity consumption.
Activated carbon	Efficient removal of several types of dyes,	Highly expensive.
Membrane filtration	Effective removal of almost all types of dyes.	Concentrated sludge production.
Ion exchange	No adsorbent loss due to regeneration.	Not effective for all dyes.
Electrochemical coagulation	Economical	High sludge production

3.1. Nanomaterials for Photocatalytic Degradation of Dyes

Semiconducting materials such as Cu_2O, ZnO, SnO_2, CeO_2, ZrO_2, Fe_2O_3, CaO, MgO, etc. have been extensively studied as photocatalysts for the degradation of a number of dyes as well as various organic and inorganic pollutants.

The use of iron-cerium mixed oxide as photocatalyst for degradation of phenol, methylene blue and congo red was reported by Gajendra Kumar and Parida et al. 2010. The compound exhibited good catalytic efficiency for removal of methylene blue and congo red. Ding Shiwen et.al.2003, have reported the synthesis of nano $TiO_2 – MnO_2$ by hydrothermal method. The compound exhibited almost complete removal of two dyes acidic red B and acidic black 234. Core-shell α- Fe2O3/Fe_3O_4 has been used for the photocatalytic degradation of methyl orange (Yang Tian et al. 2010).

Nano sized ZnO powders (R. Nagaraja et al.. 2012) were prepared by solution combustion route and its photocatalytic activity on RB dye both under UV and solar light irradiation was studied. The effect of catalyst dosage, concentration of dye, pH of the dye solution, light source on photocatalytic effect was investigated. The results reveal that the maximum decolorization (more than 95%) of dye occurred with ZnO catalyst in 8 min of stirring at basic pH under solar light irradiation. It was also found that chemical oxygen demand (COD) reduction takes place at a faster rate under solar light as compared to that of UV light. The results suggest that, the ZnO solar photocatalytic irradiation is better than the calcined ZnO/solar and UV light irradiation.

ZnO/Cu_2O compound photocatalysts were prepared (Chao Xu, et al., 2010) "soak-deoxidize-air oxidation" with different concentrations of Cu^{2+}. The photocatalytic activities of ZnO/Cu_2O compound were evaluated using a basic organic dye, methyl orange (MO). ZnO/Cu_2O composite shows higher photocatalytic activity than pure ZnO and pure Cu_2O.

The Er^{3+}:$YAlO_3$/ZnO composite (Jun Wang, et al.,2009) a new photocatalyst that could effectively utilize visible light, was prepared by ultrasonic dispersion method The Acid Red B dye as a model compound was degraded undersolar light irradiation to evaluate the hotocatalytic activity of the Er^{3+}:$YAlO_3$/ZnO composite. In addition, the effects of Er^{3+}:$YAlO_3$ content, heat-treated temperature and heat-treated time on photocatalytic activity of Er^{3+}:$YAlO_3$/ZnO composite were reviewed through the degradation of Acid Red B dye under solar light. It was found that the photocatalytic activity of Er^{3+}:$YAlO_3$/ZnO composite is much higher than that of pure ZnO powder for the similar system.

Porous α-Fe_2O_3nanorods with typical pore size of 2–4 nm were controlled prepared by a facile hydrothermal process of $Fe(NO_3)_3.9H_2O$ aqueous solution in the presence of NaOH, followed by a calcination treatment. Contrast experiments indicate that the morphology and crystalline structure of the hydrothermal products depend greatly on the quantity of NaOH. Hematite nano particles and microplates were respectively obtained under conditions without or with excess NaOH. The visible-light photocatalytic performances of the as-prepared samples were evaluated by photocatalytic decolorization of methylene blue at ambient temperature. The results indicate that the photocatalytic activity of the porous α-Fe_2O_3 nanorods is superior to hematite nanoparticles and platelets and exhibit good reusable feature. The photocatalytic process of porous structure is determined to be pseudo-first-order reaction. The optimum photocatalyst dosage is 20 mg per 100 mL of dye solution. The porous a- Fe_2O_3 nanorods are considered potential photocatalyst for practical application due to the excellent photocatalytic behavior and good reusability (Guo-Ying Zhang et al. 2012).

The hierarchical ZnO nanorod-array films were prepared by ink-jet printed barriers, selectively deposited ZnO stripes, and nanorod modified surface. The photo induced degradation of methyl orange (MO) using the hierarchical ZnO photocatalysts and Ag decorated hierarchical ZnO photocatalysts was investigated. The enhanced charge separation and increased surface area of the Ag decorated hierarchical ZnO photocatalysts lead to the improvement of the dye-decoloration efficiency (Chi-Jung Chang et al.. 2013).

Copper oxide (CuO) nanoflowers in the monoclinic phase were prepared by solution combustion method using glycine was used as fuel. CuO nanoflowers show very good photocatalytic activity. The apparent rate constant of this catalytic reaction was 0.016/min. The estimated direct band gap energy of the CuO was 3 eV. In order to demonstrate the photocatalytic activity of the pre pared CuO nanoflowers the degradation of MO under UV irradiation was investigated (Umadevi et al.. 2013).

The semiconductor photocatsalysts possess the favourable combination of electronic structures which is characterized by a filled valence band and an empty conduction band, light absorption properties, charge transport characteristics and lifetime of the excited states. In photocatalytic degradation nanoparticles are being used to catalyse the reactions. At the nano level, the particles have larger surface area to volume ratio compared to that in the bulk materials. This reduction in size results in the involvement of a large number of atoms in the catalysis process. These properties render the nanoparticles efficient catalytic activity. The surface atoms can occupy the corners and edges of the nanoparticles and hence become chemically unsaturated and also much more active. The smaller size increases the activity of the surface atoms and leads to surface reconstruction. The organic compounds (dyes) get adsorbed onto the catalyst surface and are then mineralized into carbon dioxide and water through a redox reaction brought about by hydroxyl or superoxide radicals. The photocatalytic process is known to occur at the surface or within a few monolayers around the catalyst particles. So, this chapter explains the application of ZnO, Fe_2O_3, $Zn_2Fe_2O_4$ –ZnO nano powders on photocatalytic degradation of synthetic dyes like acid orange 7 dye. Our results suggested that instead of individual nano metal oxides nano composite shows a better photocatalytic activity.

3.2. Photocatalytic Degradation of Acid Orange 7 Dye Using ZnO and α-Fe_2O_3 Nanopowders and $ZnFe_2O_4$-ZnO Nanocomposite

This section describes a comparison of the efficiencies of ZnO and α-Fe_2O_3 nanopowders and $ZnFe_2O_4$-ZnO nanocomposite as photocatalysts (prepared by solution combustion synthesis using the appropriate metal nitrate as oxidizer and oxalyldihydrazide as fuel, K. C. Patil et al. 2008). In this report Acid orange 7 is used as dye for degradation, it is an anionic azo dye with molecular formula $C_{16}H_{11}N_2NaSO_4$andmolecular mass equal to 350.32 gmol^{-1}. Chemically, AO 7 is sodium 4-[(2-hydroxy-1-naphthyl)azo]benzenesulfonate. Figure 1 shows the structure of AO 7. There are three benzene rings with an azo group present between the first and second rings. A sulphonate group is attached to the first ring while a hydroxyl group is attached to the second ring. The CAS Number of Acid Orange 7 is 633-96-5 and the CB Number is CB144029. AO7 has a colour index of 15510. The dye is available in the form of orange coloured powder and is readily soluble in water but slightly soluble in ethanol. It changes colour from amber (pH 7.4) to orange (pH 8.6 – 10.2) to red (pH 11.8) (Gupta et al. 2006; Silva et al. 2004). The absorption spectrum of 20 ppm AO7 solution is shown in Figure 2. The maximum absorbance was observed at 487 nm. In the photocatalytic degradation of AO 7 the UV-visible spectrum was recorded between 325 to 650 nm.

Acid orange 7 finds applications in paper and textile industries, particularly in the dyeing of various materials of nylon, silk and wool. In the past it was also used in tanneries. The dye is used in cosmetics, dressing materials for wounds, biofuel cells, inks, wood preservatives, hair dyes, nanoparticles and organic light emitting diodes. Acid Orange 7 is highly toxic in nature. It causes dermatitis, nausea, methemoglobinemia, irritation of the eyes, skin, mucous membrane and the upper respiratory tract. It is also carcinogenic in nature due to the presence of the azo group. (Zhang et al. 2014).

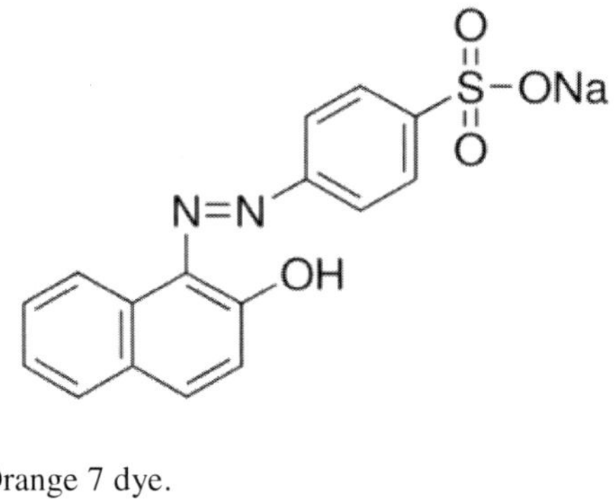

Figure 1. Structure of Acid Orange 7 dye.

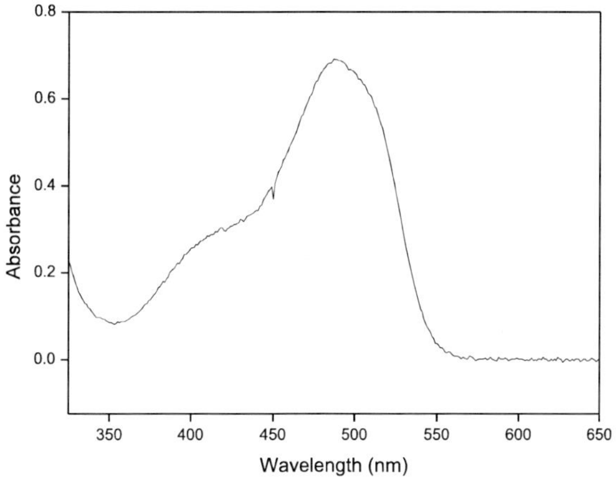

Figure 2. Absorption spectrum of 20 ppm Acid Orange 7 dye solution.

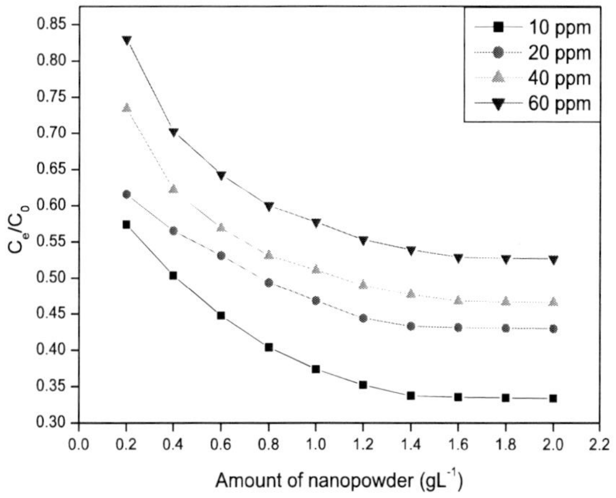

Figure 3. Effect of dosage of the photocatalyst on the photocatalytic removal of Acid Orange 7 dye by the ZnO nanopowder.

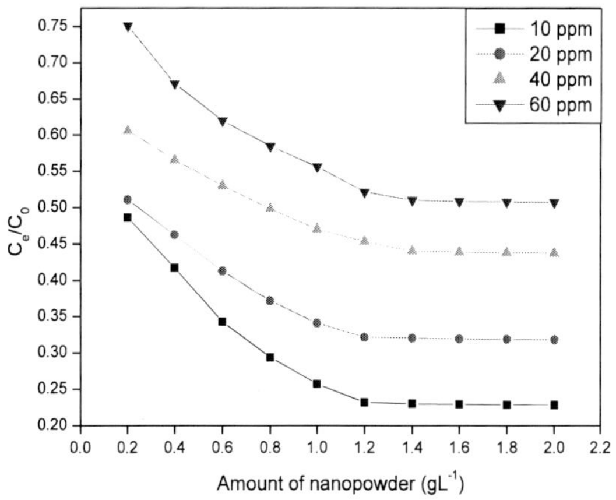

Figure 4. Effect of dosage of the photocatalyst on the photocatalytic removal of Acid Orange7 dye by the α-Fe_2O_3 nanopowder.

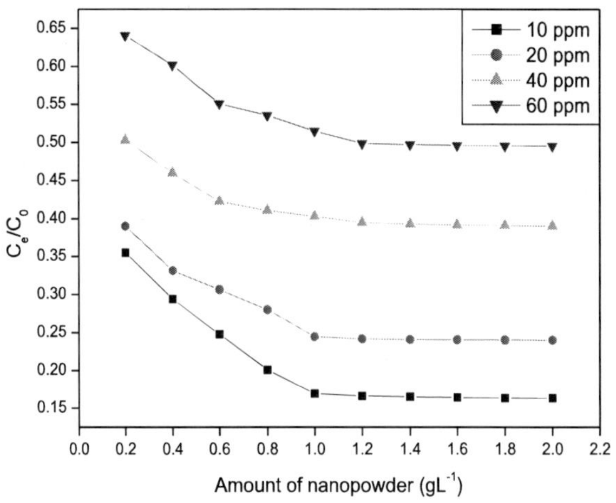

Figure 5. Effect of dosage of the photocatalyst on the photocatalytic removal of Acid Orange 7 dye by the $ZnFe_2O_4$-ZnO nanocomposite.

Figures 3 to 5 show (the plot of C_e/C_0 vs. dosage, where C_e and C_0 are respectively the equilibrium and initial concentrations of the dye, these are measured by UV-Visible spectrophotometer) the effect of dosage of the photocatalyst on the rate of photocatalytic degradation of Acid Orange 7 by the ZnO, α-Fe_2O_3 and $ZnFe_2O_4$-ZnO photocatalysts. It was noticed that in case of all the three photocatalysts, the photocatalytic degradation of AO 7 increased with increase in the dosage of the photocatalyst till an optimum value. As the dosage of the photocatalyst increased, the number of active sites available for adsorption of the dye molecules also increased. Increase in dosage of the photocatalyst also resulted in an increase in the formation of hydroxyl radicals. These two factors were responsible for the increase in the rate of photocatalytic degradation of the dye.

Beyond the optimum dosage, the increase in dye degradation was negligible. This was attributed to the fact that beyond the optimum value of the photocatalyst, the turbidity of the solution increased which resulted in a decrease in the penetration of UV light and the photo activated volume. This resulted in scattering of the UV light thereby decreasing the number of photons reaching the dye solution and consequently decreasing the formation of the hydroxyl radicals. As a consequence of this, there would be either a decrease or negligible increase in the photocatalytic activity once the optimum photocatalytic rate has been achieved. In the present study, the latter was observed (So et al. 2002; Murat et al.. 2006).

The effect of irradiation time on the photocatalytic degradation of the by the three nanopowders is depicted in Figures 6 to 8. The photocatalytic degradation of AO7 increased with increase in irradiation time up to an optimum value and then remained almost constant. In the beginning, the rate of dye degradation was higher since a large number of the active sites of the photocatalyst were available for degradation. With an increase in the irradiation time, the rate of dye degradation was less owing to the competition between the dye molecules to occupy the active sites of the photocatalyst which get saturated with lapse of time. Beyond the optimum irradiation time, the dye degradation was negligible because of the attainment of equilibrium between the adsorption and desorption of the dye molecules at the surface of the photocatalyst.

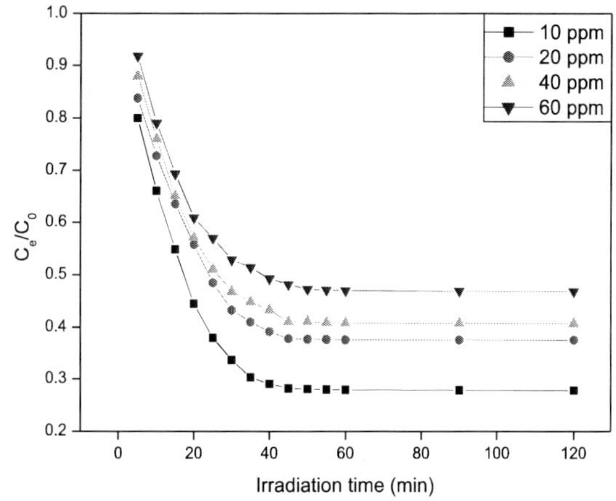

Figure 6. Effect of irradiation time on the photocatalytic removal of Acid Orange 7dye by the ZnO nanopowder.

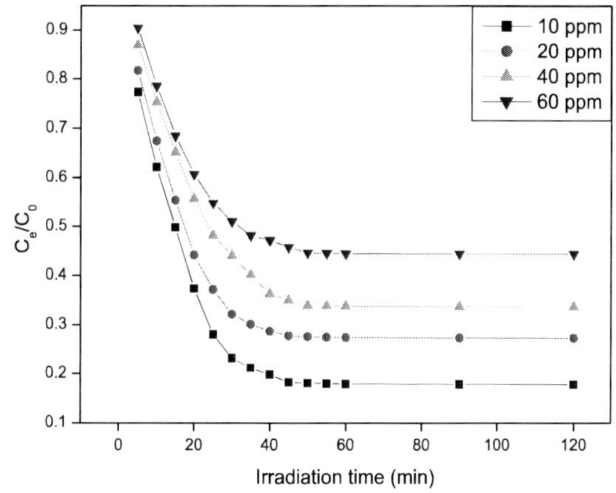

Figure 7. Effect of irradiation time on the photocatalytic removal of Acid Orange 7dye by the α-Fe_2O_3 nanopowder.

The photocatalytic degradation of AO 7 was dependent on the initial concentration of the dye solution. As the initial concentration, of the dye solution increased, the dye degradation decreased. Increase in initial concentration of the dye solution increased the number of dye molecules, but the amount of photocatalyst was fixed. This resulted in an increase in competition between the dye molecules for adsorption on the fixed number of active sites of the photocatalyst. The greater the initial dye concentration, the lesser was the photocatalytic degradation (Francoa et al. 2009; Pauporte et al. 2009). 10 ppm solution of AO7 exhibited highest dye degradation whereas the 60 ppm solution exhibited the least as is evident from Figure 9.

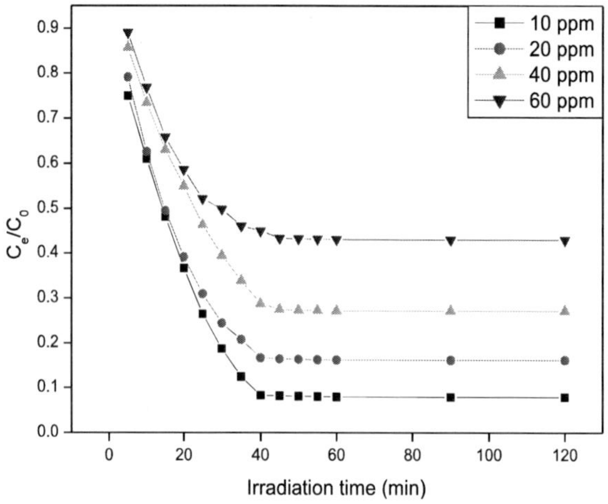

Figure 8. Effect of irradiation time on the photocatalytic removal of Acid Orange 7dye by the $ZnFe_2O_4$ - ZnO nanocomposite.

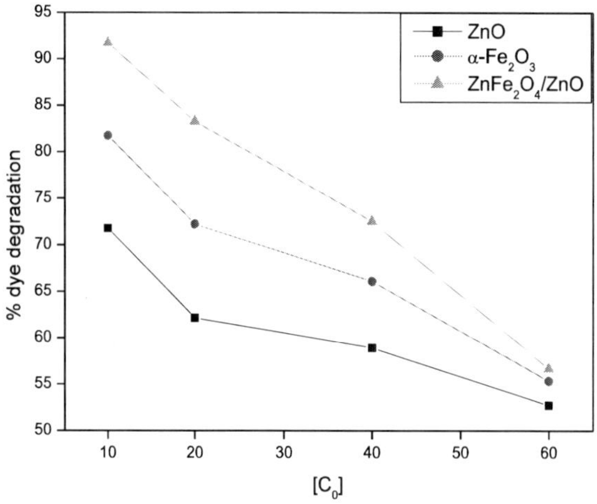

Figure 9. Effect of initial concentration of the dye solution on the photocatalytic degradation of Acid Orange 7 dye by the three photocatalysts.

3.3. Discussion on the Better Degradation of Dyes by the $ZnFe_2O_4$ -Zno Nanocomposite

It was noticed that the photocatalytic degradation of AO7 was highest in case of the $ZnFe_2O_4$ - ZnO nanocomposite followed by the α-Fe_2O_3 nanopowder while the ZnO nanopowder exhibited least photocatalytic activity. Additionally, the optimum values of dosage of the photocatalyst and irradiation time were less in case of $ZnFe_2O_4$-ZnO nanocomposite compared to the other two nanopowders. Hence, the efficiency of the three nanopowders in the photocatalytic degradation of AO7 followed the order:

$ZnFe_2O_4$-ZnO > α-Fe_2O_3> ZnO

The higher photocatalytic activity of the α-Fe_2O_3 nanopowder compared to that of ZnO nanopowder was attributed to small crystallite size of α-Fe_2O_3 (20 nm) than that of ZnO (35nm). It was also attributed to the photocorrosion property of ZnO on exposure to UV light. Table 2 gives the summary of results for the photocatalytic degradation of the AO7 by the three nanopowders.

Table 2. Summary of results for the photocatalytic degradation of Acid Orange 7 dye by the ZnO and α-Fe_2O_3 nanopowders and $ZnFe_2O_4$ - ZnO nanocomposite

Concentration of the dye solution (ppm)	Parameters	ZnO nanopowder	α-Fe_2O_3 nanopowder	$ZnFe_2O_4$-ZnO nanocomposite
10	Optimum dosage (gL^{-1})	1.4	1.2	1.0
	Optimum irradiation time (min)	45	45	40
	% dye removal	72	82	92
20	Optimum dosage (gL^{-1})	1.4	1.2	1.0
	Optimum irradiation time (min)	45	45	40
	% dye removal	62	72	83
40	Optimum dosage (gL^{-1})	1.4	1.2	1.0
	Optimum irradiation time (min)	45	45	40
	% dye removal	59	66	76
60	Optimum dosage (gL^{-1})	1.4	1.2	1.0
	Optimum irradiation time (min)	45	45	40
	% dye removal	53	55	57

It is evident from Table 2 that the $ZnFe_2O_4$-ZnO nanocomposite exhibited highest photocatalytic activity for the degradation of the azo dye Acid Orange 7 compared to the α-Fe_2O_3 and ZnO nanopowders. The highest photocatalytic activity of $ZnFe_2O_4$-ZnO nanocomposite was attributed to the synergistic mechanism between the $ZnFe_2O_4$ and ZnO phases present in it. The E_g of ZnO is greater than that of $ZnFe_2O_4$. When the $ZnFe_2O_4$-ZnO nanocomposite was exposed to UV radiation, it resulted in excitation of both the $ZnFe_2O_4$-ZnO phases leading to the production of electron-hole pairs in both the semiconductors (Figure 10). The electrons in the valence bands of both the semiconductors were promoted to their respective conduction bands. The electrons from the conduction band of $ZnFe_2O_4$ were injected into the conduction band of ZnO. This resulted in an increase in the concentration of electrons in the conduction band of ZnO. Additionally, the holes in the valence band of ZnO migrated into the valence band of $ZnFe_2O_4$. Hence, the concentration of holes in the conduction band of $ZnFe_2O_4$ increased. These interparticle charge transfer processes led to an efficient separation of the photogenerated electrons and holes. The charge-recombination was reduced to a greater extent and hence a greater concentration of the photogenerated electrons and holes was available for the degradation of the dye molecules. The electrons and holes migrated to the surface of the respective particles and participated in the redox reactions occurring at those surfaces. Hence the improved photocatalytic activity of the $ZnFe_2O_4$-ZnO nanocomposite was due to the synergistic effect involving interparticle charge transfer processes between its two phases. This synergistic effect has been noticed in case of other nanocomposite materials such as $Nb_2O_5/SrNb_2O_6$(Xing et al. 2008; Tiekun et al. 2014; Ganeshraja et al. 2015).

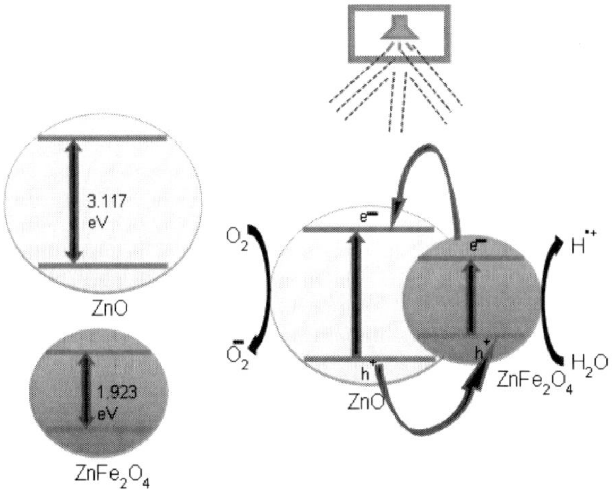

Figure 10. Mechanism of photocatalytic degradation of dyes the $ZnFe_2O_4$-ZnO nanocomposite

Conclusion

ZnO, Fe_2O_3, CeO_2 nano powders and, $Zn_2Fe_2O_4$ nano composite, are used as photcatlytic degradation of synthetic dyes like acid orange 7 dye. The degradation studies reveals that, various factors such as initial dye concentration, dosage of the photocatalyst and irradiation time effect the photocatalysis. The photocatalytic activity followed the order: ZnO < α-Fe_2O_3 < $ZnFe_2O_4$/ZnO. The mechanism of photocatalysis involves the formation of electron-hole pairs once the single nano metal oxide is exposed to radiations. However, these photogenerated electrons and holes recombine at a faster rate. This results in decrease in the photocatalytic activity and hence it is one of the major drawbacks of nano single metal oxides. This can be overcome by the use of nanocomposites where the electron-hole recombination if greatly reduced by the synergistic mechanism involving charge transfer between the constituent phases of the nanocomposite. Thus nanocomposites can serve as alternative photocatalysts for dye degradation compared to single nano metal oxides, hence nano structured materials could be used for effective removal of dyes from textile and paper effluents.

ACKNOWLEDGMENTS

The authors gratefully acknowledge the support rendered by the TEQIP Laboratory of M.S. Ramaiah Institute of Technology, Bangalore for providing the facilities to carry out the research work.

REFERENCES

Abbas J. A., Salih H. K., Falah H.H. 2008 "Photocatalytic Degradation of Textile Dyeing Wastewater Using Titanium Dioxide and Zinc Oxide" *E-Journal of Chemistry*5:219–223.

Akarsu M., Asilturk M. A., Sayilkan F., Kiraz N., Arpac E. and Sayilkan H. 2006 "A Novel Approach to the Hydrothermal Synthesis of Anatase Titania Nanoparticles and the Photocatalytic Degradation of Rhodamine B" *Turkish Journal of Chemistry* 30: 333–343.

Anjaneyulu Y., Sreedhara C. N., Samuel S. R. D. 2005 "Decolourization of Industrial Effluents – Available Methods and Emerging Technologies – A ReviewReviews" *Environmental Science and Bio-Technology* 4: 245–273.

Chang C.J., Hsu M.H., Weng Y.C., Tsay C.Y.,Lin C.K. 2013 "Hierarchical ZnOnanorod-array films with enhanced photocatalytic performance" *Thin Solid Films* 528: 167–174.

Daneshvar N., Aber S., Seyed Dorraji M. S., Khataee A. R., Rasoulifard M. H. 2007 "Preparation and Investigation of Photocatalytic Properties of ZnO Nanocrystals: Effect of Operational Parameters and Kinetic Study" *International Journal of Nuclear and Quantum Engineering* 1: 62–67.

Devipriya S., Yesodharan S.2005 "Photocatalytic Degradation of Pesticide Contaminants in Water" *Solar Energy Material and Solar Cell* 86: 309-348.

Dias S. 2004 "Adsorption of acid orange 7dye in aqueous solutions by spent brewery grains" *Separation & Purification Technology* 40: 309-315.

Francoa A., Neves M.C., Ribeiro Carrott M. M. L., Mendonca M. H.,Pereira, M. I., Monteiroa O. C. 2009 "Photocatalytic decolorization of methylene blue in the presence of TiO2/ZnS nanocomposites" *Journal of Hazardous Materials* 161: 545–550.

Fu Y., Viraraghavan T. 2001 "Fungal decolorization of dye wastewaters: a review" *Bioresource Technology* 79: 251–262.

GaneshrajaA. S., Clara A. S., Rajkumar K., Wang Y., and Anbalagan K. 2015 "Simple hydrothermal synthesis of metal oxides coupled nanocomposites: Structural, optical, magnetic and photocatalytic studies" *Appied Surface Science* 353: 553-563.

Gupta V. K., Mittal A., Gajbe V., and Mittal J. 2006 "Removal and Recovery of the Hazardous Azo Dye Acid Orange 7 through Adsorption over Waste Materials: Bottom Ash and De-Oiled Soya" *Industrial & Engineering Chemisrty Research* 45: 1446-1453.

Harrelkas F., Azizi A., Yaacoubi A., Benhammou A., Pons M. N.2009 "Treatment of textile dye effluents using coagulation–flocculation coupled with membrane processes or adsorption on powdered activated carbon" *Desalination* 235:330–339.

Jia T., Zhao J., Fu F., Deng F., Wang W., Fu Z. and Meng F. 2014 "Synthesis, Characterization, and Photocatalytic Activity of Zn-Doped SnO_2/Zn_2SnO_4Coupled Nanocomposites" *International Journal of Photoenergy* 10:1-7.

Lee R. C., Lee H.W., Song J. S., Kim W.W., Park S. 2001 "Synthesis and Ag Recovery of Nanosized ZnO Powder by Solution Combustion Process for Photocatalytic Applications" *Journal of Materials Synthesis and Processing* 9: 281–286.

Mills A., Wang J., McGrady M. 2006 "Method of Rapid Assessment of Photocatalytic Activities of Self-Cleaning Films" *Journal of Physics Chemistry B*110: 18324–18331.

Nagaraja R., Kottam N., Girija C.M., Nagabhushana B.M. 2012 "Photocatalytic degradation of Rhodamine B dye under UV/solar light using ZnO nanopowder synthesized by solution combustion route" *Powder Technology* 215-216: 91–97.

Paszczynski A., Pasti-Grigsby M. B., Goszczynski S, Crawford R. L., Crawford, D. L.,1992 "Mineralization of sulfonated azo dyes and sulfanilic acid by Phanerochaete chrysosporium and Streptomyces chromofuscus" *Applied Environmental Microbiology* 58 11: 3598–3604.

Patil K.C., Hegde M.S., Rattan T., Aruna S.T. 2008 "Chemistry of nanocrystalline oxide materials - Combustion synthesis, properties and applications" *World Scientific*, 2008:332:335.

Pauporte T., and Rathousky J. 2009 "Growth mechanism and photocatalytic properties for dye degradation of hydrophobic mesoporous ZnO/SDS films prepared by electrodeposition" *Microporous and Mesoporous Materials* 117:380–385.

Pradhan G. K. and Parida K. M. 2010 "Fabrication of iron-cerium mixed oxide: an efficient photocatalyst for dye degradation"*International Journal of Engineering, Science and Technology* 2: 53-65.

Shiwen D., Liyong W., Shaoyan Z., Qiuxiang Z., Yu D., Shujuan L., Liu Y. and Quanying K. 2003 "Hydrothermal synthesis, structure and photocatalytic property of nano-TiO_2-MnO_2" *Science in China: Series B* 46: 542-548.

Silva J. P., Sousa S., Rodrigues J., Antunes H., Porter, J. J., Goncalves I., and Ferreira-So C. M., Cheng M. Y., Yu J. C. and Wong P. K. 2002 "Degradation of azo dye Procion Red MX-5B by photocatalytic oxidation" *Chemosphere* 46: 905-912.

Tian Y., Wu D., Jia X., Yu B. and Zhan S. 2011 "Core-Shell Nanostructure ofSynthesis and Photocatalysis for Methyl Orange" *Journal of Nanomaterials* 2010:1-5.

Ullah R., Dutta J. 2008 "Photocatalytic degradation of organic dyes with manganese-doped ZnO nanoparticles" *Journal of Hazardous Material* 156:194–200.

Umadevi M. and Christy A.J. 2013 "Synthesis, characterization and photocatalytic activity of CuO nanoflowers" *Spectrochimica Acta Part A: Molecular and Biomolecular Spectroscopy* 109: 133–137.

Wang J., Xie Y., Zhang Z., Li J., Chen X., Zhang L., Xu R. and Zhang X. 2009 "Effects of substrate temperature on the structural and electrical properties of Cu(In,Ga)Se_2thin films" *Solar Energy Materials & Solar Cells* 93: 355–361.

Wesenberg D., Kyriakides I., Agathos S. N. 2003 "White-rot fungi and their enzymes for the treatment of industrial dye effluents" *Biotechnology Advances* 22: 161– 187.

Xing J., Shan Z., Li K., Bian J., Lin X., Wang W., and Huang F. 2008 "Highly Effective Silver/Semiconductor Photocatalytic Composites Prepared by a Silver Mirror Reaction" *Journal of Physical Chemistry C* 69: 23–28.

Xu C., Cao L., Su G., Liu W., Liu H., Yu Y.,Qu X. 2010 "Preparation of ZnO/Cu2O compound photocatalyst and application in treating organic dyes" *Journal ofHazardous Materials*176: 807–813.

Yan M., Mori T., Zou J., Drennan 2009 " Effect of Grain Growth on Densification and Conductivity of Ca-Doped CeO_2 Electrolyte" *Journal of American Ceramic Society* 92: 2745–2750.

Zhang G.J., Feng Y., Xu Y.Y., Gao D.Z.,Sun Y.Q. 2012 "Controlled synthesis of mesoporous α-Fe2O3 nanorods and visible light photocatalytic property" *Materials Research Bulletin* 47: 625–630.

Zhang L., Cheng Z., Guo X., Jiang X., and Liu R. 2014 "Process optimization, kinetics and equilibrium of orange G and acid orange 7 adsorptions onto chitosan/surfactant" *Journal of Molecular Liquids* 197: 353-367.

In: Photocatalysis
Editors: P. Singh, M. M. Abdullah, M. Ahmad et al.
ISBN: 978-1-53616-044-4

Chapter 11

PHOTOCATALYTIC REMOVAL OF WATER POLLUTANT BY CADMIUM SULPHIDE AND ZINC SULPHIDE SEMICONDUCTOR NANOMATERIALS

*Nida Qutub**
Department of Chemistry, Jamia Millia Islamia University, New Delhi, India

ABSTRACT

The present work is focused over the photocatalytic degradation of organic dye Acid blue-29 (AB-29) in wastewater with semiconductor (SC) cadmium sulphide (CdS) nanoparticles (NPs) and their sandwich type and core-shell type nanocomposites with zinc sulphide (ZnS) nanoparticles. CdS nanostructures exhibited enhanced photocatalytic activity. ZnS NPs alone are themselves photo-inactive in the visible region of the solar spectrum, but when combined with CdS NPs showed good photocatalytic activity as well as stability in the visible range. The core-shell type CdS and ZnS nanocomposites (*ZnS/CdS* and *CdS/ZnS*) showed greater photocatalytic activity in comparison to sandwich type CdS and ZnS nanocomposites(NCs) (*CdS-ZnS*).

Keywords: nanocomposite (NC), photocatalysis, degradation of dyes, core-shell, sandwich-type NC

---

*Corresponding Author's E-mail: drnidaqutub@gmail.com.

1. INTRODUCTION

Synthetic dyes from textile industries and other organic industrial dyestuff have emerged as the major groups of water pollutants in the world (Ma et al. 2008). Approximately 1–15% of the synthetic textile dyes are disposed of in wastewater streams during processing or manufacturing (Mahmoodi et al. 2005) (Zhu et al. 2009). The colored waste in water bodies looks displeasing and interrupts light penetration into water bodies, upsetting the ecosystem within it. Moreover, some dyes may be noxious and even mutagenic to some organisms causing direct destruction of aquatic biota leading to aesthetic pollution, perturbation of aquatic life and eutrophication (Zhu et al. 2009). The more complex problem is that some of the dyes are highly persistent and cause lethal or carcinogenic effects in aquatic organisms. Hence, complete removal of dyes from such water systems is a major environmental issue (Papic et al. 2004). The waste discharged from textile industries have great inconsistency in composition, and the dye's molecular structures are relatively stable, this makes most of the usual treatment methods (like non-destructive physical water treatment processes) inefficient for their successful removal (Ozacar & Sengil, 2003) (Sun et al. 2006) (Ki et al. 1991). A good alternative to non-destructive physical water treatment processes can be seen in terms of advanced oxidation processes (AOPs) as they can mineralize contaminants (especially organic) (Zhu et al. 2009) (Legrini et al. 1993). Among AOPs, photocatalytic degradation is one of the most proficient and economical method under easily available visible solar light (Cheng et al. 2007) (Zhiyong et al. 2008). Therefore, the use of various photocatalytic degradation processes for treatment of wastewater is a growing subject of research (Ma et al. 2008). In recent years, the focus of heterogeneous photocatalysis has gained interest in environmental applications such as treatment of wastewater and air purification. Photocatalytic purification of water can be tested using semiconductor (SC) nanomaterials as photocatalyst in UV or Visible light.

Cadmium sulphide (CdS) is a water-insoluble II-VI semiconductor material. But it is soluble in dilute mineral acids. It shows n-type intrinsic conductivity due to sulphur vacancies as a result of excess cadmium atoms (Acharya. 2009). The CdS Nanoparticles (NPs) exhibit distinctive chemical, physical and structural properties from the bulk CdS. Due to surface effects and quantum size effects, cadmium sulphide nanoparticles can display high stability, novel optical, electronic, magnetic, structural and chemical properties, thus, can be used for different important technological applications (Roduner. 2006) (El-Bially et al. 2012) (Bhattacharya & Saha. 2008) (Wang & Herron. 1990) (Banerjee et al. 2000). The energy of the band gap of CdS is 2.42eV at room temperature that allows its nanoparticles to behave remarkably in optoelectronics, photovoltaics, photonics, and photocatalysis (Lin et al. 2002) (Girginer et al. 2009) (Ma et al. 2007). CdS are significantly sensitive in visible-light, thus it can utilize solar energy efficiently. Cadmium sulphide has been widely used for photocatalytic splitting of H_2O for energy production (Shangguan and Yoshida, 2002) (Zhen et al. 2018) and for photodegradation of air and water pollutants (Yin et al. 2001) under visible light (VL) (Qutub, 2013).

Zinc sulphide (ZnS) is photoactive in the UV region as its band gap (Eg = 3.68eV) lies in UV-range. Nanosized ZnS show new physical, chemical, optical and electrical properties due to increased band gap and quantum size effect. But, as it is known that UV-region consists of only 4.5% of total solar radiation, thus it will be more beneficial if somehow ZnS is made photoactive in the visible region. The photo-activity of semiconductors can also be improved by doping, embedding or coating them with another material. When a nanocomposite has both of its constituents in close proximity to each other such that a heterojunction is formed in between them, then the nanocomposite is called as sandwich-type NC. While if in a nanocomposite there is a coating of one substance over the other (in an onion-like fashion), then it is called core-shell–type NC. In case of semiconductors when the core is enveloped by a shell of another semiconductor with a higher band gap, then core-shell is type-I, and vice-versa is core-shell type-II (Estevez-Hernandez et al. 2012) (Kim et al. 2007). In core-sell NC also, a heterojunction is developed between core and shell. In addition to this, shell act as a physical barrier in between the inner core and the outer surrounding making the nanocomposite more durable and resistant to surface chemistry, environmental changes, and photo-oxidation. ZnS has been used as a shell or capping layer in various core-shell NC's such as $CuInS_2$/ZnS (Li et al. 2009), CdSe/ZnS (Nizamoglu et al. 2007), ZnO/ZnS (Shuai and Shen, 2011), CdS/ZnS (Mandal et al. 2008) core-shell structures. In that sense, ZnS with large band gap (Eg = 3.68eV), can be a considerable semiconductor material to combine with the CdS (band gap Eg = 2.24eV) forming a continuous series of solid solutions ($Cd_{1-x}Zn_xS$, wherein the crystal structure metal atoms are replaced mutually), which can be effectively used as core-shell material for photocatalytic and biological applications and as light-emitting devices (Yang et al. 2005) (Macias-Sanchez et al. 2012) (Bahrami et al. 2012) in visible region of solar spectrum.

2. Photocatalytic Degradation Using Nanomaterials

2.1. Photocatalytic Degradation Using Cadmium Sulphide Nanoparticles

2.1.1. Synthesis of Cadmium Sulphide Nanoparticles

Synthesis of CdS nanoparticles can be done by a green synthesis using the single pot method. Simple single pot chemical precipitation method was followed in aqueous medium under ambient conditions using different combinations of chemical precursors. Where Hydrogen Sulphide (H_2S), Sodium Sulphide (Na_2S) and Ammonium Sulphide ($(NH_4)_2S$) were taken as a different source of (Sulphide) S^{2-} ions. The S^{2-} ion source has a prominent effect on nanoparticles size, the more active S^{2-} ions source will give smaller sized CdS-NPs. The presence of a stabilizing agent prevents the agglomeration of the nanoparticles. Here the study of six different CdS Nanoparticles (named R1, R2, R3, R4, R5, and R6) is presented. All the Nanoparticles are synthesized using Cadmium Nitrate ($Cd(NO_3)_2$) (0.085M) as Cd^{2+} ion source and for S^{2-} ion $(NH_4)_2S$, H_2S, and Na_2S was taken in the presence and absence of stabilizing agents. The detail reaction methods were given elsewhere (Qutub et al. 2016). It was found that the average particle sizes for all six CdS Nanoparticles were almost the same (4-10nm) when obtained by different techniques (UV-visible spectroscopy, XRD, and TEM), and the particle size decreased from R1 to R6. All the synthesized NPs displayed a considerable shift towards blue in absorption edge with respect to the bulk CdS as shown by absorption spectra. CdS-NPs exhibited quantum size effect and showed an enhancement in the band gap with the particle size reduction. Crystalline nature of the NPs was confirmed by XRD which showed hexagonal and cubic type phases. All of them showed fine elemental purity and good thermal stability (Qutub et al. 2016).

2.1.2. Procedure of Photocatalytic Degradation

The photocatalytic activity of the above-mentioned CdS nanoparticles was studied by examining the decolorization of a dye derivative Acid Blue-29 (AB-29) in the presence of visible light. The photocatalytic experiments were performed in an immersion well, double jacketed, Pyrex glass photoreactor equipped with a magnetic bar, water circulating jacket and an opening for molecular oxygen. Irradiation was carried out using a 50W halogen linear lamp (9500Lumens). An optimized catalyst dosage ($1gL^{-1}$) was added to 180mL aqueous dye solution (0.06mM) and was first magnetically stirred in dark, in presence of atmospheric oxygen for at least 20 minutes to attain adsorption-desorption equilibrium between dye and catalyst surface. After this, the first sample (0minute, 5mL) was taken out and irradiation was started. The other samples (5mL each) were collected at regular intervals during the irradiation and analyzed after centrifugation. The suspensions were continuously purged with molecular oxygen throughout each experiment. The reaction temperature was kept constant at 20±0.3°C using the refrigerated circulating liquid bath. The decolorization of AB-29 was monitored by the change in absorption using UV-visible spectroscopic analysis technique (Shimadzu UV-Vis 1601). The concentration of dye was calculated by a standard calibration curve obtained from the absorbance of the dye at different known concentrations.

The rate of degradation considerably increases with a decrease in particle sizes. Degradation rate can be calculated using the following equation:

$$\text{Degradation rate (\%)} = \left(\frac{1-C}{C_0}\right) \times 100 \qquad (1)$$

where C_0 is the initial concentration of dye (mgL^{-1}) and C is the instant concentration of dye (mgL^{-1}) in the sample at time t (Zhu et al. 2009).

Also, the degradation rate can be calculated using the formula (Singh et al. 2007) (Tariq et al. 2008):

$$- dC/dt = kC^n \tag{2}$$

k = rate constant, C = concentration of the dye, t = reaction time, n = order of reaction.

The smallest sized particle showed greatest photocatalytic activity as they possess the largest band gap and highest surface to volume ratio. Figure 1, shows the decolorization rate for the decomposition of dye AB-29 in the presence of synthesized CdS photocatalysts (R1 to R6), which revealed that the decolorization of AB-29 proceeded faster with the decrease in sizes of NP.

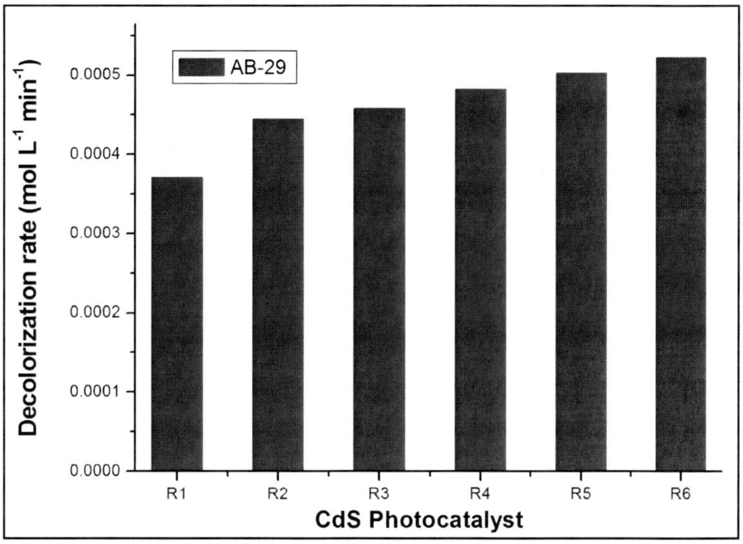

Figure 1.The decolorization rate of AB-29 in the presence of different CdS nanocatalysts (R1, R2, R3, R4, R5, and R6).

But, it was observed that the stability and reusability of the CdS NPs decreased with time after several consecutive uses in photocatalysis (Figure 2). The reason behind this could be the possible photocorrosion of CdS, i.e., the leaching of Cd^{2+} ions. Thus, certain modifications have to be done to utilize the effective photocatalytic properties of CdS, by somehow making it more stable and lesser photocorrosive.

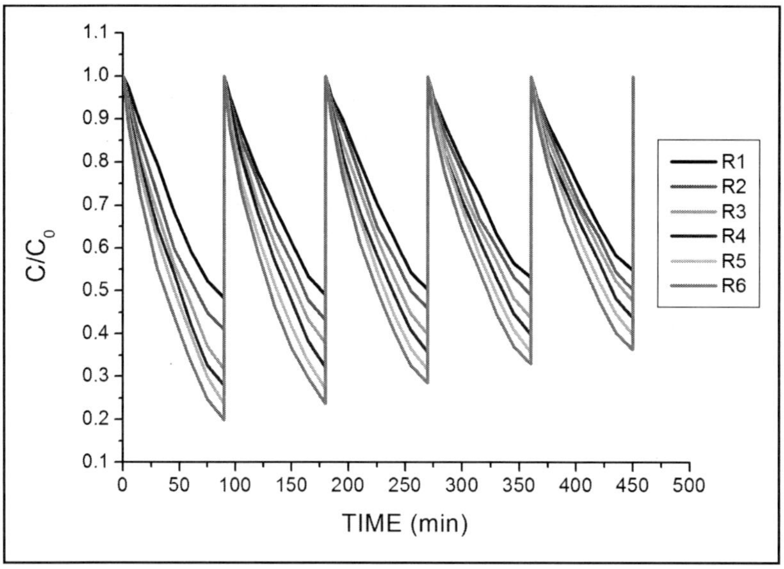

Figure 2. The relative decolorization using synthesized CdS nanocatalysts (R1, R2, R3, R4, R, and R6) under visible light irradiation for the 5 cyclic reuse after 90 minutes of reaction time.

2.1.3. Mechanism of Photocatalytic Reaction

Kinetics of a photocatalytic reaction can be expressed using the Langmuir–Hinshelwood (L–H) model (Jiang et al. 2012) (Dong et al. 2007) (Wu et al. 2006). That states when the starting concentration of dye taken is low; the photocatalytic degradation of dye with the photocatalyst can be simplified to clear pseudo-first order kinetics and is represented as

$$-dC/dt = (kKC)/(1+KC) \tag{3}$$

where k, is the reaction rate constant ($mMmin^{-1}$), t is the reaction time (min), C is the dye concentration (mM) at time "t" and "K" is the adsorption coefficient of the reactant (mM^{-1}), (Singh et al. 2007). If the concentration "C" is very small, KC will be negligible with respect to unity, so that Equation (1) can be simplified to pseudo-first-order kinetics (Li et al. 2006) (Dong et al. 2007). Thus the equation becomes:

$$-dC/dt = kKC = k_{app}C \tag{4}$$

where k_{app} is the apparent pseudo-first-order rate constant (min^{-1}).

The rate constant can also be calculated by the plot of the natural logarithm of dye concentration as a function of irradiation time (Jiang et al. 2012) (Wu et al. 2006),i.e.,

$$\ln C_0/C = k_{app}t \tag{5}$$

where C_0 is the initial concentration of dye (mM), C is the dye concentration at time t (mM), k_{app} is the apparent pseudo-first-order reaction rate constant (min^{-1}) and t is the reaction time (min). A plot of $\ln(C_0/C)$ versus t will give the slope of k_{app} (Zhu et al. 2009)(Wu et al. 2006) (Jiang et al. 2012).

The mechanism behind the photocatalysis by CdS nanoparticles can be explained on the basis of photogenerated electrons and holes obtained after excitation of CdS. Photogenerated-electrons in the HOMO jumped into LUMO, leaving behind holes in HOMO (Equation 6).

$$CdS \xrightarrow{h\nu} CdS\ (h^+ + e^-) \tag{6}$$

As a result, the photo-generated electrons/holes can act as redox centers and induce reduction/oxidation reactions respectively on the surface of the catalyst.

The photogenerated electrons in CdS nanocrystals were taken up by molecular oxygen (O_2) to give the superoxide radical anion $O_2{\bullet}-$ (Equation 7) and hydrogen peroxide H_2O_2 (Equation 8). These intermediates will then interacted to produce hydroxyl radical •OH (Equation 9). Hydroxyl radical (•OH) is itself a strong oxidizing agent capable of degrading most of the pollutants (Equation 10) (Wu et al. 2006) hence could cause complete degradation of dyes leading to the final product. On the other hand, due to the valence band potential in CdS nanocrystals, the photo-generated holes cannot oxidize hydroxyl groups to hydroxyl radicals. This would result in the photocorrosion of cadmium sulphide, forming cadmium cations (Cd^{2+}) (Wu et al. 2006) (Zyoud et al. 2010). Although photo-generated holes cannot produce hydroxyl radicals, they can, to some extent oxidized dye molecules to reactive intermediates, and further to final products (Equation 11) (Shi et al. 2012).

$$e^- + O_2 \rightarrow O_2^{\cdot -} \tag{7}$$

$$2e^- + O_2 + 2H^+ \rightarrow H_2O_2 \tag{8}$$

$$H_2O_2 + O_2^{\cdot -} \rightarrow {}^{\cdot}OH + OH^- + O_2 \tag{9}$$

$$OH + dye \rightarrow degradation\ products \tag{10}$$

$$CdS(e^- + h^+) + Dye \rightarrow CdS(e^-) + Dye^{\bullet +} \rightarrow CdS(e^-) + degradation\ products \tag{11}$$

The mechanism involved in photocatalysis using Cadmium sulphide nanoparticles can be represented as shown in Figure 3.

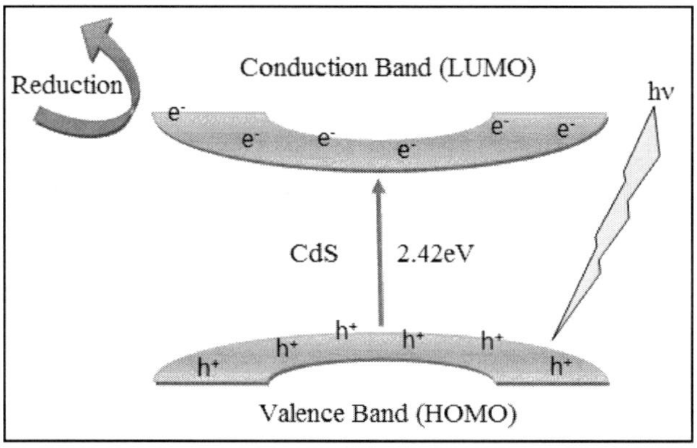

Figure 3. Schematic representation of the mechanism involved in photocatalysis by CdS NPs.

2.2. Photocatalytic Degradation Using Zinc Sulphide and Cadmium Sulphide Nanocomposites

Three different types of nanocomposites of CdS and ZnS are reported here; sandwich type CdS-ZnS, core-shell type ZnS/CdS and (CdS/ZnS) nanocomposite. The method of synthesis is given elsewhere (Qutub et al. 2015). The results revealed that among all the three CdS and ZnS NCs, core-shell nanocomposites (ZnS/CdS and CdS/ZnS) showed greater photoactivity than sandwich type NCs (CdS-ZnS). Moreover, among the core-shell types, the NC with shell material of larger band gap than the core (CdS/ZnS) displayed good photocatalytic activity in comparison to the other core-shell NC (ZnS/CdS), which has shell material with a lower band gap than the core. The decolorization rate (Figure 4) for the decomposition of AB-29 in the presence of the synthesized nano-photocatalysts (CdS-ZnS, ZnS/CdS, CdS/ZnS, ZnS, and CdS) in the visible region followed the order: (ZnS)<(CdS)<(CdS-ZnS)<(ZnS/CdS)<(CdS/ZnS).

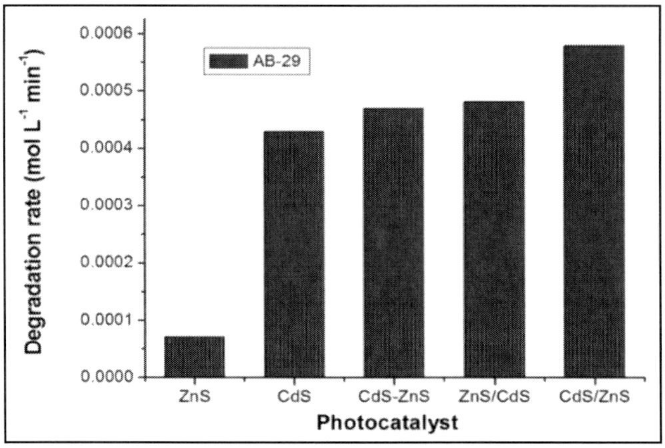

Figure 4. The decolorization rate of dye AB-29 using different nano-photocatalysts.

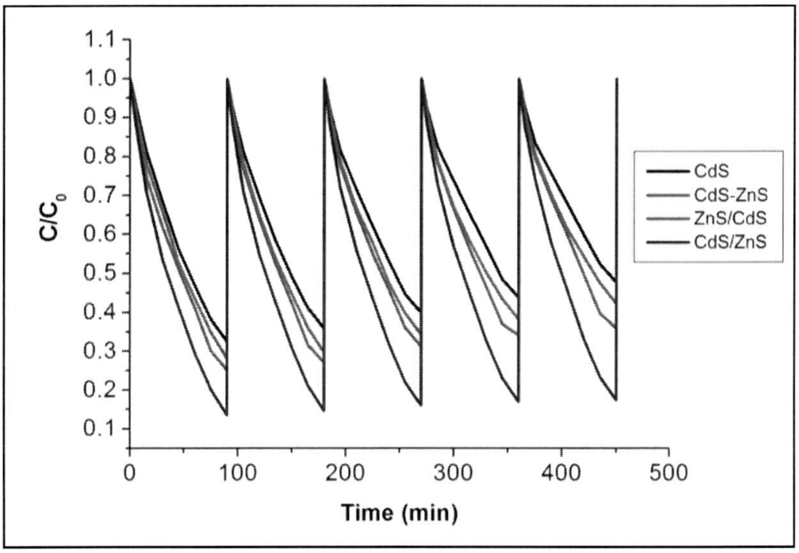

Figure 5. The decolorization rates in presence of various nano-photocatalysts (pure CdS, CdS-ZnS, ZnS/CdS, and CdS/ZnS) reused for 5 consecutive cycles after 90 minutes of reaction time each, under visible light irradiation.

The photocatalytic activity of all the nanocomposites after repetitive photo-degradation of AB-29 for five successive cycles (Figure 5) followed the order (CdS/ZnS) > (ZnS/CdS) > (CdS-ZnS). Out of core-shell (ZnS/CdS and CdS/ZnS) and sandwich (CdS-ZnS) NCs, ZnS/CdS and (CdS/ZnS) NCs showed better stability for five consecutive cycles which can be credited to the core-shell structure of (ZnS/CdS) and (CdS/ZnS) in contrast to sandwich structure of (CdS-ZnS). On the other hand (CdS/ZnS) showed maximum relative stability in the efficiency of decolorization rate in the following repetitive experiences, confirming that the CdS core was well shielded with ZnS shell. Hence (CdS/ZnS) NC has proved out to be a better photocatalyst in comparison to pure CdS and ZnS nanoparticles and (ZnS/CdS), (CdS-ZnS) nanocomposites in the visible region. (CdS/ZnS) NC showed enhanced photoactivity and improved photo-stability due to modification of surface charge and passivation of surface electronic states of CdS core by the ZnS shell, hence forming a heterojunction, thereby preventing loss of the Cd^{2+} during the degradation process.

A number of possible explanations have been proposed for the enhanced and improved photocatalytic activity of CdS&ZnS NCs. A direct relationship is seen between the photocatalytic activity and the particle size of CdS and ZnS NP's indicating that the high surface-to-volume ratio due to nano-size and the presence defect sites at CdS-ZnS interfaces might play a key role in the improved photocatalytic activity of the semiconductor nanocomposites. According to some scientists the charge separation mechanism related to nanocomposites consisting of two semiconductor photocatalysts involves the transfer of photo-generated charges from one semiconductor into the lower lying energy bands of another semiconductor (Deshpande et al. 2009), this is because, in semiconductors, the interfaces between two dissimilar materials exhibit distinct physicochemical properties due to the formation of a heterojunction. These properties are the result of the difference in the crystallographic planes present at the interfaces, which in turn can lead to in-homogeneity of atomic environments and generate certain microscale defects centers at interfacial sites (He et al. 2006). In conclusion, it can be stated that a close contact thereby creating heterojunction between CdS and ZnS nanoparticles is necessary for the enhanced photocatalytic activity (He et al. 2006). Hence, a representative diagram of the possible mechanism of photoreaction for CdS and ZnS nanocomposites showing the migration of charges through the interface can be drawn as given in Figure 6.

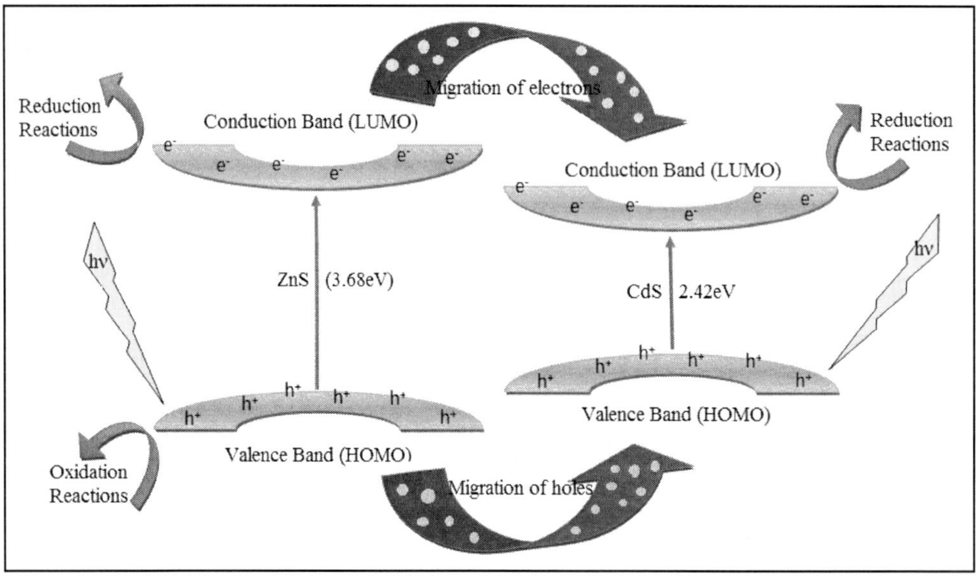

Figure 6. A representative diagram of the mechanism of photo-reaction by CdS and ZnS nanocomposite showing the migration of electrons from CB of ZnS to VB of CdS and holes from VB of ZnS to CB of CdS.

Conclusion

Thus, by decreasing the size of semiconductors to nano-range, its various properties can be enhanced. An increase in band-gap due to quantum size effect will lead to restricting photogenerated electron-hole recombination, thereby, enhancing its photo-, opto-, electronic ability. The electron-hole recombination can be further restricted by combining two different semiconductors which will lead to migration of photogenerated electrons and holes towards different materials enhancing photo-activity. In addition, the photoresponse of semiconductors that are only active in the UV range can be shifted to the visible range of solar spectrum leading to efficient use of solar light in photocatalysis.

The photo-corrosion of the catalysts causing a decrease in efficiency can be reduced by doping, capping or embedding the semiconductors with other semiconductors.

REFERENCES

Acharya K. P. 2009 "*Photocurrent Spectroscopy of CdS/Plastic, CdS/Glass, and ZnTe/GaAs Hetero-pairs Formed with Pulsed-Laser Deposition*" Bowling Green State University.

Bahrami M. N., Rezaee K. and Zobir M. 2012 "Facile Synthesis of ZnS/CdS and CdS/ZnS Core-Shell Nanoparticles Using Microwave Irradiation and Their Optical Properties" *Chalcogenide Letters* 9:379–387.

Banerjee R., Jayakrishnan R. and Ayyub P. 2000 "Effect of the Size-Induced Structural Transformation on the Band Gap in CdS Nanoparticles" *Journal of Physics: Condensed Matter* 12:10647–10654.

Bhattacharya R. and Saha S. 2008 "Growth of CdS Nanoparticles by Chemical Method and its Characterization" *Pramana* 71:187–192.

Cheng Y., Sun H., Jin W. and Xu N. 2007 "Photocatalytic Degradation of 4-Chlorophenol with Combustion Synthesized TiO_2 Under Visible Light Irradiation" *Chemical Engineering Journal* 128:127–133.

Deshpande A., Shah P., Gholap R. S. and Gupta N. M. 2009 "Interfacial and Physico-Chemical Properties of Polymer-Supported CdS·ZnS Nanocomposites and Their Role in the Visible-Light Mediated Photocatalytic Splitting of Water" *Journal of Colloid and Interface Science* 333:263–268.

Dong X., Ding W., Zhang X. and Liang X. 2007 "Mechanism and kinetics model of degradation of synthetic dyes by UV–vis/H2O2/Ferrioxalate complexes" *Dyes and Pigments* 74:470–476.

El-Bially A. B., Seoudi R., Eisa W., Shabaka A. A., Soliman S. I., El-Hamid R. K. A., and Ramadan R. A. 2012 "Preparation, Characterization and Physical Properties of CdS Nanoparticles with Different Sizes" *Journal of Applied Sciences Research* 8:676–685.

Estevez-Hernandez O., Gonzalez J., Guzman J., Santiago-Jacinto P., Rendon L., Montes E. and Reguera, E. 2012 "Mercaptopropionic Acid Capped CdS@ZnS Nanocomposites: Interface Structure and Related Optical Properties"*Science of Advanced Materials* 4:771–779.

Girginer B., Galli G., Chiellini E. and Bicak N. 2009 "Preparation of Stable CdS Nanoparticles in Aqueous Medium and their Hydrogen Generation Efficiencies in Photolysis of Water" *International Journal of Hydrogen Energy* 34:1176–1184.

He J., Ji W., Mi J., Zheng Y. and Ying J. Y. 2006 "Three-Photon Absorption in Water-Soluble ZnS Nanocrystals" *Applied Physics Letters* 88:181113–181114.

Jiang R., Zhu H., Yao J., Fu Y. and Guan Y. 2012 "Chitosan Hydrogel Films as a Template for Mild Biosynthesis of CdS Quantum Dots with Highly Efficient Photocatalytic Activity" *Applied Surface Science* 258:3513–3518.

Ki R., Stucki S. and Carcer B. 1991 "Electrochemical Waste Water Treatment Using High Overvoltage Anodes. Part I: Physical and Electrochemical Properties of SnO2 Anodes" *Journal of Applied Electrochemistry* 21:14–20.

Kim M. R., Kang Y.M., and Jang D.-J. 2007 "Synthesis and Characterization of Highly Luminescent CdS@ZnS Core-Shell Nanorods" *The Journal of Physical Chemistry C* 111: 18507–18511.

Legrini O., Oliveros E. and Braun A. M. 1993 "Photochemical Processes for Water Treatment" *Chemical Reviews* 93:671–698.

Li L., Daou T. J., Texier I., Chi T. T. K., Liem N. Q. and Reiss P. 2009 "Highly Luminescent $CuInS_2$/ZnS Core/Shell Nanocrystals: Cadmium-Free Quantum Dots for In Vivo Imaging" *Chemistry of Materials* 21:2422–2429.

Li Y., Li X., Li J. and Yin J. 2006 "Photocatalytic Degradation of Methyl Orange by TiO_2 Coated Activated Carbon and Kinetic Study" *Water Research* 40:1119–1126.

Lin C. F., Shih S.-M. and Su W. F. 2002 "CdS Nanoparticle Light-Emitting Diode on Si" *Symposium on Integrated Optoelectronic Devices*:102–110). International Society for Optics and Photonics.

Ma L. L., Sun H. Z., Zhang Y. G., Lin Y. L., Li J. L., Wang E. and Wang J. B. 2008 "Preparation, Characterization and Photocatalytic Properties of CdS Nanoparticles Dotted on the Surface of Carbon Nanotubes" *Nanotechnology* 19:115708–115709.

Ma R. M., Dai L. and Qin G. G. 2007 "Enhancement-Mode Metal-Semiconductor Field-Effect Transistors Based on Single nCdS nanowires" *Applied Physics Letters* 90:93103–93109.

Macias-Sanchez S. A., Nava R., Hernandez-Morales V., Acosta-Silva Y. J., Gomez-Herrera L., Pawelec B. and Fierro J. L. G. 2012 "$Cd_{1-x}Zn_xS$ solid solutions supported on ordered mesoporous silica (SBA-15): Structural features and photocatalytic activity under visible light" *International Journal of Hydrogen Energy* 37:9948–9958.

Mahmoodi N. M., Arami M., Limaee N. Y. and Tabrizi N. S. 2005 "Decolorization and Aromatic Ring Degradation Kinetics of Direct Red 80 by UV Oxidation in the Presence of Hydrogen Peroxide Utilizing TiO_2 as a Photocatalyst" *Chemical Engineering Journal* 112:191–196.

Mandal P., Srinivasa R. S., Talwar S. S. and Major S. S. 2008 "CdS/ZnS Core-Shell Nanoparticles in Arachidic Acid LB Films" *Applied Surface Science* 254:5028–5033.

Nizamoglu S., Ozel T., Sari E., and Demir H. V. 2007 "White Light Generation Using CdSe/ZnS Core-Shell Nanocrystals Hybridized with InGaN/GaN Light Emitting Diodes" *Nanotechnology* 18:65705–65709.

In: Photocatalysis
Editors: P. Singh, M. M. Abdullah, M. Ahmad et al.
ISBN: 978-1-53616-044-4

Chapter 12

REDUCED GRAPHENE OXIDE AS PHOTOCATALYST: NANOSTRUCTURE, SYNTHESIS AND APPLICATIONS

Sunny Khan[1], Shumaila[1], Harsh[2], M. Husain[3] and M. Zulfequar[1,*]

[1]Department of Physics, Jamia Millia Islamia, New Delhi, India

[2]Centre for Nanoscience and Nanotechnology, Jamia Millia Islamia, New Delhi, India

[3] M.J.P. Rohilkhand University, Bareilly, U. P., India

ABSTRACT

Graphene is a wonder material of 21st century with two dimensional honeycomb like structure. Its extraordinary and exciting electrical, optical and mechanical properties have made it a focal point of contemporary research in material science. Graphene finds applications in transparent electrodes, liquid crystal displays, solar cells and photo catalysis etc. This chapter focuses on to review the synthesis of graphene, its properties and its applications in photo catalysis.

Keywords: graphene, structure, reduced grapheme, photocatalysis

1. INTRODUCTION

Graphene is a two-dimensional sheet of sp^2-hybridized carbon atoms. It has an extended honeycomb network which forms the basic building block of other important

[*] Corresponding Author's E-mail: mzulfe@rediffmail.com.

allotropes like graphite, nano tubes and fullerene. When stacked layer by layer to form 3D structure it is called graphite, if rolled up it forms 1D nanotubes, and if wrapped, it takes the form of 0D fullerenes. Its extraordinary thermal, mechanical and electrical properties have been a driving force for many theoreticians and experimentalists. The long range π-conjugation in graphene is the reason behind its superior properties. The real boom in the field of graphene studies was witnessed in 2004, when Andre Geim and K. Novesolov for the first time, isolated single layer samples from graphite (Geim, 2011). It is this discovery that resulted into an explosion of interest in the field as prior to this, the existence of a stable two dimensional crystal at a finite temperature was supposed to be thermodynamically unviable (Landau, 1937). The Geim and his co-workers employed a mechanical exfoliation technique often called as Scotch –tape technique to isolate two dimensional crystals from three- dimensional graphite. These single to few layered graphene flakes were held onto a substrate where the only binding force was the van der Waals forces and eventually, the substrate was etched away to have free standing graphene layers (Meyer et al. 2007; Bolotin et al. 2008; Bolotin et al. 2008; Novoselov et al. 2005). This helped the researchers to look deep into the intrinsic properties of the same. It opened the whole new vista of interesting physics about the graphene which was long pending. The scientific fraternity has shown great interest in different properties of graphene like, the ambipolar field effect (Novoselov, 2004), anomalous qantum Hall effect (Novoselov et al. 2006; Jiang et al. 2007; Jiang et al. 2007; Zhang et al. 2005; Novoselov et al. 2007; Ozyilmaz et al. 2007) high carrier mobility (Novoselov et al. 2005; Morozov et al. 2008; Han et al. 2007) and single gas molecular detection (Schedin et al. 2007; Novoselov and Geim, 2007) which has made device applications of graphene, viable. These devices may include the likes of high speed logic devices, sensors and transparent electrodes for display and solar cells.

In spite of substantial progress in the field of graphene based device, the widespread application of graphene is yet to take place. The reason being the inability of producing graphene from the perspective of both the quality and quantity (Ruoff, 2008). The mechanical exfoliation could produce the best quality graphene, only predicament being the absence of high yield. To obtain a single layer we have to overcome the van der Waals force of attraction precisely between the first and second layer without disturbing any subsequent sheets. Therefore, alternatives to this approach have been researched so as to provide a good lead towards the conceptualisation of graphene devices. These methods include chemical routes, thermal decomposition and predominantly the chemical vapour deposition method. The chemical methods have their own limitations like lack of control over the thickness of graphene layers, presence of unwanted functional groups and harsh oxidising agents leaving a hampered graphitic structure, which could have an adverse effect over device performance. Thermal decomposition has certain degrees of reservations which include controlling the thickness of graphene layers and repeatability of large area graphene (Choi et al. 2010). The most prevalent technique these days for graphene synthesis is cvd method. It offers a good quality graphene with large scale area growth. It includes the substrate based growth of graphene using transition metals as catalyst to aid the nucleation process. Amongst several applications of graphene one of the prominent ones is graphene as a photocatalytic material.

2. Structure of Graphene

The graphene has been termed as the "mother of all graphitic forms" (Geim and Novoselov, 2007). Such an adjective could be earned because of superior properties that graphene possesses and the same have been discussed in proceeding sections. Before discussing its noteworthy properties, the structure of graphene has to be understood. The graphene could be single layer, bi-layer, few layer or multilayer. The single layer graphene is defined as a one atom thick two dimensional hexagonal sheet of carbon atoms (Hass et al. 2008) (see Figure 1). Bi- layer graphene is one which has two such single layers stacked over each other. The few layer, likewise, will have 3-10 such 2D hexagonal sheet of carbon atoms. Beyond this limit graphene would be a thick graphene which is not of much importance. These graphene sheets can stack in numerous ways in order to attain graphitic structure. There are three most common and feasible stacking arrangements viz., hexagonal or AA stacking, Bernal or AB stacking and rhombohedral or ABC stacking. Practically 80% of the single crystal graphite is found to have Bernal stacking (Haering, 1958) reason being lowest stacking energy. In a graphene layer, the carbon atoms are sp^2 hybridised in a 2 D hexagonal structure having three in-plane σ bonds. These bonds form the backbone of the structure. Other than these there are partially filled out of plane p_z orbitals having π- electrons. There is this π- electron network throughout the graphene sheet which interacts with another π- electron network of graphene sheet lying above and below it, thereby forming a weak interaction force between the layers which is termed as Van der Waals force (Schabel and Martin, 1992). This force is due to interaction between π-electron so often called as π- π interactions.

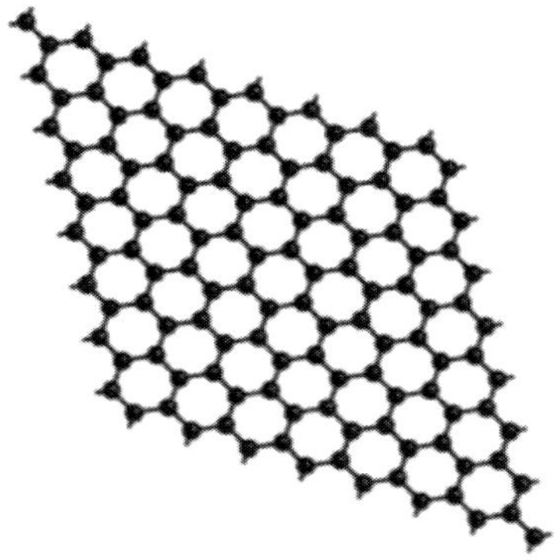

Figure 1. Single layer graphene.

3. Properties of Graphene

Theoretically the graphene layer has a thickness of 0.34nm with specific surface area of 2600 m^2/g (Chae et al. 2004). It possesses electron mobility in excess of 15000 $cm^2/V.s$ at room temperature and a thermal conductivity of 3000 W/m.K. A SLG can deal with a stress of about 1060 GPa (Dong and Chen, 2010) where the density stands merely at 2.2 g/cm^3. A single layer graphene is found to have 2.3% absorbance (Nair et al. 2008) for white light with less than 0.1% reflectance. The absorbance however linearly increases with increasing layers stacked one over another. It has been established beyond doubt that the graphene is the strongest material till date know to the human species. The single layer graphene is semi metal and a zero band gap semiconductor and possesses path breaking electronic properties. The charge carriers in graphene are known to have ballistic transport even under ambient conditions

4. Synthesis of Reduced Graphene Oxide

Way back in 1855, Brodie took a chemical route to synthesise graphite oxide by treating graphite powder with nitric acid and potassium nitrate (Brodie, 1855). The idea was to oxidize graphite in order to intercalate oxygen atoms into the graphitic layers. The same idea was used by Hummers and Offeman (Hummers and Offman, 1958) which involved oxidation of graphite powder to synthesise GO (Graphite Oxide/Graphene oxide) and thereby its reduction to obtain r-GO i.e., reduced graphene oxide. This process involves oxidative treatment of graphite powder with H_2SO_4, $KMnO_4$ and H_2O_2 so as to obtain GO after proper filtration and washing to remove unwanted metallic contents. The as grown samples are then reduced by suitable reducing agents which may include hydrazine hydrate or hexamine. With time certain modifications have been introduced to get better quality as well as quantity of r-GO. The modifications are in terms of filtration technique (Khan et al. 2016; Chen et al. 2009). In the traditional method the GO suspension which consists of acids and metal ions is washed with HCl and DI (de-ionized) water whereas the modified method splits the process into parts. The GO suspension is first filtered with HCl before being dried and then again dispersed in DI water which intern is washed with DI water itself. The yield and quality has been found to have improved by such modifications. A group led by khan et al. made use of modified Hummer's technique to grow r-GO and further improved its electrical and optical properties by attaching Ag nanoparticles to it (see Figure 2).

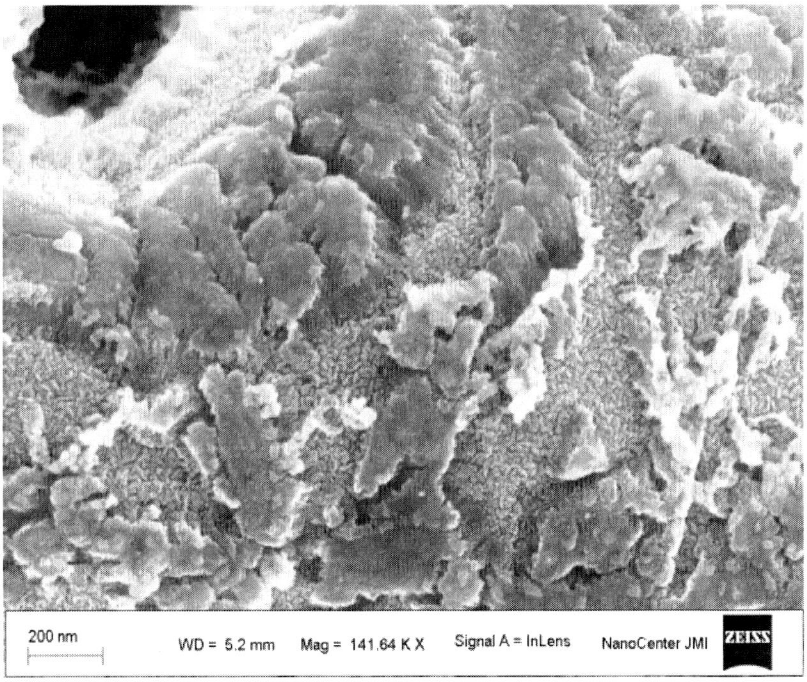

Figure 2. SEM micrograph of r-GO-Ag nano composite.

5. Photocatalysis

The impact of mankind over climate change and in turn on the ecosystem is a well known fact. However, there are certain measures which could help curb climate change and its repercussions. The shifting of interest towards a cleaner energy source namely solar energy is one such corrective measures. The photocatalysis is a part of this large family of techniques which aim towards a sustainable development of the mankind. In the earlier part of 20th century a number of reports came forth describing photocatalytic reactions but real impetus was generated in 1972 with the re-publication in English of a previous report of the splitting of water in hydrogen and oxygen at the surface of TiO2 under ultraviolet irradiation by A. Fujishima and K. Honda (Honda and Fujishima, 1972). By definition, photoctalysis stands for chemical reaction induced by the absorption of light by a solid material. The reaction takes place at the surface of this solid material which remains unchanged during and after the reaction, termed as the photocatalyst. A simple catalyst reduces the activation energy in thermodynamically favourable chemical reaction and increases the reaction rate. The photocatalyst, on the contrary can promote the reactions which are thermodynamically unfavourable because of the energy provided by light. This differentiates the catalysis with photo catalysis (Ohtani, 2010; Fuishima et al. 2008). The photocatalysis is the way forward to green chemistry leading to reduced levels of greenhouse gases.

5.1. Graphene Based Photocatalysts

One of the several applications of graphene is the synthesis of a heterogeneous graphene/semiconductor composite photoctalyst. The novel semiconductor based photocatalysts have been there for decades offering excellent applications like H_2 generation (Chen et al. 2010), water splitting (Li et al. 2014), CO_2 reduction (Marszewski et al. 2015), environmental clensing etc. Inspite of a proven efficiency of heterogeneous semiconductor based photocatalysts, there are certain predicaments, like low quantum yield, limited utilization of visible light and stability, towards having their practical applications. The wonder material graphene offers to counter such problems while leading towards better efficiency and stability.

5.2. Properties of Graphene Aiding Photocatalysis

The graphene and its derivatives enjoy an upper hand in the material science because of its properties bestowed upon it due to its unique electronic band structure, surface chemistry, molecular adsorption, activation energies etc. These very properties lend graphene an influential role in photocatalytic performance of graphene based photctalysts. These properties could be studied in the light of photocatalysis as follows.

5.2.1. Structural Properties

The photocatalytic abilities of graphene are due to the unique electronic structure of graphene emanating out of a high quality 2D crystal structure. Graphene has honeycomb lattice with sp2 hybridised carbon atoms having three in-plane σ orbitals and one out of plane (along Z axis) π orbital i.e., p_z. The crystal lattice of graphene is made up of two equivalent carbon sub lattices A and B per unit cell which leads towards a conical electronic band structure as established for the first time by Walace using DFT (Density Functional Theory) in 1947 (Wallace, 1974; Avouris, 2010). The π electrons in the adjacent p_z orbitals are associated with delocalised π (bonding) and π^* (anti-bonding) bands which correspond to highest occupied valance band and lowest unoccupied conduction band respectively (Li et al. 2016). These bands meet at six different points called as Dirac points as shown in Figure 3. The two points out of these six points labelled as K and K' correspond to linear dispersion of π and π^* states. This very feature of graphene's electronic structure makes graphene a semi-metal or a zero gap semiconductor (Geim, 2007). The unique hexagonal lattice 2D structure of graphene helps to couple with other nanostructured semiconductors so as to synthesise graphene based semiconductor photocatalysts.

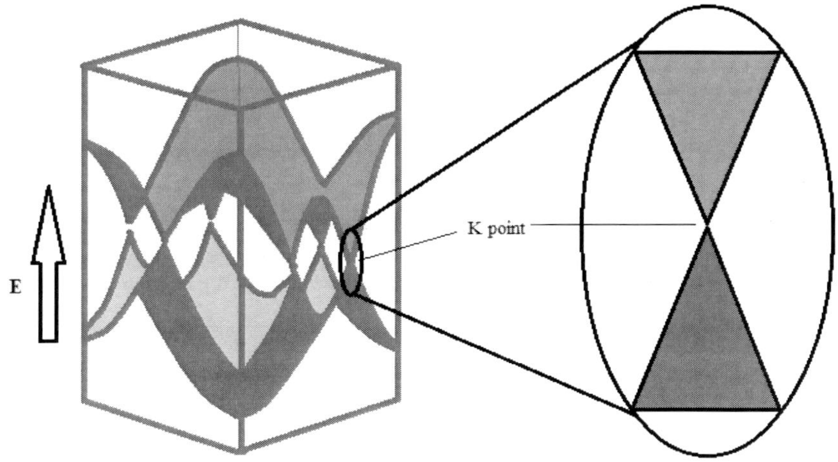

Figure 3. Electronic structure of graphene.

5.2.2. Zero Band Gap Semiconducting Properties

The ability of graphene to accept electrons enhances the photocatalytic activities of graphene based photocatalysts. The possibility to tune the work function of graphene helps reduce the contact barrier between graphene and semiconductor in graphene semiconductor composite photocatalysts. Moreover, graphene provides a good conductive support for the photo electrons involved in photocatalytic functions (Xiang and Jaroniec, 2012) by enhancing their lifetime. The large surface area offered by graphene is also a crucial aspect towards improved photocatalysis (Low et al. 2015).

5.2.3. Semiconducting Properties

The graphene has the capability of being functionalised to be n-type and p-type semi conductor (Sreeprasad and Berry, 2013). The band gap of graphene is highly tunable by means of chemical functionalization or by increasing the graphene layers. Likewise the band gap of GO (graphene oxide) is also tunable by means of changing its oxidation levels as its electronic properties depend on oxidation levels. Another important role that graphene plays as a component of the graphene-semiconductor photocatalyst is by broadening the visible light absorption range of the semiconductor. Due to the covalent chemical bonding between graphene and the semiconductor the band gap is narrowed by introduction of energy levels in between thereby increasing the visible light absorption (Yibing et al. 2013).

CONCLUSION AND PROSPECTS

Based on our brief insight into the world of graphene it can be concluded that graphene is in true sense a very fascinating and diversified materials. But to thrive for a real long time it will have to deliver its proposed applications. The success of graphene is largely dependent on two aspects first one being the large scale and controllable synthesis as for different applications, the different structures (number of layers, grain size, defects etc.) and quantities are required. Second one is the development of on- table applications, which may include energy storage devices, transparent electrodes etc.

However, it is being said that research on graphene's electronic properties has hit a plateau, it is not likely in near future that the gleam of graphene would fade away. Investigations on non-electronic properties of graphene have taken up now. In all possibilities graphene is here to stay and remain a gem in the crown of condensed matter physics.

References

Avouris P. 2010 "Graphene: Electronic and Photonic Properties and Devices." *Nano Letters* 10:4285-4294.

Bolotin K., Sikes J., Hone J., Stormer H. L and Kim P. 2008 "Temperature-Dependent Transport in Suspended Graphene." *Physics Review Letters*101:096802-4

Bolotin K., Sikes J., Jiang Z., Klima M., Fudenberg G., Hone J., Kim P. and Stormer H. L. 2008 "Ultrahigh electron mobility in suspended graphene." *Solid State Commununication*146:351-355.

Brodie B. 1855 "Note sur un Nouveau Procede pour la Purification et la Pesagregation du Graphite." ["Note on a New Process for the Purification and Pesagregation of Graphite"], *Annales de Chimie et de Physique* 45:351–353.

Castro Neto A. H., F. Guinea, and Peres N. M. R. 2006 "Drawing conclusions from graphene." *Physics World* 19:33-37.

Chae H. K., Siberio-Pérez D. Y., Kim J., Go Y. B., Eddaoudi M., Matzger A. J., O'Keeffe M. and Yagh O. M. 2004 "A route to high surface area, porosity and inclusion of large molecules in crystals." *Nature* 427:523–527.

Chen T., Zeng B., Liu J. L., Dong J. H., Liu X. Q., Wu Z., Yang X. Z. and Li Z. M. 2009 "High throughput exfoliation of graphene oxide from expanded graphite with assistance of strong oxidant in modified Hummers method." *Journal of Physics: Conference Series* 188:012051-54.

Chen X., Shen S, Guo L and Mao S. S. 2010 "Semiconductor-based photocatalytic hydrogen generation." *Chemical Reviews*110:6503-570.

Choi W., Lahiri I., Seelaboyina R. and Kang Y. S. 2010 "Synthesis of graphene and its applications: A Review." *Critical Reviews in Solid State and Materials Sciences* 35:52-71.

Dong L. X. and Chen Q. 2010 "Properties, synthesis, and characterization of graphen." *Frontier of Material Sciences in China* 4: 45-51.

Fujishima A., Zhang X. and Tryk D. A. 2008 "TiO_2 photocatalysis and related surface phenomena." *Surface Science Reports* 63:515–582

Geim A. K. 2011 "Random walk to graphene (Nobel Lecture)." *Angewandte Chemie International Edition* 50.31:6966-6985.

Geim A. K. and Novoselov K. S. 2007 "The rise of graphene." *Nature Materials* 6:183-191.

Haering R. R. 1958 "Band Structure of Rombohedral Graphite" *Canadian Journal of Physics.* 36:352-362.

Han M., Ozyilmaz B., Zhang Y. Jarillo-Herrero P. and Kim P. 2007 "Electronic transport measurements in graphene nanoribbons." *Physics Status Solidi B* 244:4134-4137.

Honda A. and Fujishima K. 1972 "Electrochemical Photolysis of Water at a Semiconductor Electrode." *Nature 238:37–38.*

Hummers W. S. and Offeman R. E. 1958 "Preparation of Graphitic Oxide." *Journal of the American Chemical Society* 80:1339-1339.

Hass, J., de Heer W. A. and Conrad E. H. 2008 "The growth and morphology of epitaxial multilayer graphene." *Journal of Physical Condensed Matter* 20:1-26.

Jiang Z., Zhang Y., Stormer H. L. and Kim P. 2007 "Quantum Hall states near the charge-neutral Dirac point in graphene." *Physics Review Letters* 99:106802-4.

Jiang Z., Zhang Y., Tan Y. W., Stormer H. L. and Kim P. 2007 "Quantum Hall effect in graphene." *Solid State Communication* 143:14-19.

Khan S., Ali J., Harsh, Mushahid H. and Zulfequar M. 2016 "Synthesis of reduced graphene oxide and enhancement of its electrical and optical properties by attaching Ag nanoparticles." *Physica E* 81:320–325.

Landau L. D. 1937 "Zu Theorio der phasenumwandlungen" ["To Theorio of the phase transformations"] *Physics Z. Sowjetunion* 11: 26-35

Li Q. Li X., Wageh S., Ghamdi A. A. A. and Yu J. 2015 "CdS/Graphene Nanocomposite Photocatalysts." *Advanced Energy Materials* 5:1500010-21.

Li X., Yu J., Low J., Fang Y., Xiao J and Chen X. 2014 "Engineering heterogeneous semiconductors for solar water splitting." *Journal of Material Chemistry A* 3:2485-2534.

Low J. Yu J. and Ho W. 2015 "Graphene-Based Photocatalysts for CO_2 Reduction to Solar Fuel." *Journal of Physical Chemistry Letters* 6:4244-4251.

Marszewski M., Cao S., Yu J. and Jaroniec M. 2015 "Semiconductor-based photocatalytic CO_2 conversion." *Materials Horizons* 2:261-278.

Meyer J. C, Geim A. K., Katsnelson M. I., Novesolov K. S., Booth T. J. and Roth S. 2007 "The structure of suspended graphene sheets." *Nature* 446:60-3.

Morozov S. V. Morozov S. V., Novoselov K. S., Katsnelson M. I., Schedin F., Elias D. C., Jaszczak J. A. and Geim A. K. 2008 "Giant intrinsic carrier mobilities in graphene and its bilayer." *Physics Review Letters*100:016602-4.

Nair R. R., Blake P., Grigorenko A. N., Novoselov K. S., Booth T. J., Stauber T., Peres N. M. R. and Geim A. K. 2008 "Fine Structure Constant Defines Visual Transparency of Graphene" *Science* 320:1308-7.

Novoselov K. and Geim A. K. 2007 "Graphene detects single molecule of toxic gas" *Advanced Materials Technology* 22:178–179.

Novoselov K. S., McCann E., Morozov S. V., Fal'ko V. I., Katsnelson M. I. and Geim A. K. 2006 "Unconventional quantum Hall effect and Berry's phase of 2π in bilayer graphene" *Nature Physics* 2:177-180.

Novoselov K. S., Geim A. K., Morozov D., Jiang D., Katsnelson M. I., Grigorieva I. V., Dubonos S. V. and Firnov A. A. 2005, "Two-dimensional gas of massless Dirac fermions in graphene" *Nature* 438:197-200.

Novoselov K. S., Geim A. K., Morozov S. V., Jiang D. and Zhang Y. 2004 "Electric field effect in atomically thin carbon films" *Science* 306:666-9.

Novoselov K. S., Jiang D., Schedin F., Booth T. L., Khotkevich V. V., Morozov S. V. and Geim A. K. 2005 "Two-dimensional atomic crystals" *Proceedings of the National Academy of Sciences of the U.S.A.* 102:10451-53.

Novoselov K. S., Jiang Z., Zhang Y., Morozov S. V, Stormer H. L., Zeitler U., Maan J. C., Boebinger G S., Kim P. and Geim A. K. 2007 "Room-Temperature Quantum HallEffect in Graphene". *Science* 315:1379.

Ohtani B. 2010 "Photocatalysis A to Z—What we know and what we do not know in a scientific sense" *Journal of Photochemistry & Photobiology, C: Photochemistry Reviews* 11:157–178.

Ozyilmaz B., Jarillo-Herrero P., Efetov D., Abanin D. A., Levitov L. S. and Kim P. 2007 "Electronic Transport and Quantum Hall Effect in Bipolar Graphene p–n–p Junctions" *Physics Review Letters* 99:166804-4.

Ruoff R. 2008 "Calling all chemists" *Nature Nanotechnolgy* 3:10-11.

Schabel M. C. and Martins J. L. 1992 "Energetics of interplanar binding in graphite" *Physical Review B* 46:7185-87.

Schedin F., Geim A. K., Morozov S. V., Hill E. W., Blake P., Katsnelson M. I. and Novoselov K. S. 2007 "Detection of individual gas molecules adsorbed on graphene" *Nature Materials* 6:652-5.

Sreeprasad T. S. and Berry V. 2013 "How do the electrical properties of graphene change with its functionalization?" *Small* 9:341-350.

Wallace P. 1947 "The band theory of graphite" *Physical Reviews* 71:622-634.

Xiang Q. J and Jaroniec M. 2012 "Graphene-based semiconductor photocatalysts" *Chemical Society Review* 41:782-96.

Xin Li, Yu J., Wageh S., Ghamdi A. A. A. and Xie J. 2016 "Graphene in Photocatalysis: A Review" *Small* 12:6640–6696.

Yibing L., Zhang H., Liu P., Wang D., Li Y. and Zhao H. "Cross-Linked g-C_3N_4/rGO Nanocomposites with Tunable Band Structure and Enhanced Visible Light Photocatalytic Activity" *Small* 9:3336-3344.

Zhang Y. B., Tan Y. W., Stormer H. L. and Kim P. 2005 "Experimental observation of the quantum Hall effect and Berry's phase in graphene" *Nature* 438:201-4.

In: Photocatalysis
Editors: P. Singh, M. M. Abdullah, M. Ahmad et al.
ISBN: 978-1-53616-044-4

Chapter 13

Recent Trends in Catalyst Based Water Splitting Technology: Research and Applications

Anjana Pandey*[*] *and Saumya Srivastava

Department of Biotechnology, Motilal Nehru National Institute of Technology Allahabad, Prayagraj, Uttar Pradesh, India

Abstract

H_2 (hydrogen) has gained importance as a clean and alternative fuel due to rise in global warming. The breakdown of water into hydrogen and oxygen is termed as water splitting. Water splitting has been observed to be occurring naturally in plants during the photosynthesis process. The other methods and equipments that have been developed for water splitting include photoelectrochemical water splitting, radiolysis, photobiological water splitting, photocatalytic water splitting, thermal decomposition of water etc. Presently current research is focused on photocatalysis for H_2 production. Artificial photosynthesis is another field of research, which replicates photosynthesis phenomena for H_2 production through water splitting. The newly developed efficient and economical water splitting methods can play a key role in hydrogen production. This chapter discusses different water splitting technologies, recent advancements in water splitting technology and equipment development and future aspects of the water splitting technology for efficient production of hydrogen as clean fuel.

[*]Corresponding Author's E-mail: apandey70@rediffmail.com.

Keywords: photoelectrochemical water splitting, radiolysis, photobiological water splitting, photocatalysis

1. Introduction

Generally, traditional fossil fuels are considered as the major source for global energy, but these current available energy supplies are expected to be consumed in coming years (Lewis and Nocera 2006; Blankenship et al. 2011). Due to the severe environmental problems that have been resulted from continuous energy demand, it is necessary to develop environmental-friendly energy conversion technologies (Chu and Majumdar 2012; Cook et al. 2010).

Release of greenhouse gases, eg. carbon dioxide has been shown to contribute significantly in global warming that have made the great interest of developing alternative and renewable energy sources across the world [http://unfccc.int/essential_background/kyoto_protocol/items/1678.php]. The most propitious technology that has been considered for energy conversion is the fuel cell, specifically hydrogen fuel (Lucia 2014; Chan 2007) because of its high energy concentration, easy storage methods and low greenhouse gas production. But, currently major source for hydrogen production is fossil fuels and biomass that leads to production and confinement of (Chiarello et al. 2010; Navarro et al. 2007) carbon dioxide. To skip this difficulty, an alternative way of hydrogen production i.e., from clean and renewable resources has been found to have maximum interest in recent years. One method is to use water splitting technique.

Generation of hydrogen and oxygen by electrolysis is simplest and efficient way of water splitting technology. The net reaction (1) for water splitting is:

$$H_2O + \text{electrical energy} \rightarrow H_2\,(g) + \tfrac{1}{2}\,O_2\,(g) \qquad (1)$$

In this whole reaction, a highly useful gas, oxygen is the only by product. Water electrolyzers have been used for production of hydrogen at the vast range from few cm^3 /min to several thousand m^3/hour satisfying approximately 4% of the total hydrogen demand across the world (Casper 1978; Divisek 1990; Dönitz et al. 1990).

Water splitting process is partitioned into two half reactions oxidation and reduction. Oxidation involves the formation of O-O bond with four protons loss resulting in slow rate of the reaction and large over potential (Kanan and Nocera 2008; Wang et al. 2013) in contrast to water reduction which occurs at a lower over potential (Landon et al. 2012).

Two types of catalysts are mainly involved in water splitting process, heterogeneous and homogeneous catalysts. Due to high activity and greater durability, oxides and hydroxides of transition metals have been placed as the main active species for the former type of catalyst. But the homogeneous catalysts have been shown more advantageous than heterogeneous catalysts because of having a wide range of active metal centres (Liu et al. 2017).

Some studies have been published in the recent years involving transition-metal catalysts (Artero et al. 2011; Du and Eisenberg 2012; Singh and Spiccia 2013; Zou and Zhang 2015; Jamesh 2016; Han et al. 2016; Kondo and Masaoka 2016). Out of which water splitting catalysts based on copper have been proved as a promising candidate because of showing high stability, reactivity and low light absorption as well as it has distinct redox properties (Liu et al. 2017).

Large amount of electricity produced by combustion of coal used to be exploited in electrolysis process, which is really a very costly affair. Therefore, for continuous development, electrical energy produced from renewable sources such as solar cells or windmills is needed (Grimes et al. 2008).

Photocatalysis (PC) and photoelectrolysis (PEC) are the two methods that exploit the solar energy for splitting of water. Photoelectrolysis differs from photocatalysis in utilizing additional electrical energy along with that photon energy. And it is found that STH efficiency i.e., solar to hydrogen conversion efficiency is 2 to 3%, much higher than photocatalytic process having STH efficiency of 0.1% (Zhou et al. 2016).

Among these numerous approaches, for conversion of solar energy to electrical or chemical energy, water splitting of semiconductor catalysts involved in solar-driven photoelectro- chemical (PEC) process has been deliberated as an efficient and promising way (Acar and Dincer 2014).

Water splitting performance of different metal oxide semiconductors like TiO_2, ZnO and Fe_2O_3 etc. have been investigated in different studies as photoanodes. They are considered as promising candidates for water splitting because of their high photocatalytic properties (Fujishima and Honda 1972; Ismail and Bahnemann 2014; Mora et al. 2018; Acar et al. 2016; Lasa et al. 2016; Li et al. 2014; Jafari et al. 2016; Ni et al. 2007; Zhao et al. 2016). Major factors which are responsible for influencing the photocatalytic quantum efficiency includes charge separation, transport of charges and surface adsorption capacity.

It is the major challenge for material science to find the effective and stable photocatalysts for overall water splitting (OWS) (Kudo and Miseki 2009; Chen et al. 2010). Much attention has been focussed on insoluble organics (e.g., g-C_3N_4 or other polymers) (Fang et al. 2017; Ghosh et al. 2015), insoluble inorganics (e.g., TiO_2, ZnO etc.), soluble organics (e.g., molecular photocatalysts) (Heyduk and Nocera 2001) and their hybrids as photocatalysts in the previous decades (Li et al. 2013; Suarez et al. 2015; Hernández et al. 2015; Maeda and Domen 2010; Tolod et al. 2017). Out of which undissolved soluble organics have been shown as a very effective method of water splitting for absorbing light and transfer of photo generated carriers because of their geometrical structures (Li et al. 2017).

Another new method of water splitting that came into light is use of pyroelectric energy. This method is considered as an alternative and novel way of hydrogen source. Pyroelectrics are capable in harvesting waste heat by converting the temperature variations into electrical energy. For generating a critical potential for initialisation of water breakdown and to increase the charge transfer for production of hydrogen, material thickness of a vast range of pyroelectric materials have been investigated (Xie et al. 2017).

2. Water Splitting Technology/Methods

2.1. Photocatalysis

Solar energy is exploited by two methods i.e., photocatalysis (PC) and photoelectrolysis (PEC) for water splitting. Due to having supreme renewable energy property on earth, use of solar energy has obtained much attention (Nunes et al. 2017; El-Khouly et al. 2017). Solar energy can accomplish all the future energy need, if harvested efficiently (Wheeler et al. 2012). Photo electrochemical (PEC) water splitting practice is one of the most favourable alternatives in imitating natural process of photosynthesis in plants by storing the solar energy in the carbohydrate form along with rearrangement of electrons in water and carbon oxide (Grätzel 2001; Montoya et al. 2017; Chu et al. 2017; Kang et al. 2015; Walter et al. 2010; Jiang et al. 2017; Roger et al. 2017). It is responsible for formation of hydrogen bond in H_2 from electromagnetic energy by utilizing solar energy.

Sodium molybdate, has the capability to be used as photocatalyst effectively for water splitting underneath band-gap irradiation. Experimental results exhibit the overall water splitting (OWS) capacity of undissolved sodium molybdate as photocatalyst, resulting in quantum yield of 0.36% at 365 nm, however, the dissolved sodium molybdate exhibits no activity. The observations of the study exhibit the capability of soluble inorganic compounds in OWS and pave the way for development of soluble inorganic compounds as catalysts in artificial photosynthesis and other accessible technologies capable of harnessing solar energy and converting it to fuel (Li et al. 2017).

The electrochemical potential necessary for splitting water into hydrogen and oxygen is recorded at 1.23V. Consequently, the potential of the H_2O/O^- redox couple is +1.23 V (Esch and Bredow 2017). In physical electro/photocatalytic systems the potential necessary for water splitting is higher than the standard value of 1.23V. Most of the arrangements have an over-voltage of 0.4 – 0.6 V (Bolton et al. 1985; Murphy et al. 2006) and thermodynamic irreversibilities ranging between 0.3 – 0.4 V (Weber and Dignam 1986). Hence the optimal bandgap for a photo catalyst lies between 1.9eV and 2.2eV.

2.2. Thermal Catalysis (Pyroelectric)

Method of using low grade waste heat for production of electrical energy is known as pyro electrolysis. Pyroelectric materials are polar materials and possess spontaneous polarization, in the absence of an applied electric field or mechanical displacement (Bowen et al. 2014). Pyroelectric materials are of great interest due to their potential to convert temperature variations from waste heat into useful electrical energy (Xie et al. 2017).

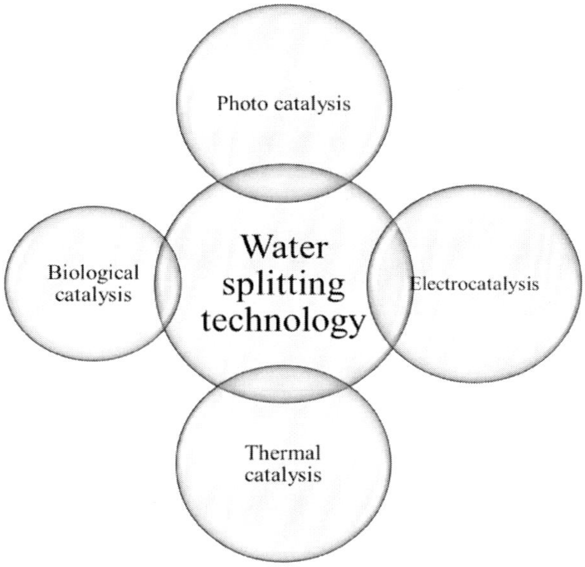

Figure 1. Different water splitting methods.

2.3. Electro Catalysis (Piezoelectric)

Numerous experiments have proved that the piezoelectric effect in combination with different electrochemical processes, permits the designing of charge-carrier conduction properties at the juncture among a strained piezoelectric material and chemical solution. Hong et al. proposed "piezo-electro-chemical" effect where mechanical energy could be transformed to hydrogen and oxygen (Hong et al. 2010). The mechanical vibration of barium titanate and zinc oxide microfibers has been studied to generate strain-induced electric charges under ultrasonic bath and the electric potential across the strained fibres was observed to be sufficient enough to trigger the redox reaction and generate hydrogen and oxygen from water (Xie et al. 2017).

It has been studied that piezoelectric materials have capability to convert mechanical energy into electric energy, with higher energy conversion efficiency (up to ~90%) as compared to other energy conversion (thermoelectric, photovoltaic) materials (Lallart et al. 2010). Mechanical vibration energy is regarded as one of the most popular energy sources in living environment (Yang et al. 2009).

Hydrogen production by means of water splitting exhibits particular interest as it utilizes water. Hence, hydrogen production using a vibrating piezoelectric material represents an alternate way to generate hydrogen energy and exhibits potential for practical applications in near future. A way has been reported for direct splitting of water for production of hydrogen with an energy conversion efficiency of ~18% using vibrating piezoelectric microfibers (Qiu et al. 2011).

2.4. Biological Catalysis

Approximately 50% of the total radiation falling in the visible region is transformed by oxidative photosynthesis into chemical energy at an efficiency of 25% (Blankenship 2002). Oxygen as product molecules are generated completely from water along with protons catalysed by the enzyme photosystem II (PSII) (Renger and Holzwarth 2005; Diner and Rappaport 2002).

A single molecule of water after splitting produces eight charged species including four electrons and four protons. When electron transfer is coupled to proton transfer, as in the oxygen evolving complex (OEC), proton transfer acts as rate-limiting factor for thermal activation.

Experimental approaches to apprehend the complex redox reaction of the oxygen evolving complex (OEC) have been escorted by different theoretical contemplations of the splitting thermodynamics. The oxygen evolving complex (OEC) should accurately couple its redox stoichiometry to its substrate (Krishtalik 1986; Krishtalik 1990).

3. Designing of Equipment

The technology employs utilization of ion exchange membranes to accumulate the ions in the solution and is regulated by an electrical potential. The three chambers namely: acid, salt and base are confined by the bipolar, cation and anion membranes. At commercial level, approximately up to 200 such cell units are assembled within a single set of electrodes to form a comprehensive water splitting unit. The salt, such as sodium sulfate, is served to the chamber amongst the anion and cation membranes. When an electrical potential is applied between the electrodes, the cations and anions migrate between the monopolar membranes and bind with the hydroxide and hydrogen ions produced at the bipolar membrane to produce the acid and base (Mani 1991). Different methods of water splitting that have been reported are summarised in Figure 2.

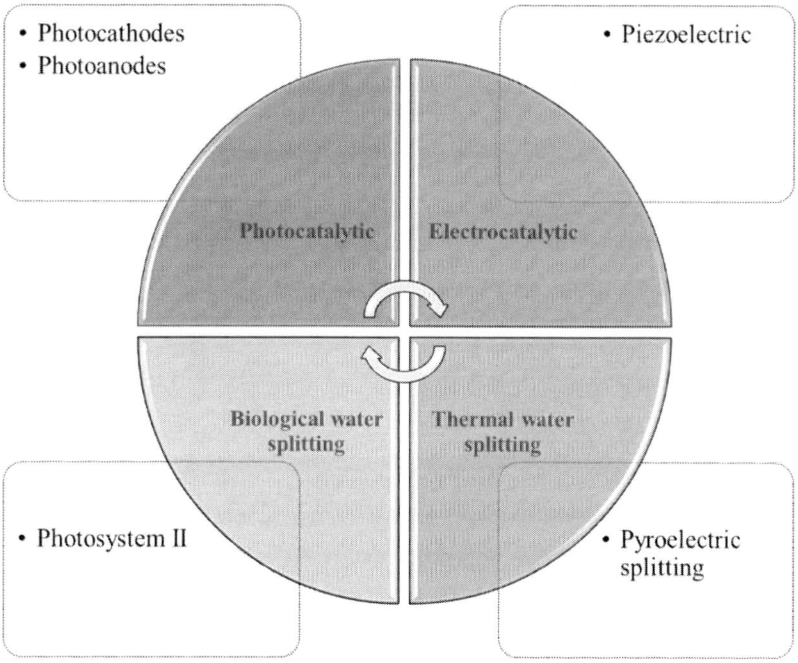

Figure 2. Different methods of water splitting.

3.1. Photocatalysis

The change in free energy for the translation of one water molecule into hydrogen and half molecule of oxygen under standard conditions is reported to be (ΔG^o) 237.2 kJ/mol, which corresponds to ΔE° value of 1.23 eV. For effective utilization of a semiconductor based device to carry out this reaction from light energy, it is necessary for the semiconductor to absorb radiant light corresponding to photon energies greater than 1.23 eV and convert the energy of radiant light into hydrogen and oxygen. This process must generate two electron and hole pairs per molecule of hydrogen (2 × 1.23 eV i.e., 2.46 eV) or four electron-hole pairs per molecule of oxygen (4 × 1.23 eV i.e., 4.92 eV). In ideal conditions, a single semiconductor device with a band gap energy large enough for splitting water, with conduction band-edge energy (Ecb) and valence band-edge energy (Evb) that spans the electrochemical potentials E° (H^+/H_2) and E° (O_2/H_2O), and have the capability to drive the hydrogen evolution reaction (HER) and oxygen evolution reaction (OER) (Kamat 2018).

3.1.1. Photocathodes for Hydrogen Evolution

Photocathodes reported for a water splitting cell are required to generate sufficient cathodic current for reduction of protons into H_2 and should be stable in aqueous condition. Additionally, for successful reduction of protons into hydrogen, the conduction band edge potential of the photocathode should be lower than the value of hydrogen redox potential. The mechanism for hydrogen evolution reaction is dependent on pH as the reaction occurs primarily through proton reduction, while at higher pH, water is predominantly reduced to produce hydroxide ions (Conway and Salomon 1964).

As reported earlier when a semiconductor is brought in contact with an electrolyte, it undergoes Fermi level equilibration with the electrochemical potential (E_{redox}) of the electrolyte (Walter et al. 2010). For a p-type semiconductor, the bands bend in a way so that the photo-generated electrons are diverted towards the interface, whereas the holes move into the bulk of the solid. Hence photoexcitation transfers the electrons from solid into the liquid phase. This cathodic current produced, to some extent, protects the semiconductor surface from oxidation. Hence the p-type semiconductors are considered to be more stable as compared their n-type counterparts. Many p-type semiconductors have been studied as photocathodes for the hydrogen evolution reaction (Levy-Clement et al. 1982). Different photocathode materials have been reported and are tabulated in along with their conversion efficiencies in Table 1.

Table 1. Different materials reported for their use in water splitting using photocatalysis

S. No.	Equipment	Material	Band gap potential (eV)	Production efficiency (%)	References
1.		GaP	2.26		(Aharon-Shalom and Heller 1982)
2.		Ru-InP	1.35	12	(Dominey et al. 1982)
3.		Rh-p-InP	Not reported	13.3	(Aharon-Shalom and Heller 1982)
4.	Photocathode	Re-p-InP	Not reported	11.4	(Aharon-Shalom and Heller 1982)
5.		GaInP2	1.83		(Aharon-Shalom and Heller 1982)
6.		p-Si	1.12	6	(Dominey et al. 1982)
7.		WSe2	Not reported	6-7	(Bicelli and Razzini 1983)
8.	Photoanode	Single band gap semiconductor cells	3	2	(Sato et al. 2005)
		metal oxides with n-type semiconducting properties	3.4	<1	(Sato et al. 2005)

3.1.2. Photoanodes for Water Splitting

An oxygen producing photoanode should be an n-type semiconductor material, so that the electric field produced by band bending, moves the holes towards the surface. The material is required to have a band gap and band-edge positions, which are suitable for their use either in single or multiple band gap systems. In addition, it is necessary that the materials are stable under water oxidization conditions. Because of the above statement, need for stability under oxidizing conditions, most of the investigated photoanode materials, are metal oxides or metal oxide anions (oxometalates), in both pure or doped forms. In ionic crystals, the valance band potential remains relatively unchanged at 3.0 V for most metal oxides and oxometalates including $SrTiO_3$, TiO_2, WO_3, ZnO and Fe_2O_3 (Scaife 1980).

After photoexcitation and separation of charges on an n-type semiconductor, minority carriers (holes) in the valance band move to the semiconductor-electrolyte interface for water oxidation. The difference between the oxygen-centred valance band at ~3.0 V and the oxygen evolution reaction potential at 1.23 V vs normal hydrogen electrode represents a challenge for the synthesis of high-performance photoanode material (Kamat 2018).

3.2. Pyroelectric Water Splitting

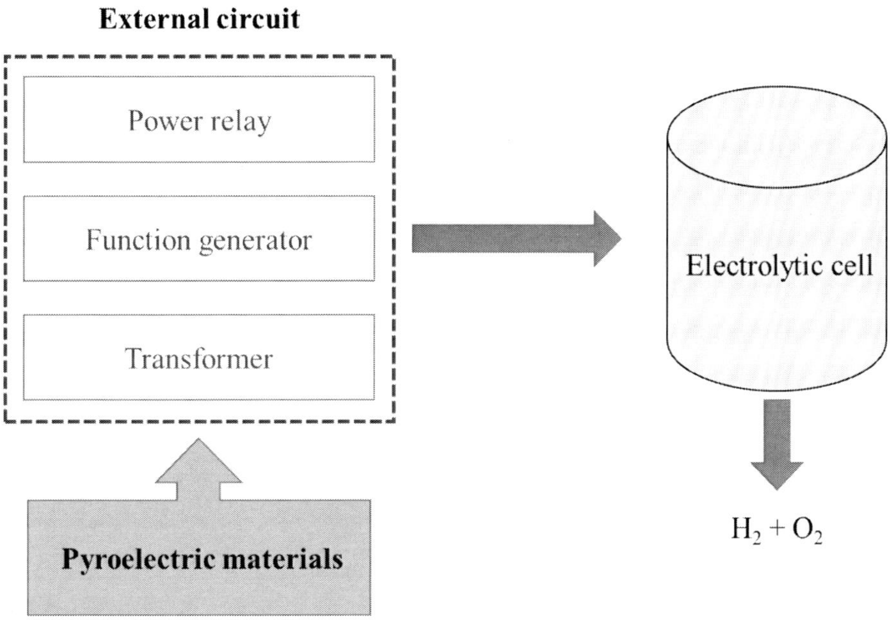

Figure 3. Schematic representation of pyroelectric water splitting.

Pyroelectric materials are polar in nature. Waste heat, is regarded as an effective source of energy, due to its omnipresence in various industrial applications along with in the environment. Pyroelectric materials are conserved to be of great interest as they possess the capability to alter temperature variations from waste heat into useful electrical energy. Pyroelectric materials exhibit impulsive polarization, in the absence of an applied electric field or mechanical displacement (Park et al. 2015) which results in existence of charge on its each surface. It is assumed that temperature of the material is responsible for polarisation level. Upon decreasing the temperature, the polarisation level of pyroelectric material also decreases due to the net dipole moment reduction which leads to the fall in surface bound charges (see Figure 3). In the conditions of open circuit, free charges stay at the surface resulting in generation of electric potential while back electric flow occurs in short circuit conditions when it is cooled (Xie et al. 2017).

3.2.1. Pyroelectric Material Selection

For generation of electrical potential, mainly three material configurations i.e., thin film, rod and particle have been considered. By the configuration of the pyroelectric material, direction of the polarisation can be assumed. Two important parameters for material selection are permittivity and pyroelectric coefficient. These parameters define the charge and voltage. It is supposed that for evolution of hydrogen, total required electrolysis potential is 1.5 V (Rand and Dell 2007). Many materials have the capability to produce large electric potential even in the presence of small temperature variations. It can be assumed that, for 1°C temperature variation, minimum thickness of less than 10 mm is required to generate 1.5 V. Materials that fulfil this condition are of low permittivity (Xie et al. 2017).

It has been investigated that by applying the temperature variation higher than 1°C, a sufficient potential changes can be generated even from the thin films. In a study, three materials viz. TGS (Triglycine sulphide) > $LiTaO_3$ (Lithium tantalite) > $LiNbO_3$ (Lithium niobate) have been found to be thinnest for generating the potential difference critically. Ferroelectric ceramics for example the PMN-PT (Lead magnesium niobate lead titanate), $BaTiO_3$ (Barium titanate), PZT-5H (lead zirconate titanate) are the bad materials as thin film because of having high permittivity. While in the case of rod or particle, materials that have been found best are TGS > PMN-PT > PZT-5H and ZnO is poorest due to its comparatively high density (Sebald et al. 2008).

3.3. Piezoelectric Water Splitting

In one study the piezoelectrochemical phenomenon was employed for generation of hydrogen and oxygen through water splitting by using piezoelectric ZnO microfibers and $BaTiO_3$ microdendrites. In this phenomenon the fibers and dendrites irradiated with ultrasonic waves resulting in a strain generated electric charge at the surface of the fibers and dendrites. With enough electric potential, strained piezoelectric fibers initiated the water splitting to produce hydrogen and oxygen (Qiu et al. 2011).

In another study a device consisting of a piezoelectric bimorph cantilever and a water electrolysis cell was developed for piezoelectrochemical hydrogen generation from water splitting. The water splitting was observed by employing a mechanical vibration of ~0.07 N and ~46.2 Hz (Zhang et al. 2012).

4. Recent Advancements in Water Splitting

4.1. Nanoscale Design of Hydrogen Evolution Sites

Certain photocatalysts have the potential to degrade water without any cocatalyst resulting in reduction of activation energy and these also provide the sites for chemical reactions. Noble transition metals including Pt and Rh are outstanding promoters for H_2 production but they also possess capability to catalyze backward reaction, resulting in water formation, thus decreasing their efficacy as cocatalysts for splitting of water. To avoid the backward reaction, transition-metal oxides are generally employed as cocatalysts for water splitting. A Nano particulate rhodium and chromium mixed oxide with a solid solution of GaN and ZnO has been reported and it signifies the role of cocatalysts in photocatalytic water splitting (Hata et al. 2008; Zong et al. 2008; Jang et al. 2008). The substantial water formation from H_2 and O_2 takes place on uncovered Rh nanoparticles during water splitting, and the water formation can be suppressed using Cr_2O_3 shells over noble metal cores. In certain conditions, though, metal oxide cocatalysts go through deprivation because of an alteration in the cocatalyst state due to exposure to the reactant. They have the potential to catalyse O_2 photoreduction, thus resulting in a decrease in activity. In another study it has been reported that the photocatalytic activity of metal oxide loaded GaN:ZnO for overall water splitting was enhanced by formation of a core/shell-like configuration with Cr_2O_3 shell (Maeda et al. 2010). Amongst the noble metals and metal oxides studied as nanoparticulate cores, Rh was observed to be most effective for increasing the activity. In another study highly dispersed Rh nanoparticles were loaded on GaN:ZnO without accumulation through adsorption of Rh nanoparticles stabilized by 3-mercapto-1-propanesulfuric acid onto GaN:ZnO, further followed by calcination under vacuum conditions at 673K for 30 min. (Sakamoto et al. 2009). Subsequently the Rh nanoparticles were glazed with a Cr_2O_3 shell, the photocatalytic activity for splitting of water was evaluated under visible light irradiation. The sample with a superior dispersion of Rh displayed three times the activity of an equivalent comprising poorly dispersed Rh nanoparticles. It is necessary to comprehend the reaction mechanism at the nanoscale for developing enhanced photocatalytic systems. The mechanism of hydrogen evolution on core/shell structured nanoparticles has been studied by using electrochemical and *in situ* spectroscopic measurements of model electrodes. The developed core/shell cocatalyst exhibits certain advantages (Maeda et al. 2010), such as

1. the possibility of utilizing different noble metals and metal oxides as a core
2. the possibility of specifically presenting active species for water splitting at different reduction sites present on the photocatalyst
3. the abolition of the requirement for activation treatment by reduction or oxidation.

Conclusion

Water splitting equipment capture the energy in one form viz. solar, thermal or mechanical and utilize it for production of clean fuel i.e., hydrogen from water. The equipment developed for this purpose till date have certain shortcomings including lower efficiency in certain cases, catalysing backward reaction to produce water from hydrogen and oxygen. Developments have been made to overcome these problems. However, the developed equipments still lack at multiple steps ranging from energy trapping efficiency, translation of the trapped energy into electrochemical cells for water splitting etc. This situation can be tackled by use of nanotechnology based equipment fabrication, which might have the potential to enhance the efficiency of the process due to unique properties of the nanodevices such as large surface area to volume ratio for catalysis, better capture of input energy and transfer of the captured energy to the electrochemical cells for water splitting. For this purpose, newer metal and metal oxide based nanofibers, dendrites and other nanoparticles are being developed which, not only increase the hydrogen yield but also help in stopping the backward reaction to produce water from hydrogen and oxygen produced.

References

Acar C and Dincer I. 2014. "Comparative assessment of hydrogen production methods from renewable and non-renewable sources." *International journal of hydrogen energy* 39:1-2.

Acar C, Dincer I and Naterer GF. 2016. "Review of photocatalytic water splitting methods for sustainable hydrogen production." *International Journal of Energy Research* 40:1449-73.

Aharon-Shalom E and Heller A. 1982. "Efficient p-InP (Rh-H alloy) and p-InP (Re-H alloy) hydrogen evolving photocathodes." *Journal of the Electrochemical Society* 129:2865-6.

Artero V, Chavarot K. M. and Fontecave M. 2011. "Splitting water with cobalt." *Angewandte Chemie International Edition* 50:7238-66.

Bicelli LP and Razzini G. 1983. "Hydriodic acid photodecomposition on layered-type transition metal dichalcogenides." *Surface technology* 20:393-403.

Blankenship RE, Tiede DM, Barber J, Brudvig GW, Fleming G, Ghirardi M, Gunner MR, Junge W, Kramer DM, Melis A and Moore TA. 2011. "Comparing photosynthetic and photovoltaic efficiencies and recognizing the potential for improvement." *Science* 332:805-9.

Blankenship RE. 2002. "Molecular Mechanisms of Photosynthesis." *Photosynthetica* 40:1-12.

Bolton JR, Strickler SJ and Connolly JS. 1985. "Limiting and realizable efficiencies of solar photolysis of water." *Nature* 316:495-500.

Bowen CR, Taylor J, LeBoulbar E, Zabek D, Chauhan A and Vaish R. 2014. "Pyroelectric materials and devices for energy harvesting applications." *Energy & Environmental Science*. 7:3836-56.

Casper MS. 1978. *"Hydrogen manufacture by electrolysis, thermal decomposition and unusual techniques."* Park Ridge, New York: Noyes Data Corporation.

Chan CC. 2007. "The state of the art of electric, hybrid, and fuel cell vehicles." *Proceedings of the IEEE* 95:704-18.

Chen X, Shen S, Guo L and Mao SS. 2010. "Semiconductor-based photocatalytic hydrogen generation." *Chemical reviews* 110:6503-70.

Chiarello GL, Aguirre MH and Selli E. 2010. "Hydrogen production by photocatalytic steam reforming of methanol on noble metal-modified TiO_2." *Journal of Catalysis* 273:182-90.

Chu S and Majumdar A. 2012. "Opportunities and challenges for a sustainable energy future." *Nature* 488:294-303.

Chu S, Li W, Yan Y, Hamann T, Shih I, Wang D and Mi Z. 2017. "Roadmap on solar water splitting: current status and future prospects" *Nano Futures* 1:022001.

Conway BE and Salomon M. 1964. "Electrochemical reaction orders: Applications to the hydrogen-and oxygen-evolution reactions." *Electrochimica Acta* 9:1599-615.

Cook TR, Dogutan DK, Reece SY, Surendranath Y, Teets TS and Nocera DG. 2010. "Solar energy supply and storage for the legacy and nonlegacy worlds." *Chemical reviews* 110:6474-502.

Diner BA and Rappaport F. 2002. "Structure, dynamics, and energetics of the primary photochemistry of photosystem II of oxygenic photosynthesis." *Annual review of plant biology* 53:551-80.

Divisek J. 1990. *"Water electrolysis in a low-and medium-temperature regime. Electrochemical hydrogen technologies-electrochemical production and combustion of hydrogen."* Oxford: Elsevier.

Dominey RN, Lewis NS, Bruce JA, Bookbinder DC and Wrighton MS. 1982. "Improvement of photoelectrochemical hydrogen generation by surface modification of p-type silicon semiconductor photocathodes." *Journal of the American Chemical Society* 104:467-82.

Dönitz W, Erdle E and Streicher R. 1990. "*High temperature electrochemical technology for hydrogen production and power generation. Electrochemical hydrogen technologies: electrochemical production and combustion of hydrogen*." Editor: H Wendt. Elsevier, Amsterdam 213.

Du P and Eisenberg R. 2012. "Catalysts made of earth-abundant elements (Co, Ni, Fe) for water splitting: recent progress and future challenges." *Energy & Environmental Science* 5:6012-21.

El-Khouly ME, El-Mohsnawy E and Fukuzumi S. 2017. "Solar energy conversion: From natural to artificial photosynthesis."*Journal of Photochemistry and Photobiology C: Photochemistry Reviews* 31:36-83.

Esch TR and Bredow T. 2017. "Band positions of Rutile surfaces and the possibility of water splitting." *Surface Science* 665:20-7.

Fang LJ, Wang XL, Li YH, Liu PF, Wang YL, Zeng HD and Yang HG. 2017. "Nickel nanoparticles coated with graphene layers as efficient co-catalyst for photocatalytic hydrogen evolution. *Applied Catalysis B: Environmental*." 200:578-84.

Fujishima A and Honda K. 1972. "Electrochemical photolysis of water at a semiconductor electrode." *Nature* 238:37-8.

Ghosh S, Kouamé NA, Ramos L, Remita S, Dazzi A, Deniset-Besseau A, Beaunier P, Goubard F, Aubert PH and Remita H. 2015. "Conducting polymer nanostructures for photocatalysis under visible light." *Nature materials* 14:505-11.

Grätzel M. 2001. "Photoelectrochemical cells."*Nature* 414:338-44.

Grimes CA, Varghese OK and Ranjan S. 2008. "Hydrogen generation by water splitting." *In Light, Water, Hydrogen* 35-113 Springer US DOI: https://doi.org/10.1007/978-0-387-68238-9_2.

Han L, Dong S and Wang E. 2016. "Transition Metal (Co, Ni, and Fe) Based Electrocatalysts for the Water Oxidation Reaction." *Advanced Materials* DOI: 10.1002/adma.201602270.

Hata H, Kobayashi Y, Bojan V, Youngblood WJ and Mallouk TE. 2008. "Direct deposition of trivalent rhodium hydroxide nanoparticles onto a semiconducting layered calcium niobate for photocatalytic hydrogen evolution." *Nano letters* 8:794-9.

Hernández S, Hidalgo D, Sacco A, Chiodoni A, Lamberti A, Cauda V, Tresso E, and Saracco G. 2015. "Comparison of photocatalytic and transport properties of TiO_2 and

ZnO nanostructures for solar-driven water splitting." *Physical Chemistry Chemical Physics* 17:7775-86.

Heyduk AF and Nocera DG. 2001. "Hydrogen produced from hydrohalic acid solutions by a two-electron mixed-valence photocatalyst." *Science* 293:1639-41.

Hong KS, Xu H, Konishi H and Li X. 2010. "Direct water splitting through vibrating piezoelectric microfibers in water." *The Journal of Physical Chemistry Letters* 1:997-1002.

Ismail AA and Bahnemann DW. 2014. "Photochemical splitting of water for hydrogen production by photocatalysis: a review." *Solar Energy Materials and Solar Cells* 128:85-101.

Jafari T, Moharreri E, Amin AS, Miao R, Song W and Suib SL. 2016. "Photocatalytic water splitting—the untamed dream: a review of recent advances." *Molecules* 21(7):900.

Jamesh MI. 2016. "Recent progress on earth abundant hydrogen evolution reaction and oxygen evolution reaction bifunctional electrocatalyst for overall water splitting in alkaline media." *Journal of Power Sources* 333:213-36.

Jang JS, Ham DJ, Lakshminarasimhan N, yong Choi W and Lee JS. 2008. "Role of platinum-like tungsten carbide as cocatalyst of CdS photocatalyst for hydrogen production under visible light irradiation." *Applied Catalysis A: General* 346:149-54.

Jiang C, Moniz SJ, Wang A, Zhang T and Tang J. 2017. "Photoelectrochemical devices for solar water splitting–materials and challenges." *Chemical Society Reviews* 46:4645-60.

Kamat PV. 2018. "Hybrid Perovskites for Multijunction Tandem Solar Cells and Solar Fuels. A Virtual Issue." *ACS Energy Letters* 3:28-9.

Kanan MW and Nocera DG. 2008. "In situ formation of an oxygen-evolving catalyst in neutral water containing phosphate and Co_2^{+}." *Science* 321:1072-5.

Kang D, Kim TW, Kubota SR, Cardiel AC, Cha HG and Choi KS. 2015. "Electrochemical synthesis of photoelectrodes and catalysts for use in solar water splitting." *Chemical reviews* 115:12839-87.

Kondo M and Masaoka S. 2016. "Water Oxidation Catalysts Constructed by Biorelevant First-row Metal Complexes." *Chemistry Letters* 45:1220-31.

Krishtalik LI. 1986. "Energetics of multielectron reactions. Photosynthetic oxygen evolution." *Biochimica et Biophysica Acta (BBA)-Bioenergetics* 849:162-71.

Krishtalik LI. 1990. "Activation energy of photosynthetic oxygen evolution: an attempt at theoretical analysis." *Bioelectrochemistry and Bioenergetics* 23:249-63.

Kudo A and Miseki Y. 2009. "Heterogeneous photocatalyst materials for water splitting." *Chemical Society Reviews* 38(1):253-78. DOI: 10.1039/B800489G.

Kyoto Protocol to the United Nations Framework Convention on Climate Change. http://unfccc.int/essential_background/kyoto_protocol/items/1678.php;[accessed 11.1.18].

Lallart M, Inman DJ and Guyomar D. 2010. "Transient performance of energy harvesting strategies under constant force magnitude excitation." *Journal of Intelligent Material Systems and Structures* 21:1279-91.

Landon J, Demeter E, İnoğlu N, Keturakis C, Wachs IE, Vasić R, Frenkel AI and Kitchin JR. 2012. "Spectroscopic characterization of mixed Fe–Ni oxide electrocatalysts for the oxygen evolution reaction in alkaline electrolytes". *Acs Catalysis* 2:1793-801.

Lasa HD, Rosales BS, Moreira J and Valades-Pelayo P. 2016. "Efficiency factors in photocatalytic reactors: Quantum yield and photochemical thermodynamic efficiency factor". *Chemical Engineering & Technology* 39:51-65.

Levy-Clement C, Heller A, Bonner WA and Parkinson BA. 1982. "Spontaneous photoelectrolysis of HBr and HI". *Journal of the Electrochemical Society* 129: 1701-5.

Lewis NS and Nocera DG. 2006. "Powering the planet: Chemical challenges in solar energy utilization. *Proceedings of the National Academy of Sciences* 103:15729-35.

Li K, An X, Park KH, Khraisheh M and Tang J. 2014. "A critical review of CO_2 photoconversion: catalysts and reactors."*Catalysis Today* 224:3-12.

Li YH, Wang Y, Zheng LR, Zhao HJ, Yang HG, and Li C. 2017. "Water-soluble inorganic photocatalyst for overall water splitting. *Applied Catalysis B: Environmental*. 209:247-52.

Li YH, Xing J, Chen ZJ, Li Z, Tian F, Zheng LR, Wang HF, Hu P, Zhao HJ and Yang HG. 2013. "Unidirectional suppression of hydrogen oxidation on oxidized platinum clusters." *Nature communications* 4:2500. DOI: 10.1038/ncomms3500.

Liu S, Lei YJ, Xin ZJ, Lu YB and Wang HY. 2017. "Water splitting based on homogeneous copper molecular catalysts." *Journal of Photochemistry and Photobiology A: Chemistry* DOI: https://doi.org/10.1016/j.jphotochem.2017.09.060.

Lucia U. 2014. "Overview on fuel cells." *Renewable and Sustainable Energy Reviews* 30:164-9.

Maeda K and Domen K. 2010. "Photocatalytic water splitting: recent progress and future challenges." *The Journal of Physical Chemistry Letters*. 1:2655-61.

Maeda K, Sakamoto N, Ikeda T, Ohtsuka H, Xiong A, Lu D, Kanehara M, Teranishi T and Domen K. 2010. "Preparation of Core–Shell Structured Nanoparticles (with a Noble-Metal or Metal Oxide Core and a Chromia Shell) and Their Application in Water Splitting by Means of Visible Light."*Chemistry-A European Journal* 16:7750-9.

Mani KN. 1991. "Electrodialysis water splitting technology." *Journal of membrane science* 58:117-38.

Montoya JH, Seitz LC, Chakthranont P, Vojvodic A, Jaramillo TF and Nørskov JK. 2017. "Materials for solar fuels and chemicals." *Nature materials*. 16:70-81.

Mora SJ, Odella E, Moore GF, Gust D, Moore TA and Moore AL. 2018. "Proton-Coupled Electron Transfer in Artificial Photosynthetic Systems". *Accounts of chemical research* DOI: 10.1021/acs.accounts.7b00491.

Murphy AB, Barnes PR, Randeniya LK, Plumb IC, Grey IE, Horne MD and Glasscock JA. 2006. "Efficiency of solar water splitting using semiconductor electrodes." *International journal of hydrogen energy* 31:1999-2017.

Navarro RM, Pena MA and Fierro JL. 2007. "Hydrogen production reactions from carbon feedstocks: fossil fuels and biomass." *Chemical reviews* 107:3952-91.

Ni M, Leung MK, Leung DY and Sumathy K. 2007. "A review and recent developments in photocatalytic water-splitting using TiO2 for hydrogen production." *Renewable and Sustainable Energy Reviews* 11:401-25.

Nunes BN, Paula LF, Costa ÍA, Machado AE, Paterno LG and Patrocinio AO. 2017. "Layer-by-layer assembled photocatalysts for environmental remediation and solar energy conversion." *Journal of Photochemistry and Photobiology C: Photochemistry Reviews* DOI: https://doi.org/10.1016/j.jphotochemrev.2017.05.002.

Park T, Na J, Kim B, Kim Y, Shin H and Kim E. 2015. "Photothermally activated pyroelectric polymer films for harvesting of solar heat with a hybrid energy cell structure." *ACS nano* 9:11830-9.

Qiu Y, Yan K, Deng H and Yang S. 2011. "Secondary branching and nitrogen doping of ZnO nanotetrapods: building a highly active network for photoelectrochemical water splitting." *Nano letters* 12:407-13.

Rand DA and Dell RM. 2007. "Hydrogen energy: challenges and prospects." *Royal Society of Chemistry* DOI: http://dx.doi.org/10.1039/9781847558022.

Renger G and Holzwarth AR. 2005. "Primary electron transfer." In *Photosystem II* 139-175. DOI: 10.1007/1-4020-4254-X.

Roger I, Shipman MA and Symes MD. 2017. "Earth-abundant catalysts for electrochemical and photoelectrochemical water splitting." *Nature Reviews Chemistry* 1:0003. DOI: 10.1038/s41570-016-0003.

Sakamoto N, Ohtsuka H, Ikeda T, Maeda K, Lu D, Kanehara M, Teramura K, Teranishi T and Domen K. 2009. "Highly dispersed noble-metal/chromia (core/shell) nanoparticles as efficient hydrogen evolution promoters for photocatalytic overall water splitting under visible light." *Nanoscale* 1:106-9.

Sato J, Saito N, Yamada Y, Maeda K, Takata T, Kondo JN, Hara M, Kobayashi H, Domen K and Inoue Y. 2005. "Ru_{O2}-loaded β-G_{e3N4} as a non-oxide photocatalyst for overall water splitting." *Journal of the American Chemical Society* 127:4150-1.

Scaife DE. 1980. "Oxide semiconductors in photoelectrochemical conversion of solar energy." *Solar Energy* 25:41-54.

Sebald G, Lefeuvre E and Guyomar D. 2008. "Pyroelectric energy conversion: optimization principles." *Ieee transactions on ultrasonics, ferroelectrics, and frequency control* 55(3). DOI: 10.1109/TUFFC.2008.680.

Singh A and Spiccia L. 2013. "Water oxidation catalysts based on abundant 1st row transition metals."*Coordination Chemistry Reviews* 257:2607-22.

Suarez CM, Hernández S and Russo N. 2015. "BiV_{O4} as photocatalyst for solar fuels production through water splitting: A short review." *Applied Catalysis A: General.* 504:158-70. DOI: https://doi.org/10.1016/j.apcata.2014.11.044.

Tolod KR, Hernández S and Russo N. 2017. "Recent advances in the BiVO4 photocatalyst for sun-driven water oxidation: Top-performing photoanodes and scale-up challenges." *Catalysts* 7(1):13.

Walter MG, Warren EL, McKone JR, Boettcher SW, Mi Q, Santori EA and Lewis NS. 2010. "Solar water splitting cells." *Chemical reviews* 110:6446-73.

Wang J, Zhong HX, Qin YL and Zhang XB. 2013. "An Efficient Three Dimensional Oxygen Evolution Electrode." *Angewandte Chemie* 125:5356-61.

Weber MF and Dignam MJ. 1986. "Splitting water with semiconducting photo-electrodes—Efficiency considerations." *International Journal of Hydrogen Energy* 11:225-32.

Wheeler DA, Wang G, Ling Y, Li Y and Zhang JZ. 2012. "Nanostructured hematite: synthesis, characterization, charge carrier dynamics, and photoelectrochemical properties." *Energy&EnvironmentalScienc* 5:6682-702.

Xie M, Dunn S, Le Boulbar E and Bowen CR. 2017. "Pyroelectric energy harvesting for water splitting." *International Journal of Hydrogen Energy* DOI: https://doi.org/10.1016/j.ijhydene.2017.02.086.

Yang R, Qin Y, Li C, Zhu G and Wang ZL. 2009. "Converting biomechanical energy into electricity by a muscle-movement-driven nanogenerator." *Nano Letters* 9:1201-5.

Zhang J, Wu Z, Jia Y, Kan J and Cheng G. 2012. "Piezoelectric bimorph cantilever for vibration-producing-hydrogen." *Sensors* 13:367-74.

Zhao Y, Hoivik N and Wang K. 2016. "Recent advance on engineering titanium dioxide nanotubes for photochemical and photoelectrochemical water splitting." *Nano Energy* 30:728-44.

Zhou H, Yan R, Zhang D and Fan T. 2016. "Challenges and perspectives in designing artificial photosynthetic systems." *Chemistry-A European Journal* 22:9870-85.

Zong X, Yan H, Wu G, Ma G, Wen F, Wang L and Li C. 2008. "Enhancement of photocatalytic H_2 evolution on CdS by loading MoS_2 as cocatalyst under visible light irradiation." *Journal of the American Chemical Society* 130:7176-7.

Zou X and Zhang Y. 2015. "Noble metal-free hydrogen evolution catalysts for water splitting." *Chemical Society Reviews* 44:5148-80.

In: Photocatalysis
Editors: P. Singh, M. M. Abdullah, M. Ahmad et al.
ISBN: 978-1-53616-044-4

Chapter 14

A BRIEF OVERVIEW ON PHYSIO-CHEMICAL ASPECT OF TiO_2 AND ITS NANO-CARBON COMPOSITES FOR ENHANCED PHOTOCATALYTIC ACTIVITY

Md. Rakibul Hasan[1], Zaira Zaman Chowdhury[1,*], Suresh Sagadevan[1,†], Rahman Faizur Rafique[2], Wan Jefry Basiro[3], Md. Abdul Khaleque[4] and Jiban Podder[5]

[1]Nanotechnology and Catalysis Research Centre (NANOCAT), Institute of Postgraduate Studies (IPS), University of Malaya, Kuala Lumpur, Malaysia

[2]Rutgers Cooperative Extension Water Resources Program, Rutgers, The State University of New Jersey, NJ, US

[3]Department of Chemistry, Faculty of Science, University of Malaya, Kuala Lumpur, Malaysia

[4]Independent University of Bangladesh, Dhaka, Bangladesh

[5]Crystal Growth Lab, Department of Physics, Bangladesh University of Engineering and Technology (BUET), Dhaka, Bangladesh

* Corresponding Author's E-mail: zaira.chowdhury76@gmail.com.

† Corresponding Author's E-mail: sureshsagadevan@gmail.com.

ABSTRACT

Application of Titania oxide (TiO_2) is one of the most promising approaches which has got immense importance in the field of energy and environmental issues. The unique properties of TiO_2 make its best candidates among the other existing photocatalyst. However, application of this photocatalyst is limited up to a certain extent owing to its structural defects. In this context, graphene (Gr) or it's oxidized (GO) or reduced form (RGO) can be incorporated inside its structure for better performance. Carbon nanotubes (CNTs) have also attracted a lot of consideration because for some applications, such as electronic devices, chemical biological sensors, and the CNTs must be of high purity. Both Graphene and CNT being environmentally friendly and highly efficient catalyst materials have been considered as the architect of future green technology. TiO_2-Carbon nanocomposites have versatile applications in H_2O splitting, pollutant degradation, CO_2 conversion, anti-bacterial and self-cleaning hydrophilicity. Application of TiO_2-Carbon nanocomposites is one of the most intriguing agenda in energy management activities all over the world. A lot of initiatives have been taken to investigate the mechanism for the catalytic activity of TiO_2-Carbon composite. This chapter highlighted the evolution of TiO_2 catalyst with time. This article briefly summarizes the significant applications of TiO_2-Carbon composites, addressing its outstanding physiochemical properties as well as future perspective for large scale applications. A brief mechanism of photo-catalysis using TiO_2-Carbon composites is also provided in a subsequent section.

Keywords: Titania oxide (TiO_2), graphene (Gr), CNTs and TiO_2-Carbon composites

INTRODUCTION

Rigorous attention has been given to the utilization of TiO_2 photocatalyst nowadays with the fast-growing advancement of science and technology. There are many other successful materials which have showed better photocatalytic activities. However, TiO_2 materials have become successful to grab the lion's share of the commercial market and in the industrial sector. From the trend, it is expected that TiO_2 will sustain this lead in the future also. It is interesting that the application of TiO_2 over the last half-century has reached a sustainable level. It has been extensively used by researchers due to its high photocatalytic activity, nontoxicity, easy availability as well as stability in the working environment (Hashimoto et al. 2005). Its specific properties make it chemically and biologically resistant and stable but highly active towards photoinduced chemical reactions and organisms. TiO_2 can resist high temperature as well as work in ambient temperature. The physiochemical properties of TiO_2 can be varied up to a greater extent with the size reduction to the nanometer scale. Until recently TiO_2 materials are prepared as nanoparticles, nanowires, nanofibers and nanotubes phases. Thus it has become a potential research element in respect to energy and environment perspective.

TiO_2 nanomaterial can be prepared via easy synthesis processes using mild conditions. Among various methods, the sol-gel method is widely used for TiO_2 nanoparticles preparation (Kuznetsova et al. 2005). In general, this process usually proceeds via acid-catalyzed hydrolysis step of titanium (IV) alkoxide followed by condensation. Micelles and inverse micelles are commonly employed to synthesize TiO_2 nanomaterials. Hydrothermal treatment is another popular preparation method for TiO_2 nanoparticles as well as nanorods. In a typical hydrothermal process for TiO_2 preparation, an acidic ethanol-water solution of titanium alkoxide is treated by hydrothermal reaction. To impose better control over the entire process, the solvothermal method was applied. The size and shape distributions and the crystallinity of the TiO_2 particles can be controlled much better in the solvothermal process. Both nanoparticles and nanorods can

Table 1. A brief summary for application of TiO_2 as photo-catalyst

Timeline	Significant Applications/ Findings observed	
Ancient times	Used in pigments, cosmetics and food coloring (Kay et al. 1993)	
Earlier 20th century	First observed flaking of paints and degradation of fabrics (Taoda 2009)	
1938	Report on Photo-bleaching of dyes by TiO_2 (Mills et al. 1994) Production of active oxygen on TiO_2 upon UV absorption	
1956	Report on "Auto-oxidation by TiO_2 as a photocatalyst". It was first termed as 'photocatalyst' rather than 'sensitizer'(Mashio et al. 1956)	
1969	Photoelectrocatalysis of H_2O using single crystal n-type TiO_2 semiconductor (Fujishima et al. 1969)	
1969	Demonstration of solar photoelectrolysis of H2O(Fujishima et al. 1972)	
1972	A report in 'Nature' on Electrochemical photolysis of water(Kawai et al. 1980)	
2011	Comparison with small band gap semiconductor (CdS, CdSe) photocatalysts. But efficiency is lower than TiO_2 (Lee et al. 2011)	
1980s	The research on H_2 production using TiO_2 lost interest due to low absorption efficiency (Kudo et al. 2009)	
1977	Cyanide decomposition in the presence of aqueous TiO_2 suspensions (Frank et al. 1977)	
Late 1980s	Immobilization of powdered TiO_2 on supports for easy handling (Kohle et al. 1996)	
Until 1990	Low light energy density of TiO_2 and narrow UV response photoactivity. no real industrial technology	--
1992	Photocatalytic cleaning material with a ceramic tile (Heller 1995)	
1992	Self-cleaning cover glass for tunnel light (Takeuchi et al. 2005)	
1995	Cu and/or Ag-doped TiO_2 coating on antibacterial ceramic tile first commercialized full-scale manufacturing by TOTO Ltd. Japan	--
Early 21st Century	TiO_2 Nanotechnology and Nanoengineering	--
2000s	Deep research on Photoinduced Hydrophilicity of TiO_2 coating and composites (Sakatani et al. 2001)	
2001	Visible-light-sensitive TiO_2-based powders and thin films were reported (Ihara et al. 2001)	
2005	Self-cleaning and anti-fogging action of TiO2 coated building material explained by photoinduced hydrophilicity(Chen et al. 2007)	
2005/06	High surface-volume ratio, enhanced light absorption rate (Tay et al. 2013)	
2013	H_2 production and clean H_2O technology, photodegradation of pollutants, photocatalytic self-cleaning, photoinduced super hydrophilicity, next-generation Li-ion batteries, antibacterial filtration membrane and wound dressing, (Zhai et al. 2013; Kamegawa et al. 2012: Devi et al. 2013)	

be formulated via this process (Li et al. 2006). Oxidation of titanium metal using oxidants or under anodization, liquid precursor delivery, physical vapor deposition (PVD) method or thermal deposition, sonochemical method, electrodeposition, microwave radiation can be named for preparation of TiO_2.

TiO_2 powders are being used from ancient times as cosmetic ingredients and pigments as they are non-toxic, cheap. But TiO_2 can be excited under a UV irradiation which can prompt the photocatalytic reaction. This photocatalytic property of TiO_2 was first reported in a scientific paper in 1920. But it was known earlier from some incidents like flaking corrosion of paints and fabric degradation under sunlight. Nowadays, it is being used for wastewater treatment as well. It is highly specific in nature that it can be effective against a very trace level of pollutants. It can be used for water splitting to produce H_2 production. TiO_2 nanomaterials are used in gas sensing and dye-sensitized solar cell (DSSC) (Xiao et al. 2012). Furthermore, it can be used for CO_2 conversion into useful fuels and so on. Up to date, TiO_2 has been used extensively by the researchers for various purposes. A brief summary for application of TiO_2 as photo-catalyst as presented in Table 1.

A lot of initiatives have been taken to investigate the mechanism for the catalytic activity of TiO_2-Carbon composite. This chapter highlighted the evolution of TiO_2 catalyst with time. Furthermore, this chapter briefly summarizes the significant applications of TiO_2-Carbon composites, addressing its outstanding physiochemical properties as well as future perspective for large scale applications. A brief mechanism of photo-catalysis using TiO_2-Carbon is also provided in a subsequent section.

PHYSIOCHEMICAL ASPECTS OF TiO_2 PHOTOCATALYST

Out of three crystal structures, i.e., anatase, brookite, and rutile, anatase phase TiO_2 show better photocatalytic activity than other crystal structures (Ding et al. 2000). The photocatalytic activity first reported for Hydrogen production where TiO_2 anatase was used as powder photocatalyst. However, the low utilization efficiency of visible light and quantum efficiency of photocatalytic reactions on TiO_2 limit its practical applications due to the relatively wide band gap (e.g., 3.2 eV for anatase; 3.0 eV for rutile) and high recombination rate of electron-hole pairs as shown in Figure 1. Although the rutile phase has a lower band gap, anatase has its advantages over rutile, including a higher reduction potential and a lower recombination rate of photogenerated carriers.

The conduction band edge of anatase is located upper by the order of 0.1V (more negative) than that of rutile as shown by Figure 2 (David et al. 2013). TiO_2 itself is an n-type semiconductor due to oxygen defects in the structure and thus non-stoichiometric. The numbers of free electrons are proportional to oxygen defects of the crystal structure, which essentially lead to the n-type characteristics of a semiconductor.

TiO_2 Degussa P25 (80% anatase, 20% rutile) is the most widely used photocatalyst powder (Khan et al. 2012). In the composite junctions, the rutile phase of TiO_2 can harvest more light and the excited electrons shuttle to anatase crystals. The interface region is where the chemistry happens apparently. Eventually, the junctions offer enhanced charge carrier separation, reduced electron-hole recombination and facilitated charge transfer to adsorbed species.

The optimum band gap for a semiconductor should be 1.35 eV to utilize solar energy efficiently (Malato et al. 2009). However, TiO_2 has low solar photoconversion efficiency. This is due to its wide band gap energy (3.2 eV for anatase phase). Thus it can only absorb light wavelength up to ~387.5 nm which apparently covers only ~5% of the solar energy. It is primarily unacceptable to utilize bare TiO_2 for either energy production or for example, water or air purification in industrial purposes because the energy density of sunlight is low and also not uniform everywhere. In addition, bare TiO_2 can utilize only the small part of sunlight (UV radiation) contained in the solar spectrum. In water photoelectrolysis, when semiconductor electrodes are used as either photoanodes or photocathodes, the theoretical bandgap should be at least 1.23 eV(Yerga et al. 2009) (i.e., the equilibrium cell potential for water electrolysis at 25°C and 1 atm). Earlier experiments showed that primarily H_2 and O_2 gases were produced in water photolysis, but they recombined to regenerate H_2O molecules eventually. In fact, the backward reactions occur in the powder photocatalyst. This is due to the closer active sites and they facilitate the reactants (H_2 and O_2) to come closer to each other. Figure 3 illustrates the mechanism of photocatalysis in terms of band energy diagram.

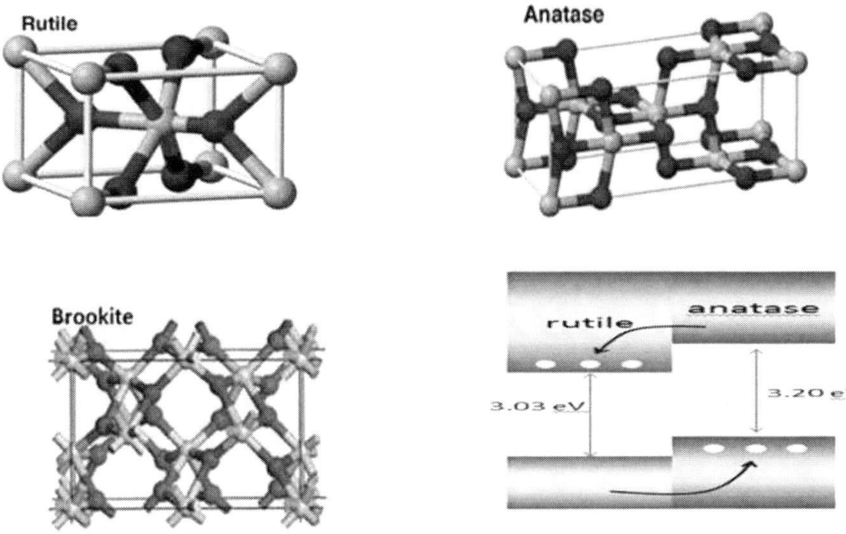

Figure 1. Structural phase of TiO_2: (a) rutile, (b) anatase, (c) brookite, (d) the relative position of the conduction band and valence band in TiO_2anatase and rutile phase.

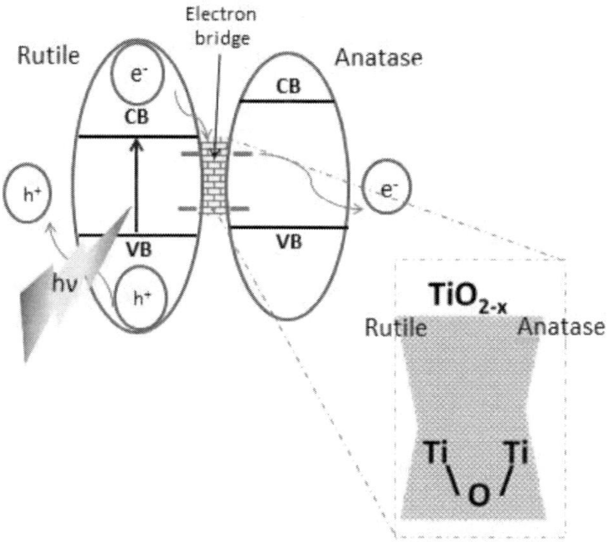

Figure 2. Solid-solid interface: the key to highly efficient and reactive photo-catalyst.

Since electron transfer reactions are involved, both oxidation and reduction occur in a photocatalytic reaction and these opposite processes should have same reaction rates for the photocatalyst to remain unchanged. In fact, a photocatalyst becomes activated with the light shine to facilitate a photocatalytic reaction. It accelerates the reaction rate by accumulating photon energy. During this time, it should not be changed physically or chemically. The important fundamental steps involved in a typical photocatalytic reaction are the absorption of the incident light, bandgap excitation, separation of the stimulated e-/h+ pairs and redox couple reactions on the semiconductor surface (Bessegato et al. 2014). Absorption of incident light occurs when the energy of the incident light is greater than the band gap energy between the valence band and conduction band. Then electrons from valence band will get sufficient energy to travel to the conduction band. Thus electron-hole separation occurs that can either open the active sites for further redox reaction on the semiconductor surface or, recombination can happen. It is noted that these electron-hole pairs are energetically unstable and recombination rates are very fast on the order of nanoseconds. Recombination occurs at boundaries and defects in large particles due to longer migration distances. Thus, it can be minimized by reducing the particle size at the nanoscale level (Chen et al. 2013). After the successful initiation of active sites, the chemical species participate in the reaction. But this surface reaction depends on the active sites and the surface area as well. The surface activity can be increased to a significant extent by using some co-catalysts. Some typical co-catalyst examples are noble metals i.e., Ru, Rh, Pt, and Pd, etc. Various modification and doping process introduce defect sites in the bulk nanostructure of TiO_2 and this phenomenon is very helpful for photoabsorption property. These sites can enhance the absorption of visible as well as IR irradiation and also they can act as trapping site for the charge carriers. A large number of lattice disorder in photocatalyst material could yield a continuous trapping band in the middle whose energy distributions differ from that of a single defect in a crystal (Anpo 2000). A lot of research works for TiO_2 modification have been done and several preparation routes have come out for the last two decades, metal-ion implanted TiO_2 (using transition metals: V, Cr, Mn, Fe, Co, Ni, Cu, Nb, Mo, Ru, Ag, Pt, Au), reduced TiO_x photocatalysts, non-metal doped-TiO_2 (B, C, N, F, S, P, I), TiO_2-semiconductor composites having lower band gap energy (e.g., CdS particles, TiO_2 with dye sensitizers (e.g., thionine)) and TiO_2 doped with upconversion luminescence agent.

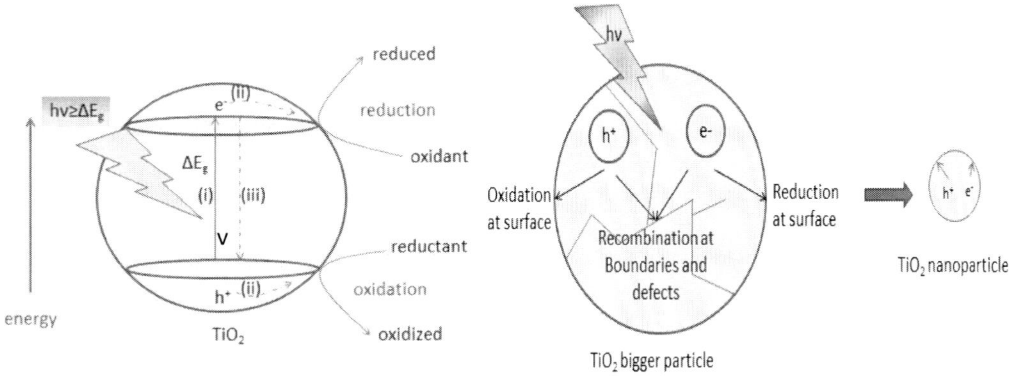

Figure 3. Excitons pathway in semiconductor photocatalysis: (i) incident photon absorption and electron-hole separation. (ii) exciton migration, (iii) surface chemical reaction at active sites and(iv) particle size effect on the recombination process.

PREPARATION OF VISIBLE ACTIVE TiO_2

Generally, TiO_2 anatase is a UV light active substance. Different techniques are available to make TiO_2 as a visible light active substance. Firstly, transition metal doping in the TiO_2 lattice can extend the absorbance range for TiO_2. Secondly, the visible light sensitivity of TiO_2 can be enhanced greatly by introducing dyes, dopants, and impurities. This is due to presented dye molecules that can operate as a visible-light-driven electron pump of the device (Miyauchi et al. 2000). Thirdly, adding organic compounds in the catalysis process can drive the photocatalytic reaction in the forward direction. One example from earlier water splitting experiment can be stated. In the case of photo-assisted water splitting, producing H_2 and O_2 may recombine in backward reaction. Organic compounds can solve this problem by altering redox potentials. Kawai and Sakata 1980 presented that, oxidation of H_2 can be minimized by adding ethanol with the platinized TiO_2 photocatalyst while photocatalytic water splitting. They observed that the organic compounds were first oxidized rather than photogenerated holes producing H_2 and this incident retards H_2 from oxidizing again. Fourthly, another highly sensitive system can be produced using visible light active heterogeneous TiO_2/WO_3 system. WO_3 has a band gap value of 2.8 eV and hence, it can be absorbed in a wavelength of around 440nm, where fluorescent light is also included. Thus the experimental results of TiO_2/WO_3 system showed enhanced hydrophilicity through photon absorption in the visible range (Lee et al. 2013). An extremely weak UV light from a fluorescent lamp was used. But still, high hydrophilicity was achieved. Thus it can be concluded that hyperactive UV sensitive TiO_2 can be utilized in a broader field including indoor conditions. Fifthly, TiO_2 doping with nitrogen also showed notable photocatalytic activities in the visible region. In fact, similar photocatalysts sensitive to visible light have already been reported in several kinds of literature. Some other significant modifications of TiO_2 metal-ion implantation and reducing of TiO_2 doping. Substitution of Ti sites by Cr, Fe or Ni in Ti lattice was also another approach (Li 2013). Besides, the formation of Ti^{+3} sites by creating oxygen vacancy in TiO_2 or self-doping method.

Designing a photocatalyst with an optimal band gap and good mechanistic properties need a deep understanding of both hypothetical and experimental experiences. For example, the conventional TiO_2, when doped with nitrogen, can absorb more photons and hence it becomes visible light active photocatalyst. On the contrary, the redox potential of TiO_2 is disturbed by the N doping process. Consequently, overall photocatalytic efficiency decreases. Therefore, whether to use nitrogen-doped TiO_2 or conventional TiO_2 depends on the light source (Beata et al. 2018). A lot of various preparation methods have been investigated to synthesize visible light active TiO_2. Unfortunately, those methods were not widely accepted until the year 2000 for their chemical instability and lack of reproducibility. The progress to maximize the usefulness of TiO_2 has been very slow, indicating that an alternative photocatalyst system is worth considering. TiO_2 nanocomposites can be a promising replacement of for those traditional photocatalyst systems. In this regard, different forms of graphene incorporated with TiO_2 crystal showed a high potential to enhance its photocatalytic activity significantly.

Physiochemical Aspects of Carbon Nanotubes (Graphene, CNTs)

CNTs can be considered as cylindrical hollow micro-crystals of graphite which have some unique properties such as for absorption process(Zhou et al. 2010; Ajayan et al. 2001). Because they have a large specific area, CNTs have attracted the interest of researchers as a new type of adsorbent. CNTs are graphitic carbon needles and have an outer diameter ranging from 4–30 nm and a length of up to 1 mm. MWCNTs are made of concentric cylinders with spacings between the adjacent layers (Ebbesen 1996) of about 3.4 Å, as shown in Figure 4(d). SWCNTs (Figure 4(d)) were discovered by Iijima 1991.

Li et al. 2013 found that oxidized CNTs can be good Cd^{2+} adsorbents and have great potential applications in environmental protection. Graphene is often used as catalysts and promoter owing to its large specific surface area, high optical transmittance and excellent electronic properties. It is the basic building block for carbon nanotubes and fullerenes. Unfortunately, this fundamental 2D material was discovered just a few years ago after all other carbon forms were known. Prior to this, fullerenes were discovered by (Kroto 1997) in 1985 and carbon nanotubes were identified by Iijima 1991. A mechanical exfoliation method of single and multi-layer graphene from mineral graphites was developed by Novoselov and Geim 2004. However, these two-dimensional honeycomb-like nanostructure consists of sp^2 bonded carbon atoms. It has low density ($2g/cm^3$) and a large surface area of $2630m^2\ g^{-1}$ with π-conjugation structure. Hence it possesses excellent electronic mobility (200,000 $cm^2V\text{-}1s^{-1}$) (Zhang et al. 2012). It is highly transparent (almost 97% light pass through it) and mechanically strong (1100 GPa modulus, fracture strength 130 GPa). These significant properties of graphene are very useful for photocatalytic application. Graphene can act as a very strong co-catalyst which can facilitate electron movement. It shows semi-metallic behavior with linear band dispersion at the Fermi level. Several techniques like hydrocarbon dissociation, segregation, multi-layer graphene preparation by carbon precipitation, ultra-high vacuum (UHV) treatment of hydrocarbon and exfoliation of graphene have been developed. But none of those methods were compatible enough for large scale preparation for the fabrication of the device. Recently, graphene oxide reduction and chemical vapor deposition (CVD) method have been developed for high-quality graphene. Though it is not difficult to produce single-layer graphene sheets, the yield percentage from various preparation methods is quite low. Besides, after the successful separation of the single layers sheets, again they have a tendency to aggregate back to bulk graphite structure gradually. This is due to the strong Van der Waals interactions. In fact, highly dispersed graphene sheets are not stable in solution phase. Later on, it was found that the structure of graphene sheets can be controlled up to a certain extent. Basically, the aggregation tendency depends on the surface properties of graphene. Liang et al. 2013)found that solvent exfoliated graphene possesses fewer defects than the graphene prepared by the oxidation-reduction route. For example, solvent-exfoliated graphene exhibited enhanced efficiency in photocatalytic reduction of carbon dioxide under visible light. This implies that the preparation method plays a vital role to get better performance from graphene. It was also found that graphene coupled with other materials to form nanocomposites are able to show better photocatalytic activity. So far, many metal and metal oxide nanoparticles such as Pd, Pt, Ag, Au, TiO_2, SnO_2, and MnO_2 have been deposited on graphene sheets. As both TiO_2 and graphene have good tunable structural properties and stability, it can be stated that the coupling of graphene with TiO_2 can improve their physio-mechanical properties up to a larger extent. Graphene has recently attracted a great deal of attention for potential applications in many fields, such as nanoelectronics, fuel-cell technology, supercapacitors and catalysts (Geim et al. 2007; Wang et al. 2008).

Hummers and Offeman 1958 developed a method to prepare graphene oxide (GO) which was based on strong chemical oxidation of graphite, followed by the sonication for exfoliation of the obtained graphite oxide. GO is strongly hydrophilic due to the distribution of epoxide and hydroxyl groups on the basal planes as well as the location of carbonyl and carboxyl groups at the edges. Due to the presence of oxygen-rich functional groups present in the GO structure, it becomes readily exfoliated in water and dispersed into single-layer sheets (Figure 4). This way it becomes stable. The possibility to obtain graphene (Gr) or reduced graphene oxide (RGO) by reduction of graphene oxide (GO) using simple chemical methods facilitates its application in the synthesis of composite materials at affordable production costs. When GO sheets are reduced to RGO, the amount of oxygen functional groups will decrease and a greater extent of the π network will be restored within the graphene structure. Thus the resultant structure will be hydrophobic. Many works have been reported to prepare covalent/non-covalent, modified, reduced graphene oxide (RGO) nanosheets to meet various needs of applications. Figure 4 shows the structure of graphene, graphene oxide, and reduced graphene oxide.

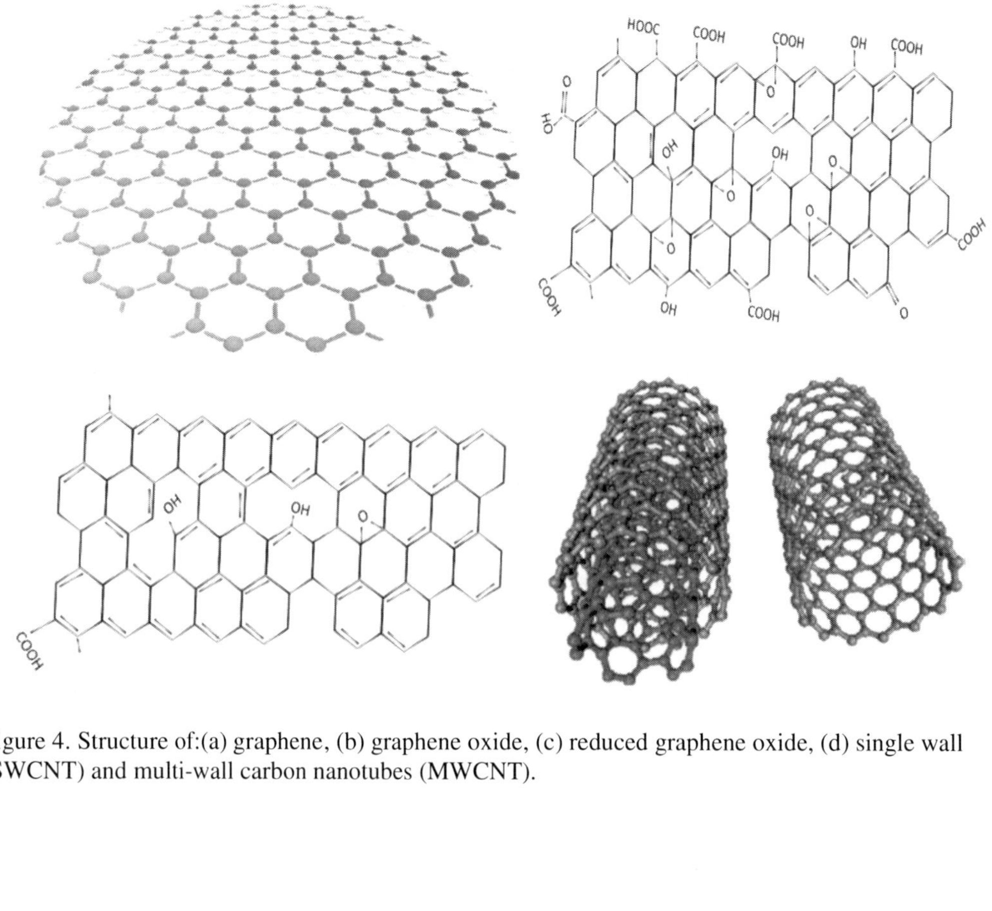

Figure 4. Structure of:(a) graphene, (b) graphene oxide, (c) reduced graphene oxide, (d) single wall (SWCNT) and multi-wall carbon nanotubes (MWCNT).

Reviews on CNT–TiO_2 composites were published by Woan et al. 2009. The tendency observed for graphene-based P25 composites was also observed for CNT–P25 composites, although the G–P25 composites exhibited slightly higher activity for degradation of methylene blue (MB) than their CNT–P25 equivalents, suggested that graphene was in essence similar to CNT regarding the obtained photocatalytic performances. In fact, the preparation method plays a very important role in the photocatalytic performance of TiO_2 Carbon-based composites.

EVOLUTION OF TiO_2–CARBON NANOCOMPOSITES AND ITS APPLICATION AS PHOTOCATALYST

Graphene and TiO_2 both are a very attractive material for photocatalytic reactions due to their high chemical stability and corrosion resistivity. Carbon nanotubes are also promising and could be used as novel sensitizers to improve photocatalytic activity and selectivity under the solar spectrum. Ultraviolet light can drive TiO_2 for photocatalytic reaction in two ways: the photo-induced redox reactions of adsorbed substances and photo-induced hydrophilic conversion of TiO_2 itself. Both the reactions have been studied well and they showed several novel photocatalytic applications of TiO_2. With the development of nanotechnology, some fundamental properties i.e., photoabsorption and charge transfer kinetics were also developed and understood well. In the case of carbon-TiO_2, it can be expected that the broad expanse of the electronic state due to carbon-Ti coupling will show higher photocatalytic activity under visible light. In general photocatalytic activity of TiO_2 depends on many factors whereby carbon nanomaterials are the most suitable materials which can assist TiO_2 to upgrade its activity. Gr-TiO_2 nanocomposites possess high adsorption capacity, extended light absorption range and enhanced charge separation and transportation properties. The functional groups on the plane of GO, such as the hydroxyl, epoxide, carbonyl, and carboxyl groups, offer the anchoring points for nanoparticles. Like another metal catalysis i.e., Ag, Au, Pt, and semiconductors (CdS, Cu_2O), Graphene coupled TiO_2 semiconductors have been widely reported. In particular, graphene sheets possess high specific surface area approximately 2600 m^2g^{-1} and easy tunable surface functionality (Salunkhe et al. 2014). These properties allow it as a host substrate for active TiO_2 nanoparticles. The surface functional groups i.e., epoxy and hydroxyl groups on graphene present favorable nucleation sites for active nanomaterials. This nucleation of heterogeneous clusters on graphene actually directs the final particles size and microstructure.

The different compositions may not have the same effect on trapping electrons or holes on the surface during the interface charge transfer because of the different active site density in the host lattice. However, a fundamental understanding of the kinetics and mechanisms of carbon-TiO_2 catalyst system is crucial from the viewpoint of optimum performance build up for advanced photocatalysis technology.

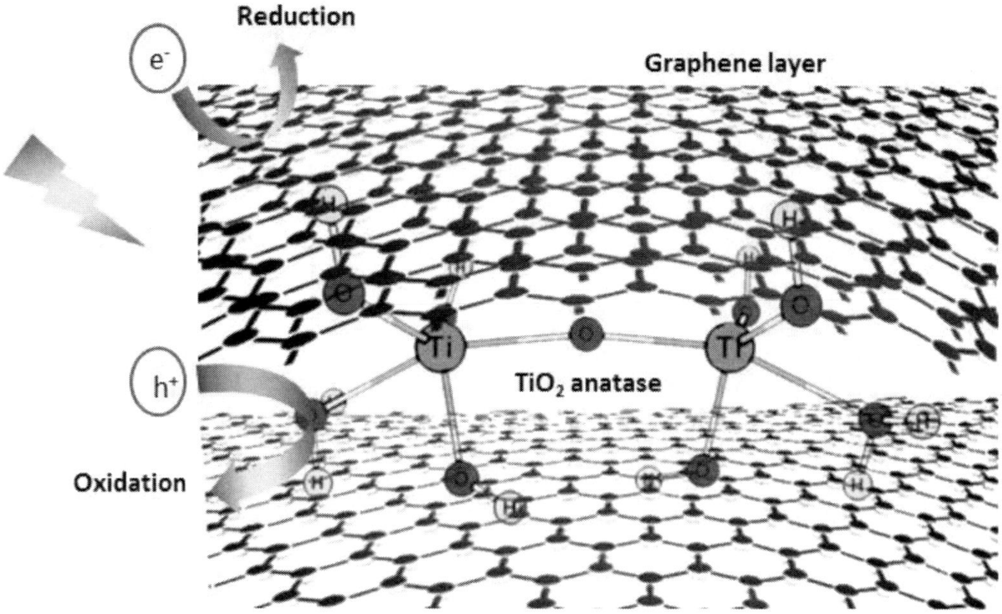

Figure 5. TiO_2/graphene layers composite as an efficient photocatalyst.

Graphene with the semiconductor particles shows efficient photocatalytic activity due to conjugative π structures which can act as 'Electron Bridge'. Graphene possesses good electrical conductivity and can reduce the charge carrier recombination to a great extent. The incorporation of graphene into TiO_2 photocatalysts can offer the characteristic properties of graphene and also new synergistic effects could be observed such as high adsorption capacity, an extended light absorption range and enhanced charge separation and transportation properties. The formation of the Ti–O–C chemical bond narrows the band gap of TiO_2 and extends the photo responding range (Basheer 2012). The transfer of excited electrons from TiO_2 to graphene suppresses the charge recombination. Electron mobility in defect-free high-quality graphene exceeds 15,000 $m^2V^{-1}s^{-1}$ at room temperature (Liu et al. 2011). These unique and excellent properties of graphene made it an essential part of an efficient photocatalyst. The coupling of TiO_2 and graphene can enhance the photocatalytic activity of TiO_2 by reducing the recombination of $h+/e-$ pairs and increasing the number of active sites. These fundamental properties actually play a role in overall photocatalytic performance. The oxidation-reduction process of graphene-TiO_2 based composite is illustrated in Figure 5. The concentration of graphene oxide in starting solution played an important role in the photoelectronic and photocatalytic performance of graphene oxide/TiO_2 composites (Štengl et al. 2013).

Graphene-based composites can capture photoinduced electrons originated from TiO_2 (or another semiconductor) particles. The synergistic photocatalytic behavior of Gr-TiO_2 is much more promising than the individual catalytic performance of these two components. However, the photocatalytic performance of any semiconductor crystal system depends on the primary factors of the components such as, surface/volume ratio, the adsorption affinity on semiconductor surface and capacity for redox couples (i.e., organic contaminants, H_2O, CO_2 etc.), e^-/h^+ pairs recombination processes in the bulk and on the semiconductor surface, intensity of the incident light and spectral distribution, crystal structures, intrinsic structural defects, stoichiometric relationship, pH, electron acceptors and the concentration of the redox species/pollutants (Al-Qaradawi et al. 2001). Several methods have been established for the preparation of Gr-TiO_2 composite sheets. Among them, simple mixing and/or sonication, sol-gel process, liquid-phase deposition, hydrothermal and solvothermal methods, electrostatic self-assembly, electrospinning fiber process technique are most common in laboratory practice (Morales-Torres et al. 2012). The first work found in the literature about the preparation of reduced graphene oxide-titania (RGO–TiO_2) composites by UV-assisted photocatalytic reduction was proposed by Williams et al. 2008. Zhou and coworkers(Zhou et al. 2011) prepared Graphene-TiO_2 composites by hydrothermal treatment. Hybridization of conjugative π structured graphene layers and TiO_2 nanomaterials were proposed and characterization results showed that the composite possessed good electrical conductivity. And hence the material was successful inefficient photoactivity. Electronic interactions were confirmed between graphene and TiO_2 by photoelectrochemical experiment. They also observed that TiO_2 inside the carbon shell of three molecular layers (~1nm thick) shows two times higher activity than that of commercial P-25 Degussa TiO_2 powder. The experiment was conducted under UV light irradiation. It can be concluded that the improved photocatalytic activity under UV irradiation is due to low charge transfer resistance and high migration efficiency of photoelectrons in between TiO_2 and graphene structures. In another experiment, TiO_2-Gr nanocomposite electrodes were prepared by (Zou et al. 2010). They used single step process via directly refluxing GO dispersion and water-soluble PTC solution. Photocurrent density was found 1.481 Acm^{-2} at graphene concentration 18.24% in the composite. The photocurrent is about three times higher than that obtained for pure TiO_2 electrode.

APPLICATION OF GR-TiO_2 COMPOSITES

Gr-TiO_2 composite materials can be classified as an environmentally friendly catalyst as well. In this section, we will show the introduction of graphene and compositional behaviour with TiO_2 in terms of photocatalysis, which can actively contribute to environmental preservation and/or improvement technologies. The most significant photocatalytic applications that have been proposed are water splitting for hydrogen generation degradation of environmental pollutants in aqueous contamination and wastewater treatment, self-cleaning activity, air purification, and carbon dioxide (CO_2^-) reduction.

PHOTOCATALYTIC WATER SPLITTING

Photocatalytic water splitting is being investigated for hydrogen production which is one of the most interesting ways to produce clean and renewable energy. The interaction of photon energy with TiO_2 nanoparticles produces electron-hole pairs. The hole oxidizes water to form oxygen whereas, the electron reduces water to form a hydrogen molecule. The inhibited electron/hole recombination, extended light absorption and favorable active sites for redox species are the key factors for efficient photocatalytic activity. Although there are several obligatory requirements for efficient photodissociation of water, for example, a minimum 1.23V potential difference is needed for water dissociation energy, TiO_2 band structure is favorable for water splitting due to the relative position of the valence band and conduction band. TiO_2 conduction band energy is adjacent to H_2 evolution energy and hence the focus goes on the valence band energy to shift it closer to O_2 formation energy. It is due to high over the potential for O_2 evolution on the TiO_2 surface. Due to the high electrical property of graphene, it shows extended electron mobility and when coupled with TiO_2, the photocatalytic activity improves greatly. For example, TiO_2/RGO composites exhibited a hydrogen evolution rate of 20mmol h-1 which is much better than the TiO_2/CNT composites (Xiaoqiang & Jimmy 2011).

Multi-walled carbon nanotube–TiO_2 composite catalysts can be used as catalysts in photocatalytic processes for water treatment. The introduction of increasing amounts of CNTs into the TiO_2 matrix prevents particles from agglomerating, thus increasing the surface area of the composite materials. A synergy effect on the photocatalytic degradation of phenol was found mostly for the reaction activated by near-UV to visible light irradiation. This improvement on the efficiency of the photocatalytic process appeared to be proportional to the shift of the UV–vis spectra of the CNT–TiO_2 composites for longer wavelengths, indicating a strong interphase interaction between carbon and semiconductor phases. This effect was explained in terms of CNTs acting as photosensitizer agents rather than adsorbents or dispersing agents. Surface defects at the surfaces of carbon nanotubes provide advantages not only for the anchoring of the TiO_2 particles but also for the electron transfer process to the semiconductor.

PHOTODEGRADATION OF POLLUTANTS

Until today, most studied photocatalytic application of Gr-TiO_2 is related to the photocatalytic degradation of environmental pollutants. Usually, phenol and phenol derivatives, organic dyes such as methyl orange, methylene blue, rhodamine B, etc. and humic substances are taken under consideration. Liu and co-workers (Liu et al. 2010)fabricated RGO wrapped TiO_2 hybrid by one-step photocatalytic reduction. They demonstrated that RGO captured dyes and photoinduced electrons during the photocatalytic degradation of organic dyes in water. Wang et al. 2012 investigated the photoinduced charge transfer between TiO_2 and graphene, using a transient photovoltage technique. After their integration with graphene, the lifetime of separated electron-hole pairs was prolonged from ~10-27s to ~10-25s. Thus, dual roles of graphene in the composite were improved by: (1) increasing the electron-hole pair separation through the electron injection from the conduction band of TiO_2 into graphene, (2) greatly retarding the recombination of electron-hole pairs in the excited TiO_2. Graphene-TiO_2 composite photocatalysts also show promising applications in the degradation of other organic molecules. Ng et al. (Du et al. 2011)fabricated homogeneous TiO_2/graphene thin film through the deposition of TiO_2/graphene suspension. During the degradation of 2,4-dichlorophenoxy acetic acid with TiO_2/graphene films, a 4-fold increase in the rate of photocatalytic degradation was achieved. Akhavan et al. 2009 investigated the influence of a graphene coating on the antibacterial activity of the TiO_2 thin film which was used as photocatalysts for the deactivation of E. coli bacteria in an aqueous solution.

Based on the above results, the enhanced photocatalytic degradation of the organic compound could be attributed to the following reasons: (1) the interactions between organic substances and the aromatic regions of graphene enhance the adsorption on photocatalysts. (2) Ti–O–C bond forms and it narrows the band gap of TiO_2 and extends the photo responding range. (3) the transfer of excited electrons from TiO_2 to graphene suppresses the charge recombination.

A few studies are available detailing SWNTs with antimicrobial activity towards Gram-positive and Gram-negative bacteria due to either a physical interaction or oxidative stress that compromises the cell membrane integrity (Qi et al. 2004) Carbon nanotubes may, therefore, be useful for inhibiting microbial attachment and biofouling formation on surfaces. However, the degree of aggregation, (Morones et al. 2005; Lyon et al. 2006) the stabilization effects by NOM (Kang et al. 2007) and the bioavailability of the nanotubes will have to be considered for these antimicrobial properties to be fully effective.

CO_2 PHOTO REDUCTION

From both thermodynamic and kinetic point of view, the CO_2 molecule is highly stable in nature and hence high energy input is needed for the activation of CO_2. First photoelectrochemical reduction of CO_2 was reported by Halmann in 1978. In1979, photocatalytic reduction of CO_2 to various organic compounds namely CH3OH, HCOOH, CH4, and HCHO was studied by Anpo et al. 1995. Various oxide and non-oxide photocatalysts such as TiO_2, ZnO, WO_3, CdS, GaP, and SiC in aqueous media were investigated and Hg and/or Xenon lamp was used as an illumination source. Anpo et al. 1995 revealed that, among various types of TiO_2 (anatase and rutile) photocatalysts, TiO_2 anatase was more selective for CH_4 production while TiO_2 rutile showed higher yield for the formation of CH_4 and CH_3OH.

CONCLUSION

A brief timeline of TiO_2 photocatalyst and its orientation with graphene were focused in this chapter. Both scientific and technological transitions about Gr-TiO_2 composite as photocatalyst were elucidated in short. Both graphene and TiO_2 became very popular materials with respect to energy and environmental aspects. Both materials have shown high resistivity and longevity, chemically inert and highly sensitive to photoactivity. They possess easily tunable structural properties and both are commercially established material for their economic excellence. Until today, individually graphene and TiO_2 are being used in versatile applications starting from environmental purification appliances to energy storage devices. According to the transition period mentioned by K. Hashimoto, TiO_2 photocatalysis became legendary in every ten years and the continuation of these transition periods are still going on. Commercially developed clean water technology by TiO_2 photo activity, enhancement of photoactivity under visible light irradiation due to Nanoengineering on Gr-TiO_2, the introduction of Gr-TiO_2 into energy management issues such as lithium ion batteries, biomedical applications such as antibacterial filtration membrane and wound dressings- all these research areas have been flourished in the early 21st century. Hopefully, in the near future, real industrial applications will be achieved regarding these research fields. It is already proved that graphene can make the poor TiO_2 material valuable photocatalyst and we believe that, this couple will be able to create new technological field based on basic scientific knowledge.

Although considerable progress on graphene modified photocatalysis has been achieved, the detailed mechanism of the enhancement is still not clear. There are only a few studies concerning the charge transfer kinetics in the graphene/TiO_2 systems. More work is necessary to clarify the charge transfer mechanism between graphene and TiO_2.

ACKNOWLEDGMENTS

The authors would like to thank High Impact Research (HIR F- 000032) for their cordial support to complete this work. The authors declare no conflict of interest.

REFERENCES

Ajayan P. M., Zhou O. Z., 2001 "Application of Carbon Nanotubes, in Carbon Nanotubes Topics" *Applied Physics* 80:381–425.

Akhavan O., Ghaderi E. J., 2009 "Photocatalytic reduction of graphene oxide nanosheets on TiO_2 thin film for photoinactivation of bacteria in solar light irradiation" *Phys. Chem. C* 113:20214-20220.

Al-Qaradawi S., Salman S. R., 2001 "Photocatalytic degradation of methyl orange as a model compound" *Journal of Photochemistry and Photobiology A: Chemistry* 148: 161.

Anpo M., 2000 "Use of visible light. Second-generation titanium dioxide photocatalysts prepared by the application of an advanced metal ion-implantation method" *Pure ApplChem* 72:1787-1792.

Anpo M., Yamashita H., Ichihash Y., Ehara Z., 1995 "Photocatalytic reduction of CO_2 with H_2O on various titanium oxide catalysts" *Journal of Electroanalytical Chemistry* 21:396.

Basheer C. 2012 "Application of Titanium Dioxide-Graphene Composite Material for Photocatalytic Degradation of Alkylphenols" *Journal of Chemistry* 456586:10.

Beata T., Magdalena W., Grzegorz Z., Guskos N., Morawski A., Colbeau J. A., Wrobel R., Nitta A., Ohtani B., 2018 "Influence of an Electronic Structure of N-TiO2 on Its Photocatalytic Activity towards Decomposition of Acetaldehyde under UV and Fluorescent Lamps Irradiation" *Catalysts* 8:85.

Bessegato G. G., Guaraldo T. T., Zanoni M. V. B., 2014 "Enhancement of Photoelectrocatalysis Efficiency by Using Nanostructured Electrodes" *Nano, Surface and Corrosion Science* doi:10.5772/58333.

Chen D., Wang J., Chen T., Shao L., 2013 "Defect annihilation at grain boundaries in alpha-Fe" *Scientific reports* 3:1450.

Chen L., Zhou Y., Tu W., Li Z., Bao C., Dai H., Zou Z., 2013 "Enhanced photovoltaic performance of a dye-sensitized solar cell using graphene–TiO_2 photoanode prepared by a novel in situ simultaneous reduction-hydrolysis technique" *Nanoscale* 5:3481-3485.

Chen X., Mao S. S., 2007 "Titanium Dioxide Nanomaterials: Synthesis, Properties, Modifications, and Applications" *Chem. Rev* 107:2891-2959.

Devi L. G., Kavitha R., 2013 "A review on non-metal ion doped titania for the photocatalytic degradation of organic pollutants under UV/solar light: Role of photogenerated charge carrier dynamics in enhancing the activity" *Applied Catalysis B: Environmental* 140-141:559-587.

Ding Z., Lu G. Q., and Greenfield P. F., 2000 "Role of the crystallite phase of TiO_2 in heterogeneous photocatalysis for phenol oxidation in water" *The Journal of Physical Chemistry B* 104:4815-4820.

Du A., Ng Y. H., Bell N. J., Zhu Z., Amal R., Smith S. C., 2011 "Hybrid Graphene/TitaniaNanocomposite: Interface Charge Transfer, Hole Doping, and Sensitization for Visible Light Response" *The Journal of Physical Chemistry Letters* 2:894-899.

Ebbesen T. W. 1996. Wetting, filling and decorating carbon nanotubes. *J. Phys. Chem. Solids* 57:951–955.

Frank S. N., Bard A. J., 1977 "Heterogeneous photocatalytic oxidation of cyanide ion in aqueous solution at TiO_2 powder" *Journal of American Chemical Society 99*:303-304.

Fujishima A., Honda K., 1972 "Electrochemical photolysis of water at a semiconductor electrode" *International journal of science 283*:37-38.

Fujishima A., Honda K., Kikuchi S., Kagaku K., Zasshi, 1969 "Photosensitized electrolytic oxidation on semiconducting n-type TiO2 electrode" *The Journal of the Society of Chemical Industry* 72:108-113.

Geim A. K., Novoselov K. S., 2007 "The rise of graphene" *Nat. Mater* 6:183-191.

Halmann M., 1978 "Photoelectrochemical reduction of aqueous carbon dioxide on p-type gallium phosphide in liquid junction solar cells" *Nature* 275:115-6.

Hashimoto K., Irie H., Fujishima A., 2005 "TiO_2 Photocatalysis: A Historical Overview and Future Prospects" *Japanese Journal of Applied Physics* 44:8269-8285.

Heller A., 1995 "Chemistry and applications of photocatalytic oxidation of thin organic films" *Acc. Chem. Res* 28:503-508.

Hummers J. W. S., Offeman R. E., 1958 "Preparation of graphitic oxide" *Journal of the American Chemical Society* 80:1339-1339.

Ihara T., Ando M., Sugihara S., 2001 "Preparation of visible light active TiO_2 photocatalysis using wet method" *Photocatalysis* 5:19.

Iijima S., 1991 "Helical microtubules of graphitic carbon" *International Journal of Science* 354:56–58.

Kamegawa T., Shimizu Y., Yamashita H., 2012 "Superhydrophobic Surfaces with Photocatalytic Self-Cleaning Properties by Nanocomposite Coating of TiO_2 and Polytetrafluoroethylene" *Advanced Materials* 24: 3697-3700.

Kang S., Pinault M., Pfefferle L. D., Elimelech M., 2007 "Single-walled carbon nanotubes exhibit strong antimicrobial activity" *Langmuir* 23:8670–8673.

Kawai T., Sakata T., 1980 "Conversion of carbohydrate into hydrogen fuel by a photocatalytic process" *International journal of science 286*:474-476.

Kay A., Gratzel M., 1993 "Artificial Photosynthesis Photosensitization of TiO_2 Solar Cells with Chlorophyll Derivatives and Related Natural Porphyrins" *J. Phys. Chem 97*: 6272-6277.

Khan A., Mir N. A., Faisal M., and Muneer M., 2012 "Titanium Dioxide-Mediated Photocatalysed Degradation of Two Herbicide Derivatives Chloridazon and Metribuzin in Aqueous Suspensions" *International Journal of Chemical Engineering* 850468:1-8.

Kohle O., Ruile, S., Grätzel M., 1996 "Ruthenium(II) Charge-Transfer Sensitizers Containing 4,4'-Dicarboxy-2,2'-bipyridine. Synthesis, Properties, and Bonding Mode of Coordinated Thio- and Selenocyanates" *American Chemical Society* 35: 4779-4787.

Kroto H., 1997 "Symmetry, space, stars and C" *Reviews of Modern Physics* 69:703.

Kudo A., Miseki Y., 2009 "Heterogeneous photocatalyst materials for water splitting" *ChemSoc Rev 38*:253-278.

Kuznetsova I. N., Blaskov V., Stambolova I., Znaidi L., and Kanaev 2005 "TiO_2 pure phase brookite with preferred orientation, synthesized as a spin-coated film" *Materials Letters* 59:3820-3823.

Lee H. U., Lee S. C., Choi S., Son B., Kim H., Lee S. M., Kim H. J., Lee J., 2013 "Influence of visible-light irradiation on physicochemical and photocatalytic properties of nitrogen-doped three-dimensional (3D) titanium dioxide" *J Hazard Mater* 258:10-18.

Lee J., Kim J., Choi W., 2011 "TiO_2 Photocatalysis for the Redox Conversion of Aquatic Pollutants" *American Chemical Society* 1071:199-222.

Li M., 2013 "The research and development of Fe doped TiO_2" *Research of Materials Science* 2:28-33.

Li X. L., Peng Q., Yi, J. X., Wang X., Li Y., 2006 "Near monodisperse TiO_2 nanoparticles and nanorods" *Chem. Eur. J* 12:2383.

Liu G., Desalegne T., Balandin A. A., 2011 "Tuning of graphene properties via controlled exposure to electron beams" *Nanotechnology, IEEE Transactions* 10: 865-870.

Liu G., Wang L., Yang H. G., Cheng H. M., Lu G. Q. M., 2010 "Titania-based photocatalysts—crystal growth, doping, and heterostructuring" *Journal of Materials Chemistry* 20:831-843.

Lyon D. Y., Adams L. K., Falkner J. C., Alvarez P. J. J., 2006 "Antibacterial activity of fullerene water suspensions: effects of preparation method and particle size" *Environmental Science and Technology* 40:4360–4366.

Malato S., Fernández-Ibáñez P, Maldonado M. I., Blanco J., Gernjak W., 2009 "Decontamination and disinfection of water by solar photocatalysis: Recent overview and trends" *Catalysis Today* 147:1-59.

Mashio F. and Kato S. 1956 "Autooxidation by TiO2 as a photocatalyst" Abtr. Book Annu. Meet.

Mills A., Belghazi A., Davies R. H., Worsley D., Morris S., 1994 "A kinetic study of the bleaching of rhodamine 6G photosensitized by titanium dioxide" *Journal of Photochemistry and Photobiology A: Chemistry 79*:131-139.

Miyauchi M., Nakajima A., Hashimoto K., Watanabe T., 2000 "A highly hydrophilic thin film under 1 μW/cm2 UV illumination" *Adv. Mater.* 12:1923-1927.

Morones J. R., Elechiguerra J. L., Camacho A. Holt K., Kouri J. B., Ramírez J. T., Yacaman M. J., 2005 "The bactericidal effect of silver nanoparticles" *Nanotechnology* 16:2346–2353.

Qi L. Z., Xu X. J., Hu C., Zou X., 2004 "Preparation and antibacterial activity of chitosan nanoparticles" *Carbohydrate Research* 339:2693–2700.

Sakatani Y., Okusako K., Koike H., Ando H., 2001 "Development of TiO_2 photocatalysis with visible light response" *Photocatalysis* 4:51.

Salunkhe R. R., Lee Y. H., Chang K. H., Li J. M., Simon P., Tang J., Torad N. L., Hu C. C., Yamauchi Y., 2014 "Nanoarchitectured Graphene-Based Supercapacitors for Next-Generation Energy-Storage Applications" *Chemistry-A European Journal* 20:13838-13852.

Scanlon D. O., Dunnill C. W., Buckeridge J., Shevlin S. A., Logsdail A. J., Woodley, S. M., Catlow C. R. A., Powell M. J., Palgrave R. G., Parkin I. P., Watson G. W., Keal T. W., Sherwood P. W., Sherwood A., Sherwood S., Alexey A., 2013 "Band alignment of rutile and anatase TiO_2. *Nature Materials* 12: 798–801.

Sergio M. T., Pastrana M., Luisa M., Figueiredo J. L., Faria J. L., Silva, Adrián M. T. 2012 "Design of graphene-based TiO_2photocatalysts—a review" *Environ Sci. Pollut. Res.* 19:3676-3687.

Štengl V., Bakardjieva S., Grygar T. M., Bludská J., Kormunda M., 2013 "TiO_2-graphene oxide nanocomposite as advanced photocatalytic materials" *Chemistry Central Journal* 7:41.

Takeuchi M., Sakamoto K., Martra G., Coluccia S., Anpo M., 2005 "Mechanism of photoinduced superhydrophilicity on the TiO_2 photocatalyst surface" *J. Phys. Chem* B 109:15422.

Taoda H., 2009 "Development and application of photocatalytic technology-Industrialization of sustainable eco-technology" *Synthesiology English edition 1*:287-295.

Tay Q., Liu X., Tang Y., Jiang Z., Sum T. C., Chen Z., 2013 "Enhanced Photocatalytic Hydrogen Production with Synergistic Two-Phase Anatase/Brookite TiO_2 Nanostructures" *The Journal of Physical Chemistry C* 117:14973-14982.

Wang P., Ao Y., Wang C., Hou J., Qian J., 2012 "Enhanced photoelectrocatalytic activity for dye degradation by graphene–titania composite film electrodes" *Journal of Hazardous Materials* 223-224:79-83.

Wang X., Zhi L. J., Mullen K., 2008 "Transparent, conductive graphene electrodes for dye-sensitized solar cells" *Nano Lett* 8:323-327.

Williams G., Seger B., Kamat P. V., 2008 "TiO_2-graphene nanocomposites. UV-assisted photocatalytic reduction of graphene oxide" *ACS Nano* 2:1487-1491.

Woan K., Pyrgiotakis G., Sigmund W., 2009 "Photocatalytic carbon-nanotube-TiO_2 composites" *Adv. Mater* 21: 2233-2239.

Xiao Y., Wu J., Yue G., Lin J., Huang M., Fan L., and Lan Z., 2012. Preparation of a three-dimensional interpenetrating network of TiO_2 nanowires for large-area flexible dye-sensitized solar cells. *RSC Advances* 2:10550-10555.

Xiaoqiang A., Jimmy C. Y., 2011 "Graphene-based photocatalytic composites" *RSC Advances* 1:1426-1434.

Yerga R. M. N., Alvarez M. C., Galvan F. D. V., Mano J. A. V. D. L., Fierro J. L. G., 2009 "Water Splitting on Semiconductor Catalysts under Visible-Light Irradiation" *Chemsuschem* 2: 471-485.

Zhai C., Zhu M., Ren F., Yao Z., Du Y., Yang P., 2013 "Enhanced photoelectrocatalytic performance of titanium dioxide/carbon cloth based photoelectrodes by graphene modification under visible-light irradiation" *Journal of Hazardous Material* 263:291-298.

Zhang X., Kumar P. S., Aravindan V., Liu H. H., Sundaramurthy J., Mhaisalkar S. G., Duong H. M., Ramakrishna S., Madhavi S., 2012 "Electrospun TiO_2–Graphene Composite Nanofibers as a Highly Durable Insertion Anode for Lithium Ion Batteries" *The Journal of Physical Chemistry C* 116:14780–14788.

Zhou D. M., Wang Y. J., Wang H. W., Wang S. Q., Cheng J. M., 2010 "Surface-modified nanoscale carbon black used as sorbents for Cu(II) and Cd(II)" *J. Hazard. Mater* 174: 34–39.

Zhou K., Zhu Y., Yang X., Jiang X. and Li C., 2011 "Preparation of Graphene-TiO_2 Composites with Enhanced Photocatalytic Activity" *New Journal of Chemistry* 2:353-359.

Zou F., Yu Y., Cao N., Wu L. and Zhi J. 2010 "A novel approach for synthesis of TiO_2 graphene nanocomposites and their photoelectrical properties" *Scripta Materialia* 64:621-624.

ABOUT THE EDITORS

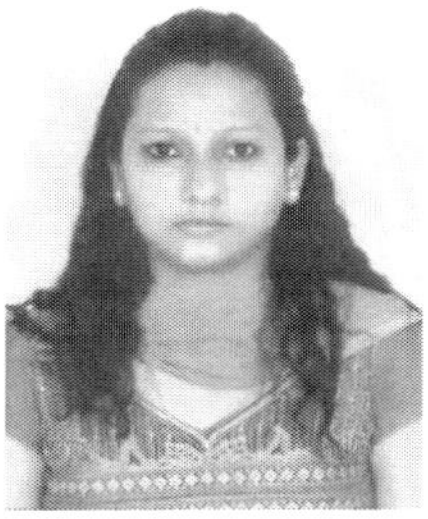

Dr. Preeti Singh is a Post Doctorate Fellow, UGC in Bio/Polymers Research Laboratory, Department of Chemistry, Jamia Millia Islamia, New Delhi. She was awarded her PhD from the Department of Physics, Faculty of Natural Sciences, Jamia Millia Islamia, New Delhi, India in Physics (Material Science). She has several research publications in the area of crystal growth and their defects, synthesis of nanomaterials and their applications in Photocatalysis and Sensors. She has authored 22 research articles and three book chapters.

Dr. M. M. Abdullah is assistant professor in the department of physics, Najran University, Najran, Saudi Arabia. In addition, he is actively involved in scientific research in Promising Center for Sensors and Electronic Devices (PCSED) and in Advanced Materials and NanoResearch Center, Najran University. He had completed his doctoral degree from Jamia Millia Islamia, New Delhi, India. He did his PhD work in collaboration with National Physical Laboratory, New Delhi, India. His doctoral thesis encompasses the preparation, characterization, and dielectric application of single- and poly-crystalline chalcogenide compounds. At present, the research interest of Dr. Abdullah is mainly focused on the understanding and evaluation of material's properties, performances, and applications. His recent area of research work is to prepare metal oxide nanostructures and their functional applications in sensors, catalysis, and capacitors. He has more than 30 research articles published in SCI journals. He had already completed several funded projects and is currently involved in many other funded projects.

Dr. Mudasir Ahmad is currently a Post Doc faculty at School of Natural and Applied Sciences Northwestern Polytechnical University Xi'an, PR China. His current research is focus on Synthesis of Noval Organic/Inorganic Carbon Nanotubes for various industrial applications. In 2018, He has completed his PhD from the Department of Chemistry Jamia Millia Islamia, New Delhi and worked as SRF fellow from University Grants Commission (UGC) of India. He works closely with several programs, State Key Program of National Natural Science of China (51433008) the National Natural Science China (21704084) and the Fundamental Research Funds for the Central Universities (3102017jc01001).

Dr. Saiqa Ikram is working as an Associate Professor, in the Department of Chemistry, Jamia Millia Islamia (A Central University by an Act of Parliament) Delhi, India. She is a PhD in the Faculty of Technology, Delhi University (JRF & SRF); followed by Post-doc from CBME, IIT, Delhi.

Her pioneering research is in the modification of polymers for sustainable developments. The core polymers are "Chitosan & Cellulose" modified as bio-nanocomposites for therapeutics and compostable materials for wastewater treatment. She had received research funding from DBT, UGC, JMI Innovative Grant, US-AID & MHRD-SPARC with University of Queensland, Australia. Under these fundings, "Best Innovative Project Award" also been felicitated by US-AID & TERI University in October 2017 (TERI University Research Grant for Innovative Projects) funded by USAID. This research brought publications in Journals: Elsevier, Springer, RSC, and ACS, including the book *Chitosan: Derivatives, Composites & Applications* for Wiley-Scrivener -April 2018 & *Biocomposites, Biomedical & Environmental Applications*-November 2017 for Pan Standford Publishing. She has been published in more than 70 publications, international publications, five International (edited) books along with more than ten chapters with other publishers. She had supervised 08 PhD scholars and six more continuing.

INDEX

A

B

C

D

E

G

H

I

M

N

O

P

R

S

T

U

W

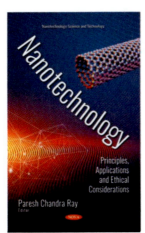

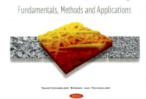

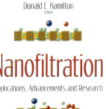

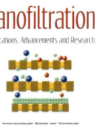